Introduction to Sports Biomechanics

LIB/ND/001

WP 2215641 0

Other titles available from E & FN Spon

Kinanthropometry and Exercise Physiology Laboratory Manual
Tests, procedures and data
Edited by Roger Eston and Thomas Reilly

Drugs in Sport
2nd edition
David Mottram

Foods, Nutrition and Sports Performance
Edited by Clyde Williams and John R Devlin

Notational Analysis
Mike Hughes and Ian Franks

Physiology of Sports
Thomas Reilly, Peter Snell, Clyde Williams and Nils Secher

Science and Racket Sports II
Edited by Adrian Lees, Mike Hughes, Thomas Reilly and Ian Maynard

Science and Soccer
Thomas Reilly

Sport, Leisure and Ergonomics
Edited by Greg Atkinson and Thomas Reilly

Visual Perception and Action in Sport
Mark Williams, Keith Davids and John Williams

Journal of Sports Sciences
Editor: Roger Bartlett

For more information about these and other titles published by E & FN Spon, please contact:

The Marketing Department, E & FN Spon, 11 New Fetter Lane, London EC4P 4EE
Tel: 0171 583 9855; Fax: 0171 842 2303; or visit our web site at **www.efnspon.com**

Introduction to Sports Biomechanics

Roger Bartlett

UNIVERSITY OF WOLVERHAMPTON
LEARNING RESOURCES
Acc No. 2215641
CLASS 612.76 BAR
CONTROL 0419208402
DATE 21 JUL 2000
SITE WL

E & FN SPON
An Imprint of Routledge
London and New York

First published 1997
by E & FN Spon, an imprint of Chapman & Hall

Reprinted 1997, 1999
by E & FN Spon, an imprint of Routledge
11 New Fetter Lane, London EC4P 4EE
29 West 35th Street, New York, NY 10001

© 1997 Roger Bartlett

Typeset in 10½/12pt Sabon by
Saxon Graphics Ltd, Derby
Printed in Great Britain by
Alden Press, Oxford

All rights reserved. No part of this book may be reprinted or reproduced or utilized in any form or by electronic, mechanical, or other means, now known or hereafter invented, including photocopying and recording, or in any information storage or retrieval system, without permission in writing from the publishers.

British Library Cataloguing in Publication Data
A catalogue record for this book is available from the British Library

Library of Congress Cataloguing in Publication Data
Catalog Card Number: 96–67515

ISBN 0–419–20840–2

∞ Printed on permanent acid-free text paper, manufactured in accordance with ANSI/NISO Z39.48–1992 and ANSI/NISO Z39.48–1994 (Permanence of Paper)

To my dearest Mel

Contents

Preface

Sports biomechanics uses the scientific methods of mechanics to study the effects of various forces on the sports performer. It is concerned, in particular, with the human neuromusculoskeletal system. It also considers aspects of the behaviour of sports implements, footwear and surfaces where these affect performance or injury. It is a scientific discipline that is relevant to all students of the exercise and sport sciences, to intending physical education teachers and to all those interested in sports performance and injury. This book is intended as the first of two volumes covering aspects of sports biomechanics from first-year undergraduate to postgraduate level. The content of this volume mostly applies to first- and second-year undergraduate level. The focus is largely practical in approach, reflecting the nature of the discipline, and Part Two deals in detail with the main measuring techniques that sports biomechanists use to study the movements of the sports performer. This is not to suggest that sports biomechanics is simply about measuring things. It has a sound theoretical basis, as is evident in Part One, which covers the anatomical and mechanical foundations of biomechanics. Wherever possible, this is approached from a practical sport viewpoint. Being based on mechanics, there is a strong mathematical element in biomechanics, which often deters students without a mathematical background. Where the author considers that basic mathematical equations add to the clarity of the material, these have been included. However, extensive mathematical development of the topics covered has been avoided and the non-mathematical reader should find the great majority of the material easily accessible.

The production of any textbook relies on the cooperation of many people other than the author. I should like to acknowledge the invaluable, carefully considered comments of Carl Payton on all the chapters of the book and of Neil Fowler on Chapter 1. All those who acted as models for the photographic illustrations are gratefully acknowledged: Phil Gates, Vicky Goosey, Caroline Jordan, Mike Lauder, Calvin Morriss and Keith Tolfrey, as is the assistance of Dimitrios Tsirakos with the biomechanics laboratory photographs. I am also grateful to Vasilios Baltzopoulos and Simon Coleman for pointing out a few errors in the first printing. Thanks are also due to Steve Watts and Russell Fisher, who took the anatomical and laboratory photographs, to Terry Bolam for help with other photographic illustrations, and to Tim Bowen for his advice on various aspects of the software packages used for the illustrations.The book could not have been produced without the support of Professor Les Burwitz, nor without Ian Campbell taking over my administrative responsibilities for a considerable period. I am also grateful to those publishers and authors who allowed me to reproduce their illustrations, and to the British Association of Sport and Exercise Sciences (BASES) for allowing me to adapt some sections of their biomechanics guidelines (Bartlett, R.M. (ed) (1992) *Biomechanical Analysis of Performance in Sport*, BASES, Leeds). Last, and by no means least, my deepest gratitude to dearest Melanie, without whose encouragement I would never have started on the book and whose love and humour sustained me throughout.

Roger M. Bartlett
Alsager

Permissions

Figure 1.10 reproduced with kind permission from J.V Basmajian.

Figure 1.18 adapted with kind permission from Marieb, E. (1992) *Human Anatomy and Physiology*, Second Edition, Benjamin Cummings Publishing Company.

Figure 2.20 reprinted from Yeadon, M.R. (1993) The biomechanics of twisting somersaults. *Journal of Sports Sciences,* **21**, 187-225, with kind permission from Chapman & Hall, 2-6 Boundary Row, London, SE1 8HN, UK.

Figure 3.14 reprinted from Coulton, J. (1977) *Women's Gymnastics*, with kind permission from EP of Wakefield.

Figures 3.21 and 3.22 reprinted from Hay, J.G. (1993) *Biomechanics of Sports Techniques,* with kind permission from Allyn & Bacon, Needham Heights, MA 02194-2310, USA.

Figure 8.1a reprinted from Winter, D.A. (1990) *Biomechanics and Motor Control of Human Movement,* with kind permission from John Wiley & Sons Inc, 605 Third Avenue, New York, NY 10158 0012, USA.

Figure 8.1b reprinted from Cochran, A. and Stobbs, J. (1968) *The Search for the Perfect Swing,* Heinemann, London, with kind permission from the Golf Society of Great Britain, South View, Thurleston Kingsbridge, Devon, TQ7 3NT, UK.

Figure 8.2 reprinted from Robertson, G. and Sprigings, E. (1987) Kinematics in *Standardizing Biomechanical Testing in Sport* (eds D.A. Dainty and R.W. Norman), pp. 9-20, with kind permission from Human Kinetics Publishers Inc, Box 5076, Champaign, IL 61820, USA.

Figure 8.5 reprinted with kind permission from Movement Techniques Ltd, Unit 5, The Technology Centre, Epinal Way, Loughborough, LE11 0QE, UK.

Figure 8.6 reprinted from Chao, E.Y.S. (1980) Justification of triaxial goniometer for the measurement of joint rotation. *Journal of Biomechanics,* **13**, 989-1006, with kind permission from Elsevier Science Ltd, The Boulevard, Langford Lane, Kidlington, OX5 1GB, UK.

Figure 8.8 reprinted from Bishop, P.J. (1993) Protective equipment: biomechanical evaluation, in *Sports Injuries: Basic Principles of Prevention and Care* (ed P.A.F.H. Renström), pp. 355-373, with kind permission from Blackwell Science Ltd, Osney Mead, Oxford, OX2 0EL, UK.

Part One
Foundations of Biomechanics

Introduction

Biomechanics is the study of both the structure and function of biological systems using the methods of mechanics. In sports biomechanics, the interest focuses, clearly, on the sports performer. However, interest also extends to the behaviour of inanimate structures that influence performance, such as sports implements, footwear and surfaces. The foundations of sports biomechanics are mechanics and the anatomy of the musculoskeletal system. It is the key material of these disciplines that forms the content of this part of the book, related wherever possible to examples from sport. Mechanics is closely related to mathematics and, therefore, an understanding of the mathematical concepts that underpin biomechanics is essential. These include aspects of calculus and vector algebra. These concepts will be introduced in the following chapters along with, where necessary, basic mathematical equations. However extensive mathematical development of the topics covered has been avoided.

Chapter 1 covers the anatomy of the human musculoskeletal system. Attention is focused on the anatomical principles that relate to movement in sport and exercise. This includes consideration of the planes and axes of movement and the principal movements in those planes. The functions of the skeleton, types of bone, processes of bone growth and fracture, and typical surface features of bone are covered. Attention is also given to the tissue structures involved in the joints of the body, the process of joint formation, joint stability and mobility and the identification of the features and classes of synovial joints. The features and structure of skeletal muscles are considered along with how they develop and the ways in which muscles are structurally and functionally classified. The chapter concludes with a consideration of the types and mechanics of muscular contraction, how tension is produced in muscle and how the total force exerted by a muscle can be resolved into components depending on the angle of pull.

In Chapter 2 the kinematic principles which are important for the study of movement in sport and exercise are covered. Kinematics is the branch of mechanics that deals with the geometry of movement without reference to the forces that cause the movement. Chapter 2 considers the types of motion and the model appropriate to each, vectors and scalars, and vector addition, subtraction and resolution. The importance of differentiation and integration in sports biomechanics is addressed with reference to practical examples, and graphical differentiation and integration are covered. The importance of being able to interpret graphical presentations of one linear kinematic variable (displacement, velocity or acceleration) in terms of the other two is stressed. Projectile motion is considered and equations are presented to calculate the maximum vertical displacement, flight time, range and optimum projection angle of a simple projectile for specified values of the three projection parameters. Deviations of the optimal angle for the sports performer from the optimal projection angle are explained. Finally, simple rotational kinematics are introduced, including vector multiplication and the calculation of the velocities and accelerations caused by rotation.

Chapters 3 and 4 both deal with the kinetics of movements in sport. Kinetics is the branch of mechanics that covers the action of forces in producing or changing motion. In Chapter 3, the basics of linear and rotational kinetics are addressed. Linear kinetics are considered, including the definition of force and its SI unit, the identification of the various external forces acting in sport, the laws of linear kinetics and related concepts such as linear momentum, and the ways in which force systems can be classified. The segmentation method for calculating the position of the whole body centre of mass of the sports performer is explained. Some important forces are considered in more detail. The ways in which friction and traction influence movements in sport and exercise are addressed, including reducing and increasing friction and traction. An appreciation is provided of the factors which govern impact, both direct and oblique, of sports objects, and the centre of percussion is introduced and related to sports objects and performers. The vitally important topic of rotational kinetics is covered, including the laws of rotational kinetics and related concepts such as angular momentum and the ways in which rotation is acquired and controlled in sports motions. The chapter concludes with a very brief introduction to spatial (three-dimensional) rotation.

In Chapter 4 the forces involved in moving through a fluid and the energetics of both linear and rotational motion are covered. This includes motion through a fluid, which always characterizes sport, and the forces which affect that motion. The principles and simple equations of fluid flow are outlined. The important forms of fluid dynamic drag are introduced, as well as the way in which differential boundary layer separation can cause a sideways force, as exemplified by cricket-ball swing. The mechanisms of lift generation on sports objects are also explained. The chapter also covers the principles of thermodynamics and their application to the energetics of the sports performer. The non-steady-flow energy equation is introduced and how this could be simplified for sport and exercise is considered. The limitations of the concept of efficiency are outlined. Finally, an understanding is provided of the models of inter- and intrasegmental energy transfers and their application to movement in sport and exercise.

Anatomical principles 1

This chapter is designed to provide an understanding of the anatomical principles which relate to movement in sport and exercise. After reading this chapter you should be able to:

- define the planes and axes of movement, and name and describe all of the principal movements in those planes;
- identify the functions of the skeleton and give examples of each type of bone;
- describe the processes of bone growth and fracture, and typical surface features of bone;
- understand the tissue structures involved in the joints of the body, the process of joint formation and the factors contributing to joint stability and mobility;
- identify the features of synovial joints and give examples of each class of these joints;
- understand the features and structure of skeletal muscles;
- classify muscles both structurally and functionally;
- describe the types and mechanics of muscle contraction;
- appreciate how tension is produced in muscle;
- understand how the total force exerted by a muscle can be resolved into components depending on the angle of pull.

1.1 Movements of the human musculoskeletal system

1.1.1 INTRODUCTION

In order to specify clearly and unambiguously the movements of the human musculoskeletal system in sport, exercise and other activities, it is necessary to define an appropriate scientific terminology. While 'bending knees' and 'raising arms' may be acceptable in everyday language, the latter is ambiguous and the former is often deemed to be scientifically unacceptable.

A description of human movement requires the definition of a reference position or posture, from which these movements are specified. Two are used (Luttgens, Deutsch and Hamilton, 1992).

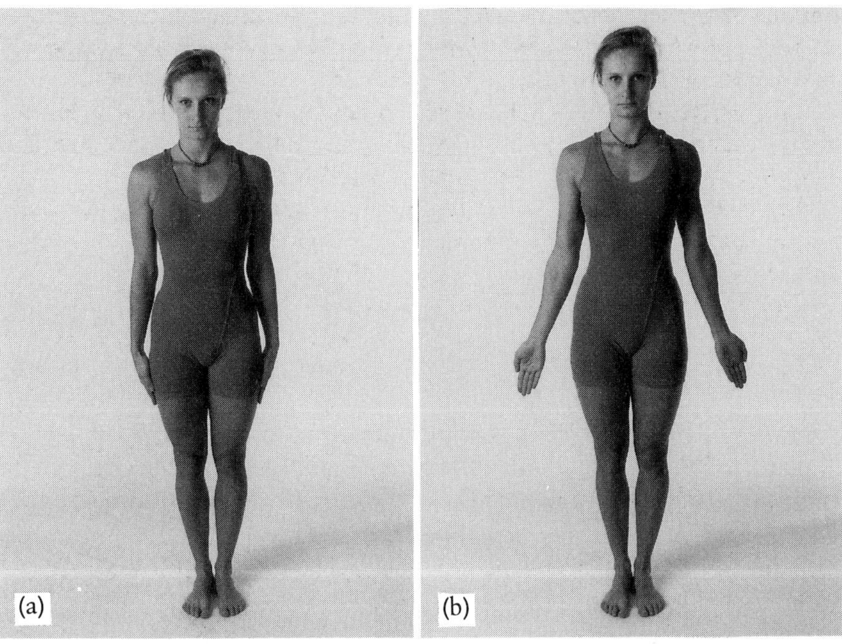

Figure 1.1 Reference positions: (a) fundamental; (b) anatomical.

Fundamental position (Figure 1.1(a)).

This is a familiar position as it closely resembles the military 'stand to attention'. Figure 1.1(a) clearly explains this. With the exception of the forearms and hands, the fundamental and anatomical reference positions are the same. The forearm is in its neutral position, neither pronated nor supinated (see below).

Anatomical position (Figure 1.1(b)). Here, the forearm has been rotated from the neutral position of Figure 1.1(a) so that the palm of the hand faces forwards. Movements of the hand and fingers are defined from this position, movements of the forearm (radioulnar joints) are defined from the fundamental position and movements at other joints can be defined from either. As the name implies, this is anatomically the basic reference position.

1.1.2 PLANES OF MOVEMENT

Various terms are used to describe the three mutually perpendicular (orthogonal) intersecting planes in which many, though not all, joint movements occur (Figure 1.2(a),(b),(c)).

Obviously, many such orthogonal systems can be described, depending on their common point of intersection. This is most conveniently defined as either the centre of the joint being studied or the centre of mass of the whole human body. In the latter case, the planes are known as cardinal planes, as depicted in Figure 1.2.

The sagittal plane is a vertical plane passing from posterior (rear) to anterior (front) dividing the body into left and right halves (Figure 1.2(a)) (it is also known as the anteroposterior plane).

The frontal plane is also vertical and passes from left to right, dividing the body into posterior and anterior halves (Figure 1.2(b)) (it is also known as the coronal plane).

The horizontal plane divides the body into superior (top) and inferior (bottom) halves (Figure 1.2(c)) (it is also known as the transverse plane).

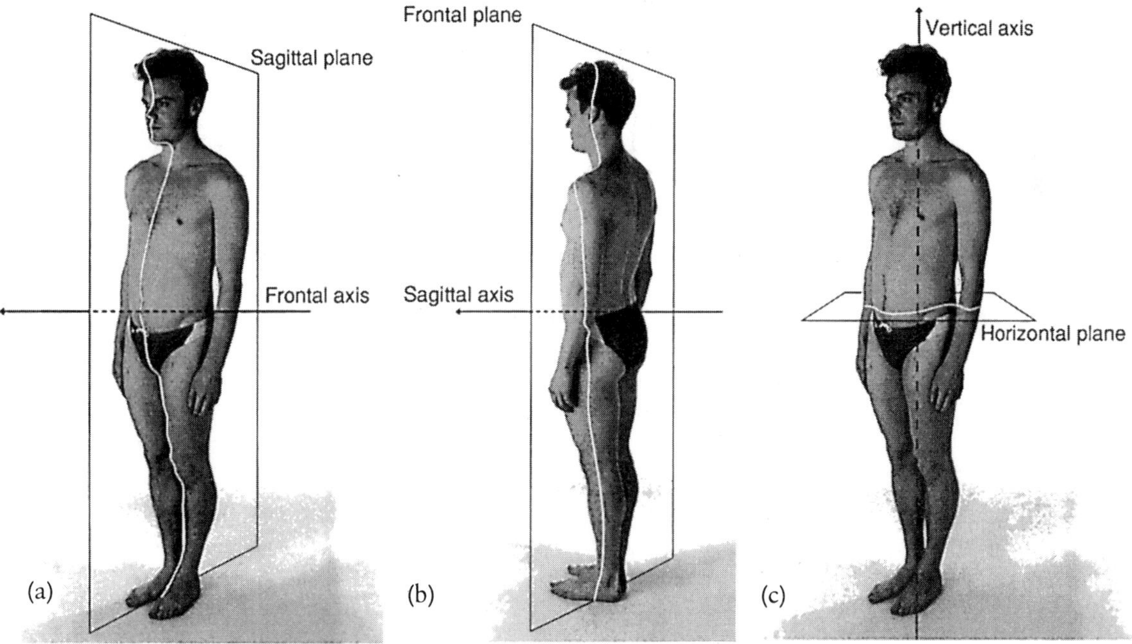

Figure 1.2 Cardinal planes and axes of movement: (a) sagittal plane and frontal axis; (b) frontal plane and sagittal axis; (c) horizontal plane and vertical axis.

1.1.3 AXES OF MOVEMENT

Movements at the joints of the musculoskeletal system are largely rotational, and take place about a line perpendicular to the plane in which they occur. This line is known as an axis of rotation. Three axes can be defined by the intersection of pairs of the above planes of movement (Figure 1.2).

The sagittal axis (Figure 1.2(b)) passes horizontally from posterior to anterior and is formed by the intersection of the sagittal and horizontal planes.

The frontal axis (Figure 1.2(a)) passes horizontally from left to right and is formed by the intersection of the frontal and horizontal planes.

The vertical (longitudinal) axis (Figure 1.2(c)) passes vertically from inferior to superior and is formed by the intersection of the sagittal and frontal planes.

1.1.4 MOVEMENTS IN THE SAGITTAL PLANE ABOUT THE FRONTAL AXIS

Flexion (Figure 1.3) is a movement away from the middle of the body in which the angle between the two body segments decreases (a 'bending' movement).

The movement is usually to the anterior (except for the knee, ankle and toe). The term **hyperflexion** is sometimes used to describe flexion of the upper arm beyond the vertical. It is cumbersome and is completely unnecessary when, as is usually the case in biomechanics, the range of movement is quantified.

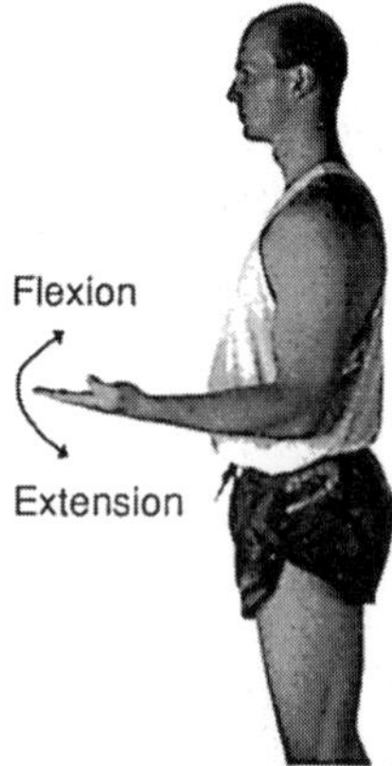

Figure 1.3 Movements in the sagittal plane about the frontal axis: flexion and extension.

Extension (Figure 1.3) is the return movement from flexion. Continuation of extension beyond the reference position is termed **hyperextension.** The return movement from a hyperextended position is often referred to as flexion. (In strict anatomical terms, the latter movement should be described as reduction of hyperextension.)

Dorsiflexion and plantar flexion are normally used to define extension (foot moving towards the anterior surface of the leg) and flexion (foot moving towards the posterior surface of the leg) of the ankle joint respectively.

1.1.5 MOVEMENTS IN THE FRONTAL PLANE ABOUT THE SAGITTAL AXIS

Abduction (Figure 1.4(a)) is a sideways movement away from the middle of the body or, for the fingers, away from the middle finger.

Radial flexion or **radial deviation** denotes the movement of the middle finger away from the middle of the body and can also be used for the other fingers. The term **hyperabduction** is sometimes used to describe abduction of the upper arm beyond the vertical.

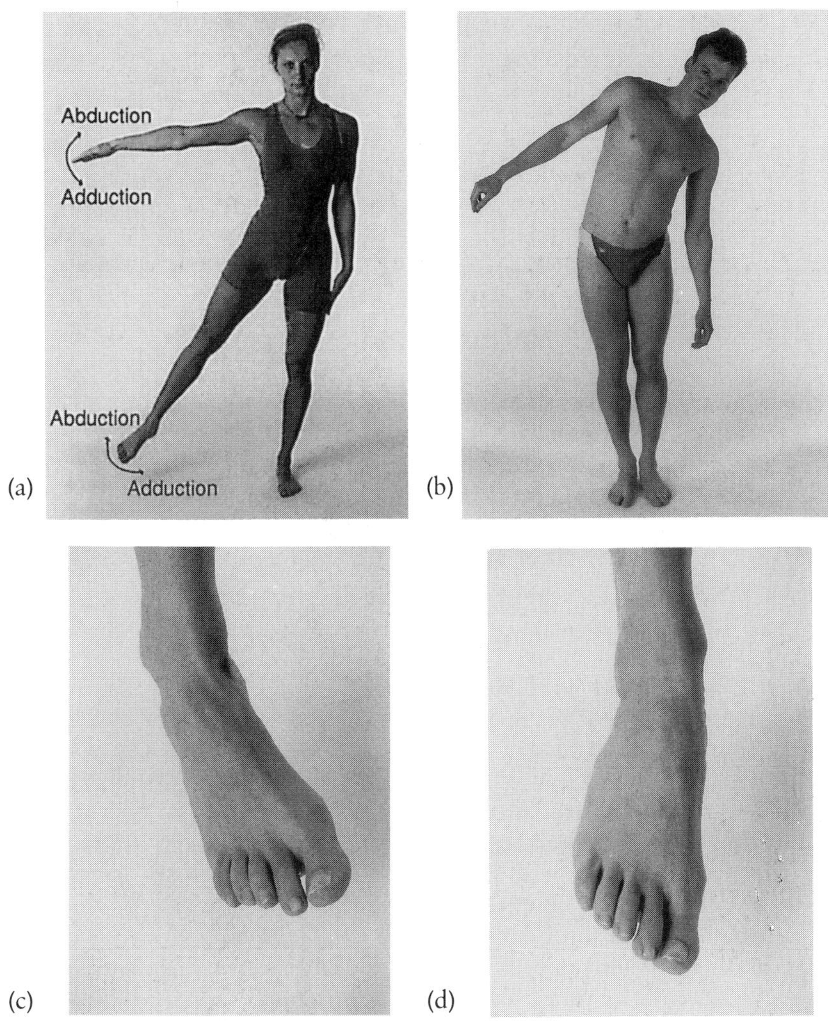

Figure 1.4 Movements in the frontal plane about the sagittal axis: (a) abduction and adduction of the arm and leg; (b) lateral flexion; (c) supination; (d) pronation.

Adduction (Figure 1.4(a)) is the return movement from abduction towards the middle of the body or, for the fingers, towards the middle finger. **Ulnar flexion** or **ulnar deviation** denotes the movement of the middle finger towards the middle of the body and can also be used for the other fingers. Continuation of adduction beyond the reference position is termed **hyperadduction.** This is only possible when combined with some sagittal plane motion. The return movement from a hyperadduction position is often called abduction. (In strict anatomical terms, the latter movement should be referred to as reduction of hyperadduction.)

Lateral flexion to the right or to the left (Figure 1.4(b)) is the term used to define the sideways bending of the trunk to the right or left and, normally, the return movement from the opposite side.

Eversion and inversion are used to refer to the raising of the lateral and medial border of the foot respectively with respect to the other border. (Note that medial and lateral denote positions respectively nearer to and further from the middle of the body.) Eversion cannot occur without the foot tending to be displaced into a toe-out (abducted) position and, likewise, inversion tends to be accompanied by adduction. The terms **supination** and **pronation of the foot** (Figure 1.4(c),(d)) are widely used in describing and evaluating running gait, and may already be familiar to you for this reason. Pronation of the foot involves a combination of eversion and abduction (Figure 1.4(d)), along with dorsiflexion of the ankle. Supination involves inversion and adduction (Figure 1.4(c)) and plantar flexion. These terms should not be confused with pronation and supination of the forearm (see below). When the foot is bearing weight, as in running, its abduction and adduction movements are restricted by friction between the shoe and the ground. Medial and lateral rotation of the lower leg are then more pronounced than in the non-weight-bearing positions of Figure 1.4(c),(d).

1.1.6 MOVEMENTS IN THE HORIZONTAL PLANE ABOUT THE VERTICAL AXIS

Lateral (outward) and medial (inward) rotation (Figure 1.5(a)) describe the movements of the leg or arm in this plane. Lateral and medial rotation of the forearm are referred to respectively as supination and pronation.

Rotation to the left and rotation to the right are the rather obvious terms for horizontal plane movements of the head, neck and trunk.

Horizontal flexion and extension (Figure 1.5(b)) define the rotation of the arm about the shoulder (or the leg about the hip), from a position of 90° abduction. Movements from any position in the horizontal plane towards the anterior are usually called horizontal flexion and those towards the posterior horizontal extension. (In strict anatomical terms, these movements are named from the 90° abducted position, with the return movements towards that position being reduction of horizontal flexion and extension respectively.)

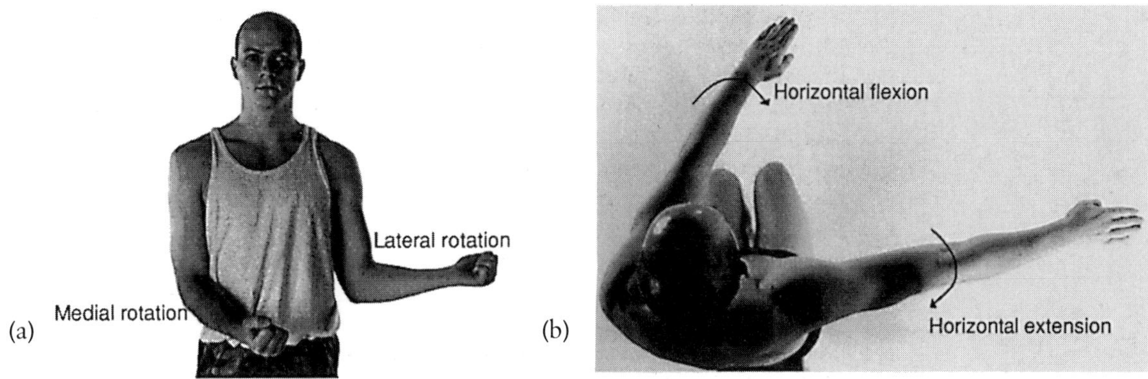

Figure 1.5 Movements in the horizontal plane about the vertical axis: (a) lateral and medial rotation; (b) horizontal flexion and extension.

1.1.7 OTHER MOVEMENTS

The movements of the thumb (Figure 1.6(a–e) appear to some to be confusingly named. **Abduction** (Figure 1.6(a)) and **adduction** are used to define movements away from and towards the palm of the hand in the sagittal plane, with **hyperadduction** (Figure 1.6(b)) as the movement beyond the starting position. **Extension** (Figure 1.6(c)) and **flexion** (Figure 1.6(d)) refer to frontal plane movements away from and towards the index finger, with **hyperflexion** (Figure 1.6(e)) as the movement beyond the starting position.

The movement of the arm or leg to describe a cone is termed **circumduction** and is a combination of flexion and extension with abduction and adduction. Several attempts have been made to define movements in other, diagonal planes but none has been adopted universally.

The movements of the shoulder girdle are shown in Figure 1.7, along with the humerus movements with which they are usually associated. **Elevation** (Figure 1.7(a)) and **depression** are upward and downward linear movements of the scapula. They are generally accompanied by some upward (Figure 1.7(b)) and downward **scapular rotation** respectively, movements approximately in the frontal plane. These rotations are defined by the turning of the distal end of the scapula (that further from the middle of the body) with respect to the proximal end (that nearer the middle of the body). **Abduction** (protraction) and **adduction** (retraction) describe the movements of the scapula away from (Figure 1.7(c)) and towards the vertebral column. These are not simply movements in the frontal plane but also have anterior and posterior components owing to the curvature of the thorax. **Posterior** (upward) and **anterior** (downward) **tilt** describe the movement of the inferior angle (lower tip) of the scapula away from (Figure 1.7(d)) or towards the thorax.

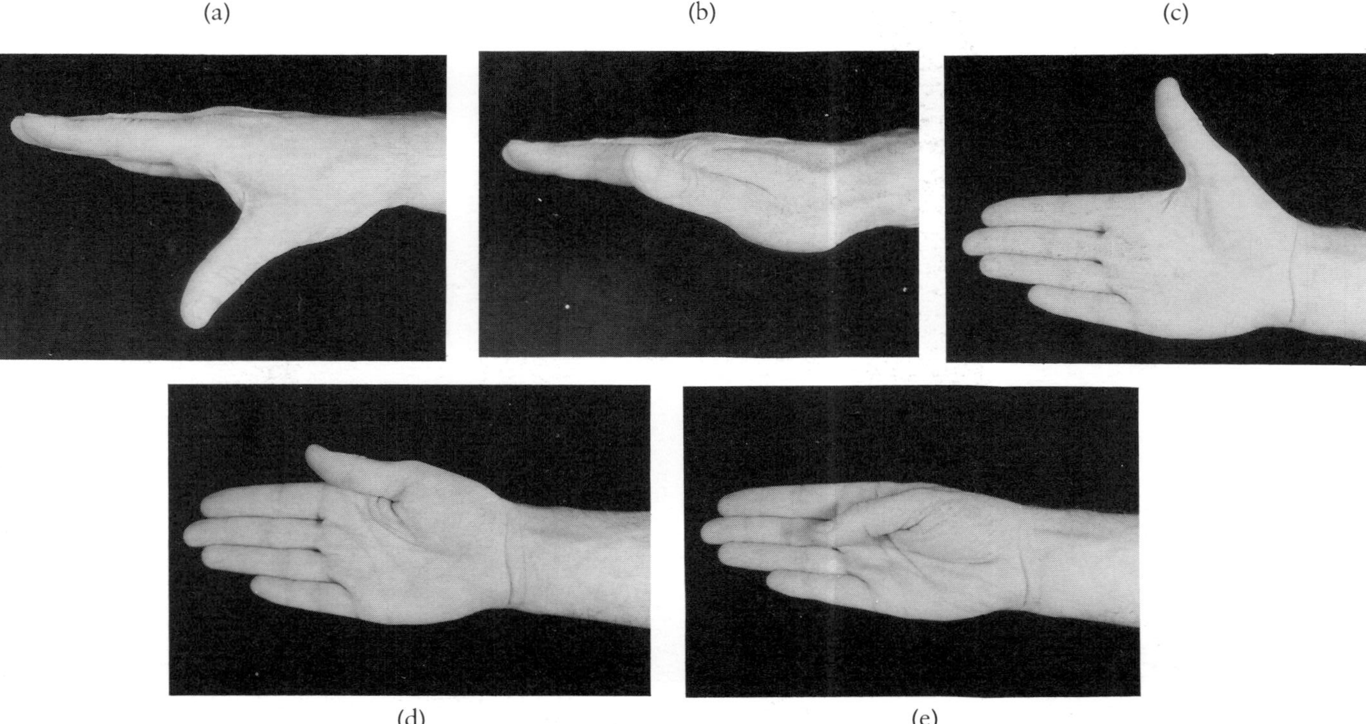

Figure 1.6 Movements of the thumb: (a) abduction; (b) hyperadduction; (c) extension; (d) flexion; (e) hyperflexion.

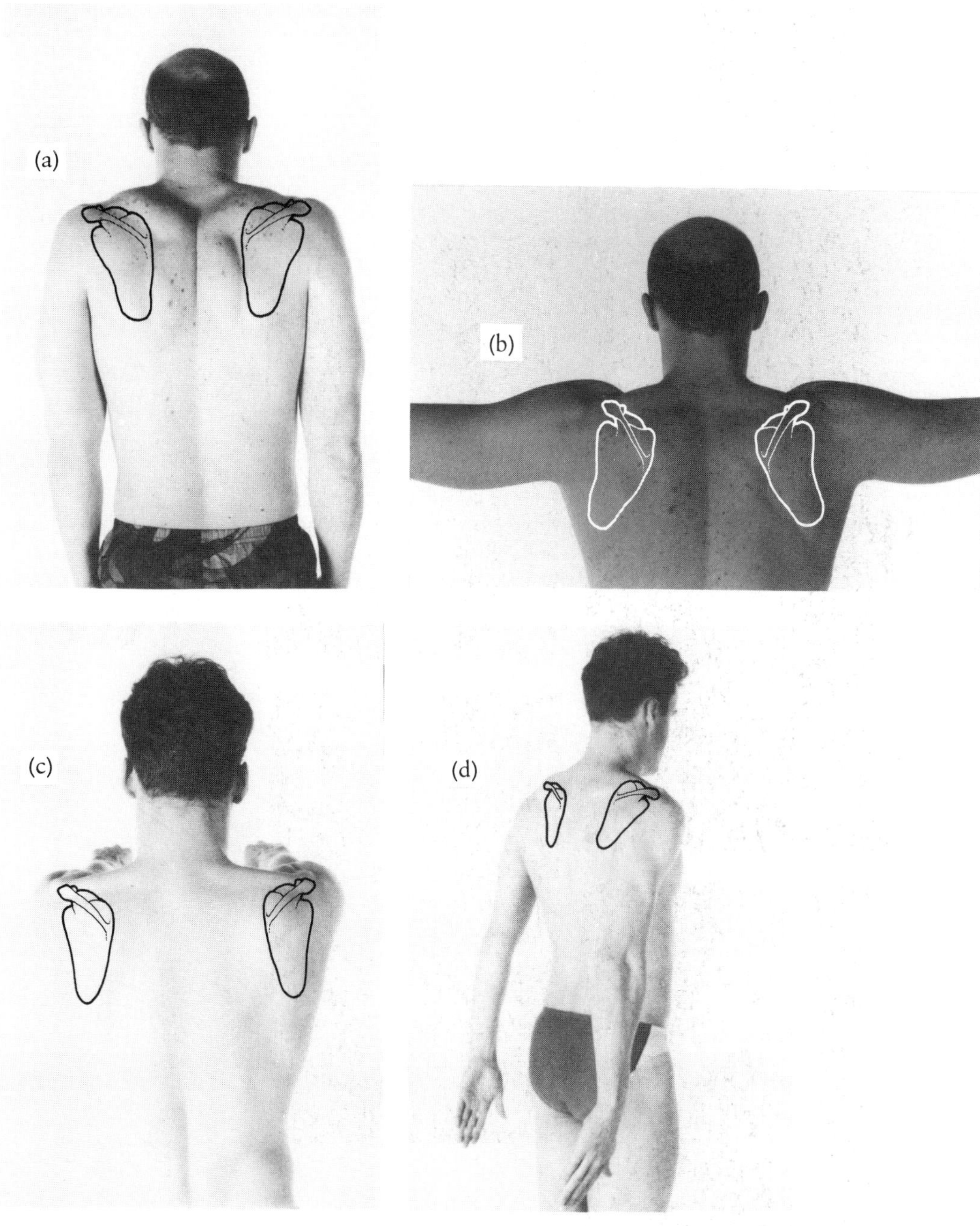

Figure 1.7 Shoulder girdle movements: (a) elevation; (b) upward rotation; (c) abduction; (d) upward tilt.

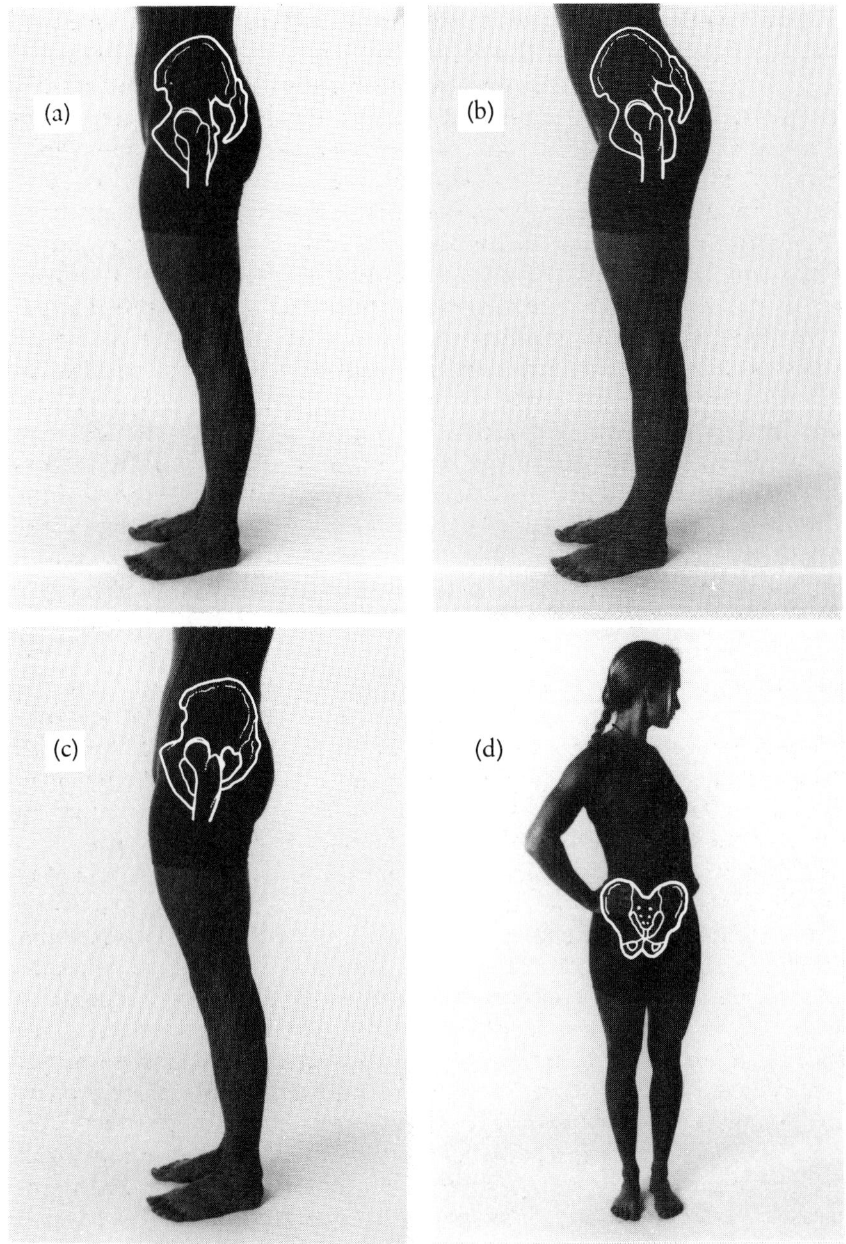

Figure 1.8 Pelvic girdle movements: (a) neutral sagittal plane position; (b) forward tilt; (c) backward tilt; (d) rotation (to the left).

Changes in the position of the pelvis are brought about by the motions at the lumbosacral joint (that between the lowest lumbar vertebra and the sacrum) and the hip joints. Movements at these joints

(Figure 1.8) permit the pelvis to tilt forwards, backwards and sideways and to rotate horizontally (Luttgens, Deutsch and Hamilton, 1992).

The pelvic tilts are due to opposite movements at the hip and sacroiliac joints. The normally rigid structure of the pelvic girdle makes independent movement of the right side in relation to the left impossible. **Forward tilt** (from Figure 1.8(a) to Figure 1.8(b)) involves increased inclination in the sagittal plane about the frontal axis. This results from lumbosacral hyperextension and, in the standing position, hip flexion. The lower part of the pelvic girdle where the pubic bones join (the symphysis pubis) turns downwards and the posterior surface of the sacrum turns upwards. **Backward tilt** (from Figure 1.8(a) to Figure 1.8(c)) involves decreased inclination in the sagittal plane about the frontal axis. This results from lumbosacral flexion (reduction of hyperextension) and, in the standing position, hip extension. The symphysis pubis moves forwards and upwards and the posterior surface of the sacrum turns somewhat downwards. **Lateral tilt** describes the movement of the pelvis in the frontal plane about the sagittal axis in such a manner that one iliac crest is lowered and the other is raised. This can be demonstrated by standing on one foot with the other slightly raised directly upwards off the ground, keeping the leg straight. The tilt is named in terms of the side of the pelvis which moves downwards. Thus in a lateral tilt of the pelvis to the left, the left iliac crest is lowered and the right is raised. This is a combination of right lateral flexion of the lumbosacral joint, abduction of the left hip and adduction of the right. **Rotation** or **lateral twist** (Figure 1.8(d)) is the rotation of the pelvis in the horizontal plane about a vertical axis. The movement is named in terms of the direction towards which the front of the pelvis turns.

Alternative names for some of the above movements will be found in the literature. Furthermore, although the vast majority of biomechanics texts define the movements of the human musculoskeletal system about the above three axes, many movements at joints with more than one axis of rotation are more complex than the above definitions suggest. It should also be noted that the planes and axes above, and the movements within and about them, are defined for the human musculoskeletal system, not in absolute space. For a swimmer, for example, frontal plane movements occur in a horizontal plane in absolute space. The three cardinal axes move with the performer and this is obviously sensible from the viewpoint of explaining movements in sport, exercise and other areas of biomechanical investigation of human motion.

1.2 The skeleton and its bones

1.2.1 INTRODUCTION

In this section, the functions of the human skeleton and the form, nature and composition of its bones will be considered. There are 206 bones in the human skeleton, of which 177 engage in voluntary movement.

The functions of the skeleton are:

- to protect vital organs such as the brain, heart and lungs;
- to provide rigidity for the body;
- to provide muscle attachments whereby the bones function as levers, allowing the muscles to move them about the joints;
- to enable the manufacture of blood cells;
- to provide a storehouse for mineral metabolism.

The skeleton (Figure 1.9) is often divided into the **axial skeleton** (skull, lower jaw, vertebrae, ribs, sternum, sacrum and coccyx), which is mainly protective, and the **appendicular skeleton** (pectoral girdle and upper extremities, pelvic girdle and lower extremities), which functions in movement.

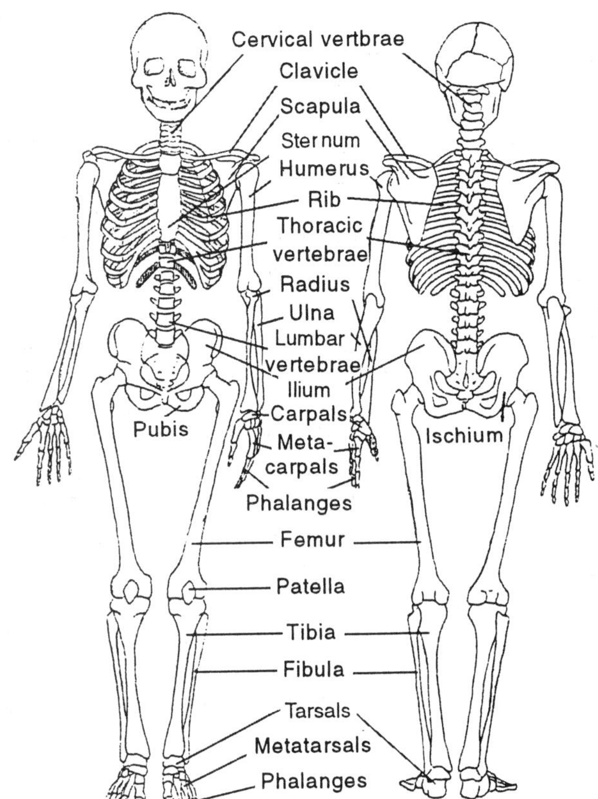

Figure 1.9 The skeleton.

1.2.2 BONE COMPOSITION AND STRUCTURE

Bone contains 25–30% water. The rest of its contents are a protein, collagen (25–30% of the bone's dry weight), and minerals (65–70% of the bone's dry weight), mostly calcium phosphate and carbonate. The collagen gives bone its resistance to tensile loading (that tending to stretch the bone along its longitudinal axis). The minerals provide resistance to compression (the opposite of tension). The minerals also give bone its properties of hardness and rigidity while the protein content contributes ductility (flexibility; ability to deform) and toughness, i.e. resistance to shock loading (Nordin and Frankel, 1989).

Bone is a highly specialized form of connective tissue. Like all such tissues it consists of a multitude of cells (bone cells are called osteocytes), which contribute only a small fraction of the bone's weight, within an organic, extracellular matrix of collagen fibrils. The tissue fluid and a gelatinous ground substance of protein polysaccharides are interspersed among the collagen fibrils. This ground substance serves to cement layers of mineralized collagen fibrils. All the organic material is impregnated with minerals, the distinguishing feature of bone. The structure of bone at the microscopic level will not be considered here (see, for example, Nordin and Frankel, 1989).

Macroscopically, there are two types of bone (osseous) tissue: **cortical** or **compact bone** and **trabecular** or **cancellous bone**. The first forms the outer shell (cortex) of a bone and has a dense structure. The second has a loose latticework structure of trabeculae or cancelli (Figure 1.10), with the interstices between the trabeculae being filled with red marrow, in which red blood cells form.

Cancellous bone tissue is arranged in concentric layers (lamellae) and its osteocytes are supplied with nutrients from blood vessels passing through the red marrow. The lamellar pattern and material composition of the two bone types are similar. Different porosity is the principal distinguishing feature, and the distinction between the two types might be considered somewhat arbitrary (Nordin and Frankel, 1989).

Biomechanically, the two types of bone should be considered as one material with a wide range of densities and porosities. It can be classified as a composite material in which the strong but brittle mineral element is embedded in a weaker but more ductile one consisting of the collagen and ground substance. Like many similar but inorganic composites (for example, glass- or carbon-reinforced fibres, important in sports equipment), this structure gives a material whose strength to weight ratio exceeds that of either of its constituents.

With the exception of the articular surfaces, bones are wholly covered with a membrane (the **periosteum**). The periosteum possesses a strong outer layer of collagenous fibres and a deep layer which produces cells called osteoblasts, which participate in the growth and repair of the bone. The periosteum also contains capillaries, which nourish the bone, and it has a nerve supply. It is sensitive to injury and is the source of much of the pain from fracture, bone bruises and shin splints. Muscles generally attach to the periosteum, not directly to the bone, and the periosteum attaches to the bone by a series of root-like processes.

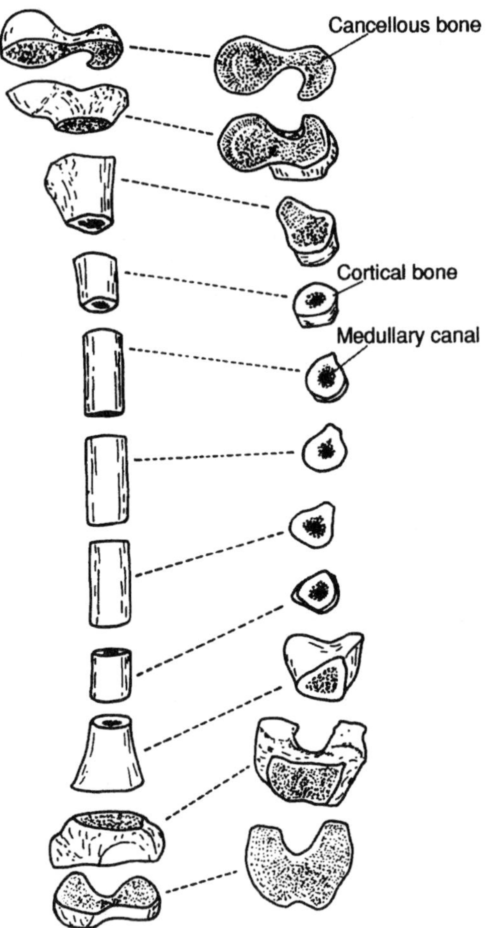

Figure 1.10 Changes in bone composition within a long bone (reproduced with kind permission from J.V. Basmajian).

1.2.3 TYPES OF BONE

Bones can be classified according to their geometrical characteristics and functional requirements, as follows.

Long bones (Figures 1.10, 1.11) occur mostly in the appendicular skeleton and function for weight-bearing and movement. They consist of a long, central shaft (body; diaphysis), the central cavity of which consists of the medullary canal, which is filled with fatty yellow marrow. At the expanded ends of the bone, the compact shell is very thin and the trabeculae are arranged along the lines of force transmission. In the same region, the periosteum is replaced by smooth, hyaline articular cartilage. This has no blood supply and is the residue of the cartilage from which the bone formed. Examples of long bones are the humerus, radius and ulna of the upper limb, the femur, tibia and fibula of the lower limb, and the phalanges (Figure 1.9).

Short bones are composed of cancellous tissue and are irregular in shape, small, chunky and roughly cubical, such as the tarsal and carpal bones.

Flat bones are basically a sandwich of richly veined spongy bone within two layers of compact bone. They serve as extensive flat areas for muscle attachment and, except for the scapulae, enclose body cavities. Examples are the sternum, ribs and ilium.

Irregular bones are adapted to special purposes and include the vertebrae, sacrum, coccyx, ischium and pubis.

Sesamoid bones form in certain tendons; the most important example is the patella.

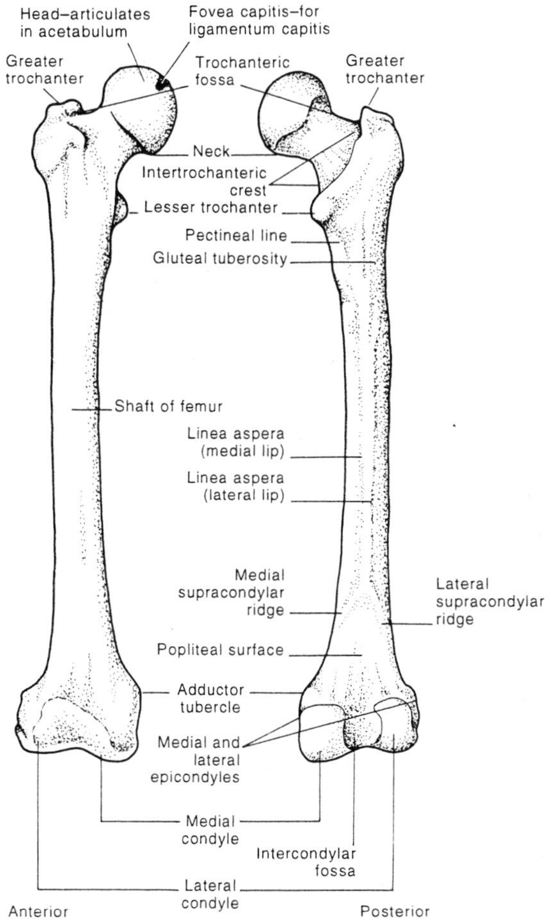

Figure 1.11 Some surface features of bone (adapted from Crouch, 1989).

1.2.4 THE BONE SURFACE

The surface of a bone is rich in markings, some examples of which are shown in Figure 1.11, which show its history.

Some of these markings can be seen at the skin surface while many others can be easily palpated (felt). These are often the anatomical landmarks used to estimate the axes of rotation of the body's joints, which are important for the biomechanical investigation of human movements in sport and exercise.

Lumps known as tuberosities or tubercles (the latter are smaller than the former) and projections (processes) show attachments of strong fibrous cords, such as tendons. Ridges and lines (linea) indicate attachments of broad sheets of fibrous tissue (aponeuroses or intermuscular septa). Grooves (furrows or sulci), holes (foramina), notches and depressions (fossae) often suggest important structures, for example grooves for tendons.

A projection from a bone was defined above as a **process**, while a rounded prominence at the end of a bone is termed a **condyle**, the projecting part of which is sometimes known as an **epicondyle**. Special names may be given to other bony prominences: the greater trochanter on the lateral aspect of the femur (Figure 1.11) and the medial malleolus on the medial aspect of the distal end of the tibia are examples.

1.2.5 BONE GROWTH

Bone grows by a process known as ossification, which involves the deposition of mineral salts in the organic, collagenous matrix. This is preceded by the laying down of the collagenous matrix, either in existing connective tissue (**intermembranous ossification**) or in hyaline cartilage (**intercartilaginous ossification**). The latter occurs for short bones and intermembranous ossification occurs, for example, for the clavicle and skull bones. Long bones, in which both mechanisms are involved, will be used to illustrate the process of bone growth (Figure 1.12).

Bone begins to grow in the embryo. In long bones (and for other intercartilaginous ossification), calcium phosphate is deposited in hyaline cartilage, a process known as calcification. Cells called osteoblasts appear around the middle of the shaft of calcified cartilage and result in the replacement of cartilage by bone. In this, the formative period of growth (5–12 weeks embryo age), ossification proceeds in all directions from the primary ossification centre or **diaphysis** (Figure 1.12(a)). Simultaneously, the bone collar ossifies intermembranously within the periosteum, forming the outside of the bone.

By birth, ossification has almost reached the ends of the cartilage and at each end a new, secondary ossification centre, or **epiphysis**, appears (Figure 1.12(b)), from which bone development now proceeds. This is the characteristic feature of the growth period, during which the bone develops to adult length. The ends of the bone are separated from the main shaft of bone by the epiphyseal plates of cartilage (Figure 1.12(c)). The bone lengthens, as new bone forms on the shaft (diaphyseal) side of the epiphyseal plate, new cartilage arises on the epiphyseal surface and the epiphyseal plate moves away from the diaphysis. The most recently formed bone at the end of the diaphysis is called the metaphysis. Diametral growth proceeds by the periosteum producing concentric rings of bone while some bone is reabsorbed in the medullary cavity (or canal), thus increasing its diameter.

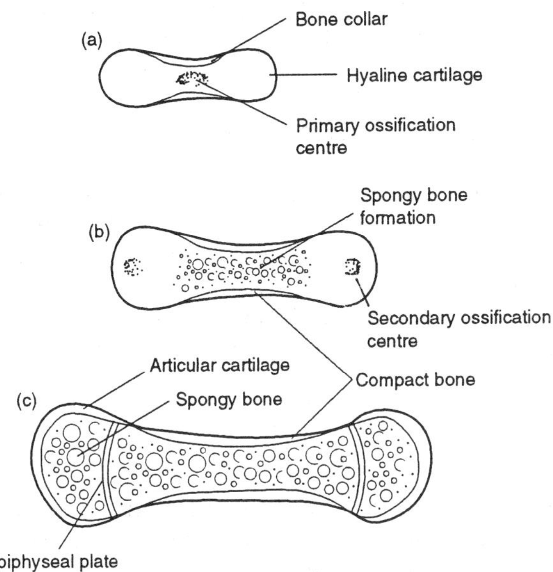

Figure 1.12 The growth of a long bone.

During the consolidation period (14th–25th year), the entire epiphyseal plate eventually ossifies and the diaphysis (shaft) and epiphysis fuse. An elevated ridge, the epiphyseal line, is left on the bone surface. Approximate ages of epiphyseal closures vary from 7–8 years for the inferior rami of the pubis and ischium to 25 years for the bones of the spinal column and thorax (Rasch and Burke, 1978). Before epiphyseal closure, the strength of the capsule and ligaments around a joint may be two to five times greater than the strength of the metaphyseal–epiphyseal junction and therefore injury is more likely at this site. Fractures or dislocations involving epiphyses may lead to cessation of bone growth. This can be used as an argument against stressful contact sports until bone maturity is achieved (about 17th–19th year). A given bone will ossify and cease growing between 1 and 3 years earlier in the female than in the male.

Bone ceases to grow or wastes (atrophies) when no stresses are applied to it, as in a period of prolonged rest (for example in a plaster cast). The reduction of circulation may be as important in this process as the absence of mechanical stress. One result is a thinning of, and reduction in, the number of trabeculae, reducing the strength of the cancellous tissue (Nordin and Frankel, 1989).

1.2.6 BONE FRACTURE

The factors affecting bone fracture, such as the mechanical properties of bone and the forces to which bones in the skeleton are subjected during sport or exercise, will not be discussed in detail here. The loading of living bone is complex because of the combined nature of the forces, or loads, applied and because of the irregular geometric structure of bones. For example, during the activities of walking and jogging, the tensile and compressive stresses along the tibia are combined with transverse shear stresses, caused by torsional loading associated with lateral and medial rotation of the tibia. Interestingly, although the tensile stresses are, as would be expected, much larger for jogging than walking, the shear stresses are greater for the latter activity (Nordin and Frankel, 1989). Most fractures are produced by such a combination of several loading modes.

After fracture, bone repair is effected by two types of cell, **osteoblasts** and **osteoclasts**. Osteoblasts deposit strands of fibrous tissue, between which ground substance is later laid down, and osteoclasts dissolve or break up damaged or dead bone. Initially, when the broken ends of the bone are brought into contact, they are surrounded by a mass of blood. This is partly absorbed and partly converted, firstly into fibrous tissue then into bone. The mass around the fractured ends is called the **callus**. This forms a thickening, for a period of months, that will gradually be smoothed away, unless the ends of the bone have not been correctly aligned. In that case, the callus will persist and form an area where high mechanical stresses may occur, rendering the region susceptible to further fracture.

1.3 The joints of the body

1.3.1 INTRODUCTION

Joints (or articulations) occur between the bones or cartilage of the skeleton. They allow free movement of the various parts of the body or more restricted movements, for example during growth or childbirth. Other tissues that may be present in the joints of the body are dense, fibrous connective tissue (including ligaments) and synovial membrane. The nature and biomechanical functions of these and other structures associated with joints will be considered in the following sections.

1.3.2 DENSE CONNECTIVE TISSUE

In these tissues, extracellular material predominates, with the cells (fibroblasts) being relatively few (about 20% of the total tissue volume). Of the remaining matrix, about 70% is water and 30% solids. Of the solid matter, over 75% is the white, unyielding protein collagen (Nordin and Frankel, 1989). The remainder is ground substance and yellow elastin. The latter is responsible for the elastic behaviour of some ligaments. These tissues are viscoelastic (their mechanical properties depend on the rate at which they are loaded), and tendons and ligaments have unique properties for the functions they fulfil. This group of tissues includes the following.

Tendons and aponeuroses. Tendons are tough cords of closely packed collagen fibres. Aponeuroses are sheet-like tendons. The functions of both are to attach muscles to bones and to transmit tensile forces between them so movement can occur. The tendon allows the muscle belly to be at the best distance from a joint without an excessive length of muscle belly (Nordin and Frankel, 1989). The ratio of tendon length to overall muscle length has been termed the tendinous ratio (MacConaill and Basmajian, 1977).

Deep fascia and intermuscular septa. The former exist between muscles, envelop individual muscles and bind muscles in groups. Intermuscular septa separate groups of muscles passing from the deep fascia to the bone.

Ligaments resemble tendons but contain varying amounts of elastic fibres (elastin) depending on their function. The structure which surrounds the joint, the joint capsule (see Figure 1.13), is ligamentous. These tissues connect bone to bone, guide the motion of the joint, enhance its stability and prevent excessive motion (Nordin and Frankel, 1989). They prohibit movements in undesired planes or limit the range of normal movement. In the former case the ligament will contain mostly white, non-elastic fibres of collagen, making it more susceptible to partial or total tears, known as sprains. In the latter case the ligament will contain a large proportion of elastic fibres of elastin, which become increasingly taut as movement progresses. Ligaments that are solely a thickening of the fibrous capsule of a joint are called **capsular ligaments.** Those which are situated near a joint but are not part of the capsule are known as **accessory** (or **collateral**) **ligaments.** Ligaments have no contractile ability, being simply passive structures having a fibre direction that indicates their restraining function. If subjected to constant tensile stress they can suffer permanent lengthening leading to joint laxity, for example pes planus ('fallen arches') in the foot.

1.3.3 CARTILAGE

Cartilage consists of cells that are widely separated within a fibrous, organic matrix. This takes two forms, the first of which may appear to be structureless (but is not) as in rigid **hyaline cartilage.** The second contains greater numbers of connective tissue fibres in the matrix, as in **fibrocartilage.** The cells of cartilage (chondrocytes) account for less than 10% of the tissue volume. The organic matrix consists of collagen fibres (10–30% net weight) within a concentrated solution of protoglycans (3–10% net weight). The remainder of the tissue (60–87%) is made up of water, inorganic salts and small quantities of matrix proteins, glycoproteins and lipids (Nordin and Frankel, 1989). Like most biological materials, articular cartilage is viscoelastic and its mechanical properties are dependent on the direction of loading (anisotropy). Hyaline cartilage is avascular, i.e. it has no blood supply, and relies on other tissues for its nutrients.

1.3.4 BURSAE, TENDON SHEATHS AND SYNOVIAL MEMBRANE

Tendon sheaths are cylindrical sacs containing two layers of connective tissue, one fixed to the tendon, the other to the surrounding structures. They possess a cavity lined with synovial membrane, which secretes synovial fluid into the cavity to reduce friction. The synovial membrane that lines the cartilage of synovial joints (see below) has a similar function in lubricating the joint. Simple synovial sacs or bursae protect other soft tissue structures.

1.3.5 FORMATION OF JOINTS

In the embryo, limbs develop from the sides of the body as buds filled with a tissue known as **mesenchyme.** In the region between two developing bones the mesenchyme persists to form the **primitive joint plate.** This can develop in several ways (Figure 1.13).

If the primitive joint plate remains largely unchanged and fibrous (Figure 1.13(a)) then the joint will be fibrous or **ligamentous** (syndesmosis). In fibrous joints the edges of the bone are joined by thin layers of the fibrous periosteum, as in the suture joints of the skull, where movement is undesirable. With age these joints disappear as the bones fuse. These are **immovable** (synarthrosis) **joints.** Ligamentous joints occur between two bones. The bones can be close together, as in the interosseus talofibular ligament, which allows only a little give, or further apart, as in the broad and flexible interosseus membrane of fibrous tissue between the ulna and radius, which permits free movement. These are not true joints and are classed as slightly movable (amphiarthrosis).

If the primitive joint plate becomes cartilage and then shows no further development, a **cartilaginous joint** (synchondrosis) results (Figure 1.13(c)). This consists of hyaline cartilage as in the joint between the sternum and first rib. If, however, a fibrous link becomes fibrocartilage, then a freer fibrocartilage joint results, as in the intervertebral discs. Cartilaginous joints are classed as slightly movable (amphiarthrosis).

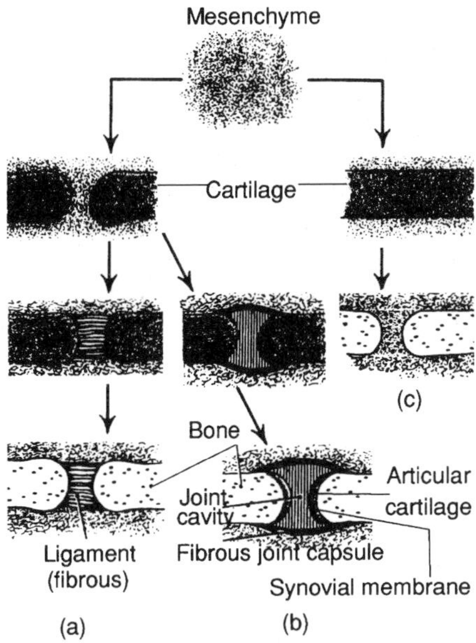

Figure 1.13 Schematic representation of the development of various joints: (a) fibrous; (b) synovial; (c) cartilaginous (adapted from Basmajian, 1976).

In the third classification of joint (**freely movable**, diarthrosis) the interior of the joint plate disappears to be replaced by a joint cavity surrounded by a sleeve of fibrous tissue, the ligamentous joint capsule, which unites the bones (Figure 1.13(b)). Friction between the bones is minimized by a residue of smooth hyaline cartilage. Although cartilage has been traditionally regarded as a shock absorber, this role is now considered unlikely. The function of the cartilage in such joints is mainly to help to reduce stresses between the contacting surfaces, by widely distributing the joint loads, and to allow movement with minimal friction (Nordin and Frankel, 1989). The inner surface of the capsule is lined with a delicate membrane (**synovial membrane**) the cells of which exude a lubricating fluid (**synovial fluid**). This fluid converts potentially compressive solid stresses into equally distributed hydrostatic ones, and nourishes the bloodless hyaline cartilage. **Synovial joints** (Figure 1.13(b)) are the most important in human movement and are characterized by having: a potential joint cavity; a lubricated articular cartilage; a capsule of fibrous (ligamentous) tissue; and a lining of synovial membrane (Basmajian, 1976). The changing relationship of the bones to each other during movement creates spaces filled by synovial folds and fringes attached to the synovial membrane. When filled with fat cells these are called fat pads. In certain synovial joints (such as the sternoclavicular and distal radioulnar), fibrocartilaginous discs occur that wholly or partially divide the joint in two and represent the persistence of the primitive joint plate.

1.3.6 CLASSIFICATION OF SYNOVIAL JOINTS

Most sports biomechanics texts use the following classification of synovial joints (Figure 1.14).

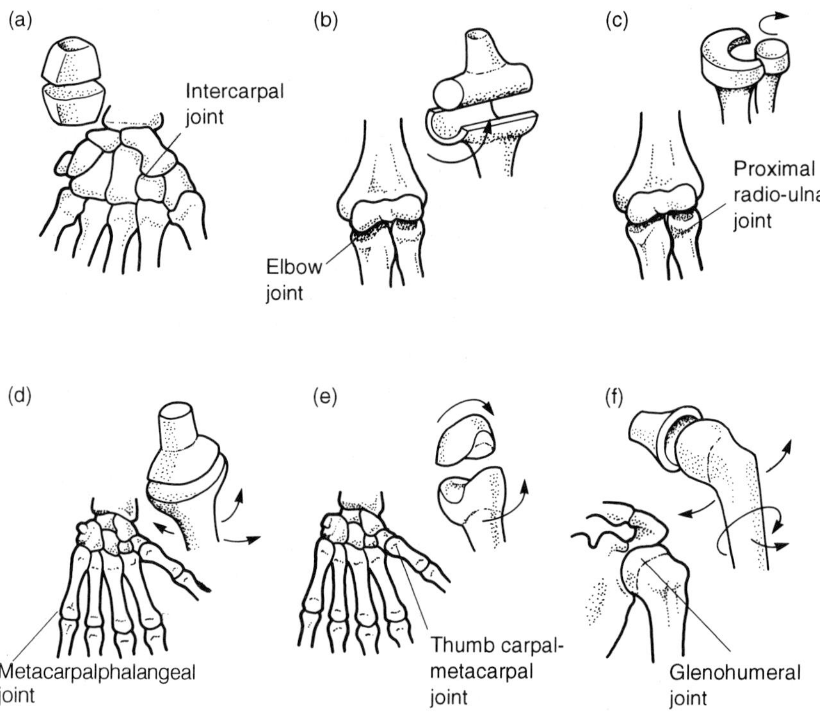

Figure 1.14 Classification of synovial joints: (a) plane joint; (b) hinge joint; (c) pivot joint; (d) condyloid joint; (e) saddle joint; (f) ball and socket joint.

Plane joints (gliding; irregular; arthrodial) are joints in which only slight, gliding movements occur. These joints have an irregular shape. Examples are the intercarpal (Figure 1.14(a)) and intertarsal joints, acromioclavicular joint and the heads and tubercles of the ribs. The joints are classed in the literature both as **non-axial** (because they glide more or less on a plane surface) and **multiaxial.** The latter term is presumably used because the surfaces are not plane but have a large radius of curvature and therefore an effective centre of rotation some considerable distance from the bone. The former description seems more useful. Although individual joint movements are small, combinations of several such joints, as in the carpal region of the hand, can result in significant motion.

Hinge joints (ginglymus) are joints in which the concave surface of one bone glides partially around the convex surface of the other.

Examples are the elbow (Figure 1.14(b)) and ankle (talocrural) joints and the interphalangeal joints. The knee is not a simple hinge joint, although it appears that way when bearing weight. These joints are uniaxial, permitting only the movements of flexion and extension.

Pivot joints (trochoid) are joints in which one bone rotates about another. This may involve the bones fitting together at one end and one rotating about a peglike pivot in the other, as in the atlanto-axial joint between the first and second cervical vertebrae. The class is also used to cover two long bones lying side by side, as in the proximal radioulnar joint (Figure 1.14(c)). These joints are uniaxial, permitting rotation about a vertical axis in a horizontal plane.

Condyloid joints are classed as biaxial joints, permitting flexion–extension and abduction–adduction (and, therefore, circumduction). The class is normally used to cover two slightly different types of joint. One of these has a spheroidal surface that articulates with a spheroidal depression as in the metacarpophalangeal or 'knuckle' joints ('condyloid' means 'knuckle-like'). These joints are potentially triaxial but lack the musculature to perform rotation about a vertical axis. The other type (sometimes classified separately as **ellipsoidal joints**) are similar in most respects except that the articulating surfaces are ellipsoidal rather than spheroidal, as in the wrist joint (Figure 1.14(d)).

Saddle joints consist of two articulating saddle-shaped surfaces, as in the thumb carpometacarpal joint (Figure 1.14(e)). They are biaxial joints, with the same movements as other biaxial joints but with greater range.

Ball and socket joints (spheroidal; enarthrosis) have the spheroidal head of one bone fitting into the cup-like cavity of the other, as in the hip and glenohumeral (shoulder) joints. The latter is shown in Figure 1.14(f). They are triaxial joints, permitting movements in all three planes.

1.3.7 JOINT STABILITY AND MOBILITY

The stability of a joint is the converse of its mobility and is an expression of the joint's resistance to displacement. It depends on:

- the shape of the bony structure, including the type of joint and the shape of the bones: this is a major stability factor in some joints, such as the elbow and hip, but of far less importance in others, for example the knee and shoulder (glenohumeral joint);
- the ligamentous arrangement, including the joint capsule (see above), which is crucial in, for example, the knee joint;
- the arrangement of fascia, tendons and aponeuroses;
- atmospheric pressure, providing it exceeds the pressure within the joint, as in the hip;

- muscular contraction – depending on the relative positions of the bones at a joint, muscles may have a force component capable of pulling the bone into the joint; this is particularly important when the bony structure is not inherently stable, as in the shoulder joint.

Joint mobility or flexibility is widely held to be desirable for sportspeople. It is usually claimed to reduce injury. While this is probably true, excessive mobility can sacrifice important stability and predispose to injury (Bloomfield, Ackland and Elliott, 1994). It is also claimed that improved mobility enhances performance and while this is impeccably logical it is not well substantiated. Mobility is highly joint-specific and is affected by body build, heredity, age, sex, fitness and exercise levels. Participants in sport and exercise are usually more flexible than non-participants owing to the use of joints through greater ranges, avoiding adaptive shortening of muscles. Widely differing values are reported in the scientific literature for normal joint ranges of movement. The discrepancies may be attributable to unreliable instrumentation, lack of standard experimental protocols and interindividual differences (Rasch and Burke, 1978).

1.4 Muscles

1.4.1 INTRODUCTION

Muscles are structures that convert chemical energy into mechanical work and heat energy. In studying sport and exercise movements biomechanically, the muscles of interest are the skeletal muscles, used for moving and for posture. This type of muscle has striated muscle fibres of alternating light and dark bands. Muscles are extensible, that is they can stretch or extend, and elastic, such that they can resume their resting length after extending. They possess excitability and contractility. **Excitability** means that they respond to a stimulus. The stimulus is chemical and the response is the generation of an electrical signal, the action potential, along the plasma membrane. **Contractility** refers to the unique ability of muscle to shorten and hence produce movement.

This section will consider the structure, function and classification of skeletal muscle. The major skeletal muscles are shown in Figure 1.15. The attachment points of skeletal muscles to bone (the origins and insertions) and their agonist functions are listed, for example, in Rasch and Burke (1978) and Marieb (1991).

1.4.2 MUSCLE STRUCTURE

Each muscle fibre is a highly specialized, complex, cylindrical cell. The cell is elongated and multinucleated, 0.01–0.1 mm in diameter and seldom more than a few centimetres long (Alexander, 1992). The cytoplasm of the cell is known as the **sarcoplasm**. This contains large amounts of stored glycogen and a protein, **myoglobin**, which is capable of binding oxygen and is unique to muscle cells. Each fibre contains a large number of smaller, parallel elements, called **myofibrils**. These run the length of the cell and are the contractile elements of skeletal muscle cells. The sarcoplasm is surrounded by a delicate plasma membrane called the **sarcolemma**.

The sarcolemma is attached at its rounded ends to the **endomysium**, the fibrous tissue surrounding each fibre. Units of 100–150 muscle fibres are bound in a coarse, collagenic fibrous tissue, the **perimysium** to form a fascicle.

The fascicles can be much longer than individual muscle fibres, for example around 25 cm long for the hamstring muscles (Alexander, 1992). Several fascicles are bound, in turn, into larger units enclosed in a covering of yet coarser, dense fibrous tissue, the **epimysium** (or deep fascia), to form muscle. The epimysium separates the muscle from its neighbours and facilitates frictionless movement.

Each muscle fibre is innervated by cranial or spinal nerves and is under voluntary control. The terminal branch of the nerve fibre ends at the neuromuscular junction or motor end-plate, which touches the muscle fibre and transmits the nerve impulse to the sarcoplasm. Each muscle is entered from the central nervous system by nerves that contain both motor and sensory fibres, the former of which are known as **motor neurons**. As each motor neuron enters the muscle, it branches into a number of terminals, each of which forms a neuromuscular junction with a single muscle fibre. The term **motor unit** is used to refer to a motor neuron and all the muscle fibres that it innervates, and these can be spread over a wide area of the muscle (Nigg and Herzog, 1994). The motor unit can be considered the fundamental functional unit of neuromuscular control (Enoka, 1994). The number of fibres per motor unit is sometimes termed the **innervation ratio**. This is small (less than 10) for muscles requiring very fine control and large (greater than 1000) for the weight-bearing muscles of the lower extremity.

The contractile cells are concentrated in the soft, fleshy central part of the muscle, called the **belly**. Towards the ends of the muscle the contractile cells finish but the perimysium and epimysium continue to the bony attachment. If distant, the connective tissue merges to a cordlike **tendon** or flat **aponeurosis**, in which the fibres are plaited in order to equally distribute the muscle force over the attachment area. If the belly continues almost to the bone, then individual sheaths of connective tissue attach on to the bone over a large area.

The strength and thickness of the muscle sheath or fascia (see above) varies with location. Protected deep muscles possess little whereas relatively exposed superficial muscles, especially near the distal end of a limb, have a thick sheath with additional protective fascia. The sheaths

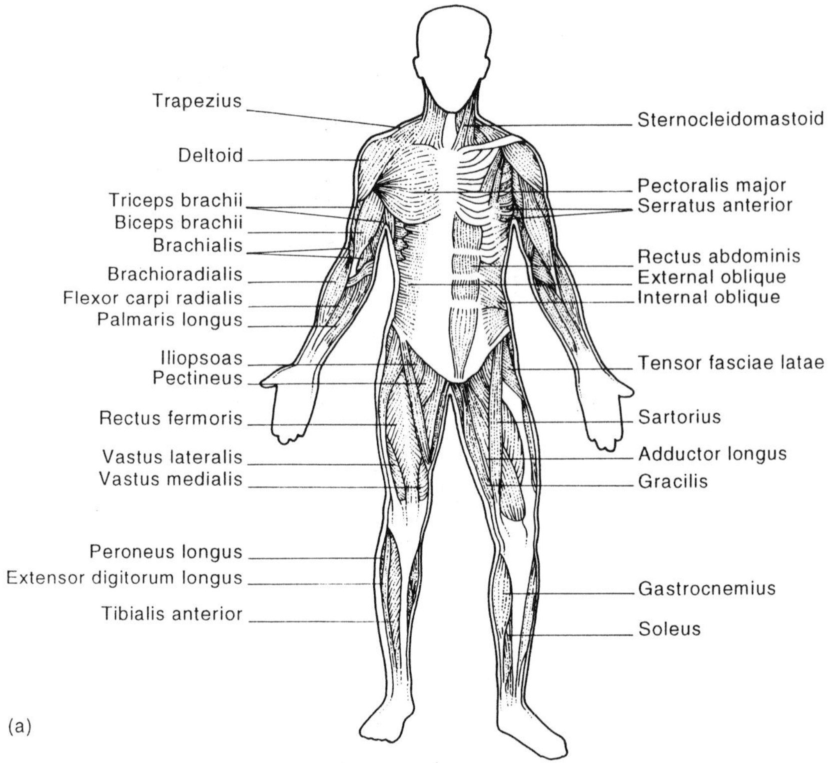

Figure 1.15 Principal skeletal muscles: (a) anterior view; (b) posterior view.

form a tough structural framework for the semi-fluid muscle tissue. They return to their original length even if stretched by 40% of their resting length. Groups of muscles are compartmentalized from others by intermuscular septa (see above), usually attached to the bone and to the deep fascia that surrounds the muscles.

The proximal attachment of a muscle, that nearer the middle of the body, is known as the **origin** and the distal attachment as the **insertion.** Skeletal muscles account for approximately 40–45% of the weight of an adult. From a sport or exercise point of view, skeletal muscles exist as about 75 pairs (Rasch and Burke, 1978). The way in which muscles develop is not relevant to the remainder of this book and can be found in standard anatomical texts, such as Crouch (1989).

1.4.3 NAMING MUSCLES

In the scientific literature, muscles are nearly always known by their Latin names. The normal rules of Latin grammar apply. The full name is musculus (often omitted or abbreviated to m. or M.) followed by

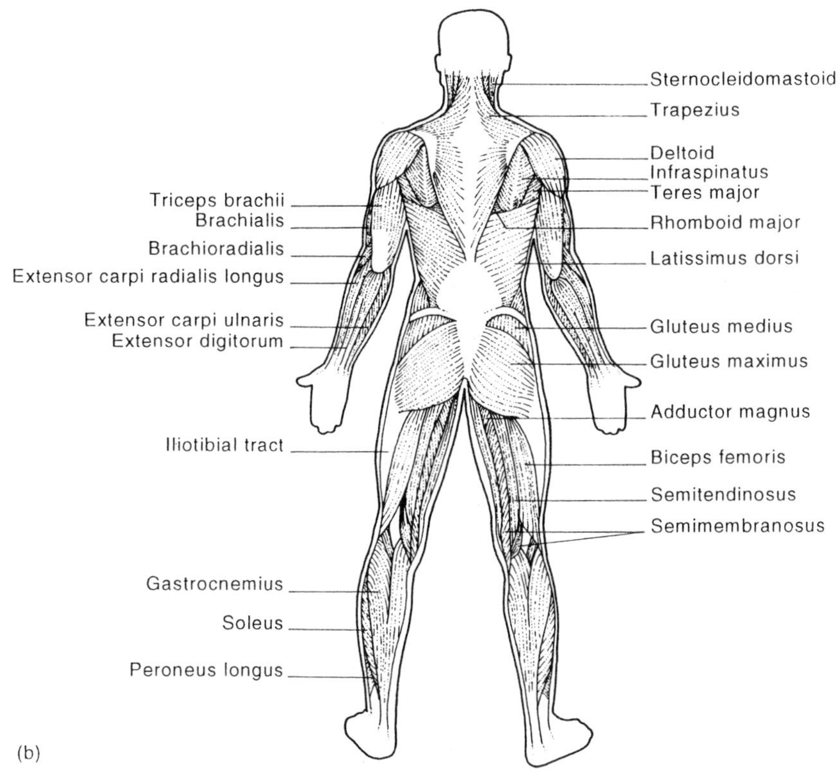

(b)

adjectives or genitives of nouns. The name may refer to role, location, size or shape of the muscle, such as:

- M. latissimus dorsi – the broadest muscle of the back;
- M. flexor digitorum profundus – the deep flexor of the fingers;
- M. trapezius – the trapezius (referring to its trapezoidal shape).

To form plurals note -us becomes -i , -is becomes -es , -or becomes -ores as in flexores. Only the first noun becomes plural. Some English names are accepted, such as the anterior deltoid; obvious English translations of the Latin names, for example the deep flexor of the fingers (see above) are not commonly encountered in most scientific literature.

Muscles are often described in terms of their role, such as the flexors of the knee and the abductors of the humerus. Most muscles have more than one role: multijoint muscles have roles at more than one joint.

1.4.4 STRUCTURAL CLASSIFICATION OF MUSCLES

The internal structure or arrangement of the muscle fascicles is related to both the force of contraction and the range of movement and therefore serves as a logical way of classifying muscles. There are two basic types each of which is further subdivided.

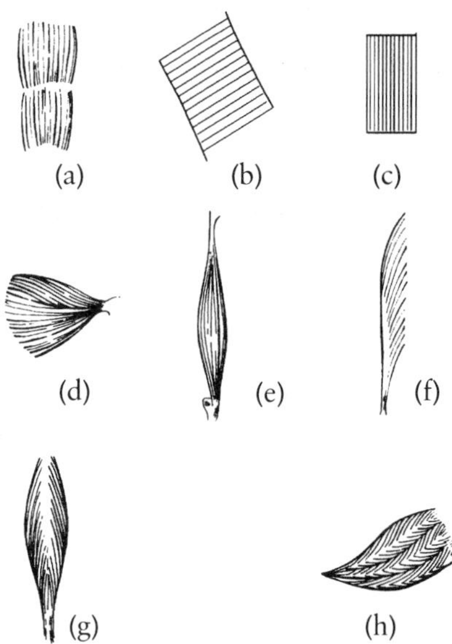

Figure 1.16 Structural classification of muscles. Collinear muscles: (a) longitudinal; (b) quadrate rhomboidal; (c) quadrate rectangular; (d) fan-shaped; (e) fusiform. Pennate muscles: (f) unipennate; (g) bipennate; (h) multipennate (adapted from Hay and Reid, 1988).

Collinear muscles (Figure 1.16(a)–(e))

The muscle fascicles are more or less parallel. A collinear muscle is capable of shortening by about one-third to one-half of its belly's length. Such muscles have a large range of movement, which is limited by the fraction of the muscle length that is tendinous. These muscles are very common in the extremities.

- **Longitudinal muscles** consist of long, straplike fascicles parallel to the long axis (Figure 1.16(a)), for example rectus abdominis, sartorius.
- **Quadrate muscles** are four-sided, usually flat, with parallel fascicles (Figure 1.16(b),(c)). They may be rhomboidal (Figure 1.16(b)), such as the rhomboideus major, or rectangular (Figure 1.16(c)) as, for example, pronator quadratus.
- **Fan-shaped (or radiate) muscles** (Figure 1.16(d)) are relatively flat with almost parallel fascicles. A good example is pectoralis major.

- **Fusiform muscles** are usually rounded, tapering at either end (Figure 1.16(e)), and include brachialis and brachioradialis.

Pennate (penniform) muscles (Figure 1.16(f)–(h))

These have shorter fascicles, which are angled away from an elongated tendon. This arrangement allows more fibres to be recruited, which provides a stronger, more powerful muscle but at the expense of range of movement and speed of the limb moved. They form 75% of the body's muscles, mostly in the large muscle groups, including the powerful muscles of the lower extremity.

- **(Uni)pennate muscles** lie to one side of the tendon, extending diagonally as a series of short, parallel fascicles (Figure 1.16(f)), for example tibialis posterior.
- **Bipennate muscles** (Figure 1.16(g)) have a long central tendon, with fascicles in diagonal pairs on either side. This group includes rectus femoris and flexor hallucis longus.
- **Multipennate muscles** converge to several tendons, giving a herringbone effect (Figure 1.16(h)), for example deltoideus.

1.4.5 TYPES OF MUSCLE CONTRACTION

The term 'muscle contraction' refers to the development of tension within the muscle. The term is a little confusing, as contraction means becoming smaller in much English usage. Some sport and exercise scientists would prefer the term 'action' to be used instead, but this has yet to be widely adopted. There are three types of muscle contraction:

- **isometric (static) contraction**, in which the muscle develops tension with no change in overall muscle length, as when holding a dumbbell stationary in a biceps curl;
- **concentric contraction**, in which the muscle shortens as tension is developed, as when the dumbbell is raised in a biceps curl;
- **eccentric contraction**, in which the muscle develops tension while it lengthens, as in the lowering movement in a biceps curl.

Both concentric and eccentric contractions can, theoretically, be isotonic (constant tension) or isokinetic (constant speed). They normally occur in a manner in which neither tension nor speed is constant.

1.4.6 GROUP ACTION OF MUSCLES

The following axioms relating to muscular contraction should be noted (Rasch and Burke, 1978).

- A muscle fibre can only develop tension within itself.
- When a fibre or muscle develops tension both ends tend to move; whether these movements occur depends on resistance and the activity of other muscles.
- When a muscle develops tension, it tends to perform all of its possible actions at all joints it crosses.

The above axioms suggest that, to bring about the movements of the human body, muscles act together rather than individually with each playing a specific role – this is one important feature of coordinated movement. In group action, muscles are classified according to their role.

Agonists are the muscles that directly bring about a movement by contracting concentrically. The group is sometimes divided into **prime movers**, which always contract to cause the movement, and **assistant movers**, which only contract against resistance or at high speed. Brachialis and biceps brachii are prime movers for elbow flexion while brachioradialis is generally considered to be an assistant mover.

Antagonists are muscles that cause the opposite movement from that of specified agonists. Their normal role in group action is to relax when the agonists contract, although there are many exceptions to this. At the elbow, the triceps brachii is antagonistic to brachialis and biceps brachii.

Stabilizers contract statically to fix one bone against the pull of the agonist(s) so that the bone at the other end can move effectively. Muscles that contract statically to prevent movements caused by gravity are called **supporting muscles**, such as the abdominal muscles in push ups.

Neutralizers prevent undesired actions of the agonists when the agonists have more than one function. They may do this by acting in pairs, as mutual neutralizers, when they enhance the required action and cancel the undesired ones. For example, the flexor carpi radialis flexes and **abducts** the wrist while the flexor carpi ulnaris flexes and **adducts** the wrist: acting together they produce only flexion. Such muscles are also called **helping synergists**. Neutralizers may also contract statically to prevent an undesired action of agonist(s) that cross more than one joint (multijoint muscles). The flexion of the fingers while the wrist remains in its anatomical position involves static action of the wrist extensors to prevent the finger flexors from flexing the wrist. Such muscles are also called **true synergists**.

1.5 The mechanics of muscular contraction

1.5.1 INTRODUCTION

This section will consider the mechanism of muscle contraction at a structural level, the mechanics of contraction and factors affecting the production of tension in muscle. For details of the microstructure of muscle fibres and the mechanism of muscular contraction see, for example, Marieb, 1991.

1.5.2 A FUNCTIONAL MODEL OF SKELETAL MUSCLE

A schematic model of skeletal muscle is often used to represent its functionally different parts. The model used (Figure 1.17) has contractile, series elastic and parallel elastic elements. The **contractile element** is made up of the myofibril protein filaments of actin and myosin and their associated coupling mechanism.

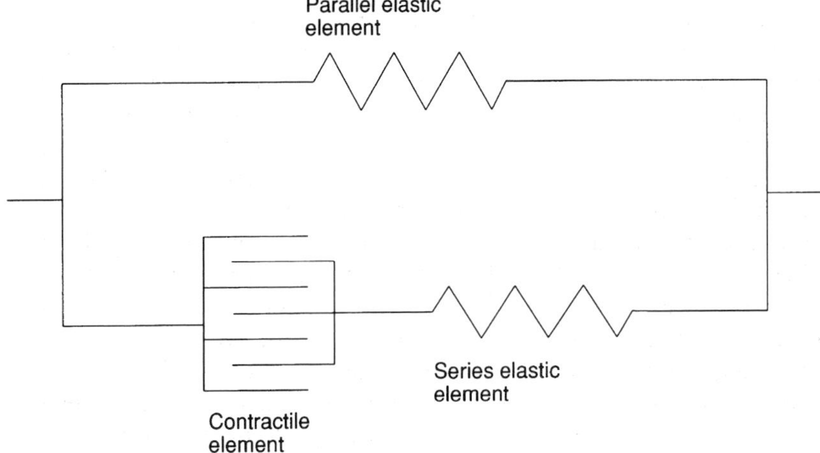

Figure 1.17 Model of skeletal muscle.

The **series elastic element** lies in series with the contractile element and transmits the tension produced by the contractile element to the attachment points of the muscle. The tendons account for by far the major part of this series elasticity with elastic structures within the muscle cells contributing the remainder (Hatze, 1981). The **parallel elastic element** comprises the epimysium, perimysium, endomysium and sarcolemma. The elastic elements store elastic energy when they are stretched and release this energy when the muscle recoils. The series elastic element is more important than the parallel one in this respect.

The elastic elements are important as they keep the muscle ready for contraction and ensure the smooth production and transmission of tension during contraction. They also ensure the return of the contractile elements to their resting position after contraction. They may also help to prevent the passive overstretching of the contractile elements when relaxed, thus reducing the risk of injury (Nordin and Frankel, 1989).

The series and parallel elastic elements are viscoelastic rather than simply elastic. This viscous property enables them to absorb energy at a rate proportional to that at which force is applied and to dissipate energy at a rate which is time-dependent. Nordin and Frankel (1989) cite the example of toe-touching, in which the initial stretch is elastic followed by a further elongation of the muscle–tendon unit owing to its viscous nature.

1.5.3 THE MECHANICS OF MUSCULAR CONTRACTION

This section will consider the gross mechanical response of a muscle to various electrical (neural) stimuli. Much of this information is derived from *in vitro*, electrical stimulation of the frog gastrocnemius. However, the assumption will be made that similar responses occur *in vivo* for the stimulation of human muscle by motor nerves. Although each muscle fibre can only respond in an all-or-none manner, a muscle contains large numbers of fibres and may contract with various force and time characteristics.

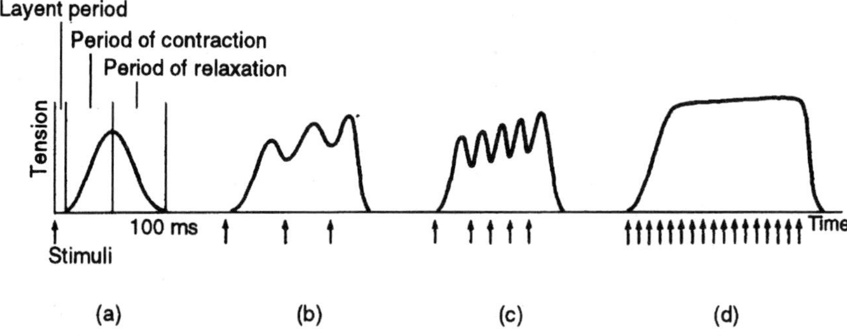

Figure 1.18 Muscle responses: (a) muscle twitch; (b) wave summation; (c) incomplete and (d) complete tetanus (adapted from Marieb, Copyright 1992 by Benjamin Cummings Publishing Company).

The muscle twitch

The muscle twitch is the mechanical response of a muscle to a single, brief, low-level stimulus. The muscle contracts and then relaxes (Figure 1.18(a)).

Following stimulation, there is a short period of a few milliseconds when excitation–contraction coupling occurs and no tension is developed. This can be considered as the time to take up the slack in the elastic elements (Nordin and Frankel, 1989) and is known as the **latency** (or

latent) period. The **contraction time** (or **period of contraction**) is the time from onset of tension development to the peak tension (Figure 1.18(a)) and lasts from 10 to 100 ms depending on the make-up of the muscle fibres. If the tension developed exceeds the resisting load, the muscle will shorten. During the following **relaxation time** (or **period of relaxation**), the tension drops to zero. If the muscle had shortened, it now returns to its initial length.

The muscle twitch is normally a laboratory rather than an *in vivo* event. In most human movement, contractions are long and smooth and variations of the degree of response are referred to as **graded responses.** These are regulated by two neural control mechanisms. The first involves increasing the rapidity of stimulation to produce wave summation (increasing the stimulation rate). The second involves recruitment of increasingly large numbers of motor units to produce multiple motor unit summation (increasing motor unit recruitment) (Hatze, 1981; Marieb, 1991).

Wave summation and tetanus

The duration of an action potential (a few milliseconds) is very short compared with the following twitch. It is, therefore, possible for a series of action potentials (an **action potential train**) to be initiated before the muscle has completely relaxed. As the muscle is still partially contracted, the tension developed as a result of the second stimulus produces greater shortening than the first (Figure 1.18(b),(c)). The contractions are summed and the phenomenon is termed **wave summation** (Nordin and Frankel, 1989). Increasing the stimulation rate will result in greater tension development as the relaxation time decreases until it eventually disappears. When this occurs, a smooth, sustained contraction results (Figure 1.18(d)) called **tetanus**; this is the normal form of muscle contraction in the body. It should be noted that prolonged tetanus leads to an eventual inability to maintain the contraction and a decline in the tension to zero. This condition is termed **muscle fatigue** (Marieb, 1991).

Multiple motor unit summation

The wide gradation of contractions within muscles is achieved mainly by the differing activities in their various motor units, in terms both of stimulation rate and in the number of units recruited. The repeated, asynchronous twitching of all the recruited motor units leads to brief summations or longer subtetanic or tetanic contractions of the whole muscle. For precise but weak movements only a few motor units will be recruited, while far greater numbers will be recruited for forceful contractions. The smallest motor units with the fewest muscle fibres (normally type I, see below) are recruited first. The larger motor units (normally type IIA, then type IIB, see below) are activated only if needed. Both wave summation and multiple motor unit summation are factors in producing the smooth movements of skeletal muscle. Multiple motor unit summation is primarily responsible for the control of the force of contraction.

Treppe

The initial contractions in the muscle are relatively weak, only half as strong as those which occur later. The tension development then has a staircase pattern called **treppe**, which is related to the suddenly increased availability of calcium ions. This, plus the increased enzymatic activity, increase in conduction velocity across the sarcolemma, and increased elasticity of the elastic elements (all consequent on the rise in muscle temperature), leads to the pattern of increasingly strong contractions with successive stimuli. The effect could be postulated as a reason for warming up before an event (Marieb, 1991), but this view is not universally accepted.

1.5.4 DEVELOPMENT OF TENSION IN A MUSCLE

The tension developed in a muscle depends upon:

- the number of fibres recruited (see above);
- the relative size of the muscle – the tension is proportional to the physiological cross-sectional area of the muscle; about 0.3 N force can be exerted per square millimetre of cross-sectional area (Alexander, 1992);
- the temperature of the muscle and muscle fatigue;
- the degree of prestretch of the muscle – a muscle which develops tension after being stretched (the stretch–shortening cycle) performs more work because of elastic energy storage and other mechanisms (see, for example, Enoka, 1994); the energy is stored mostly in the series elastic elements but also in the parallel elastic ones;
- its mechanical properties are expressed by the length–tension, force–velocity and tension–time relationships (see below).

It should be noted that there are distinct differences in the rates of contraction, tension development and susceptibility to fatigue of individual muscle fibres. The major factor here is the muscle fibre type. Thus slow-twitch, oxidative type I (red) fibres are suited for prolonged, low-intensity effort as they are fatigue-resistant because of their aerobic metabolism. However, being small and contracting only slowly, they produce little tension. Fast-twitch, glycolytic type IIB (white) fibres are of large diameter and contract quickly. They produce high tension but for only a short time, as they fatigue quickly because of their anaerobic, lactic metabolism. Fast-twitch, oxidative-glycolytic type IIA (red) fibres are intermediate between the other two, being moderately resistant to fatigue because of their mainly aerobic metabolism. These fibres are able to develop high tension but are susceptible to fatigue at high rates of activity.

The length–tension relationship

For a single muscle fibre, the tension developed when it is stimulated to contract depends on its length at that time. Maximum tension occurs at about the resting length of the fibre, because the actin and myosin filaments overlap along their entire length, maximizing the number of cross-bridges attached. If the fibre is stretched, the sarcomeres lengthen and the number of cross-bridges attached to the thin, actin filaments decreases. Conversely, the shortening of the sarcomeres to below resting length results in the overlapping of actin filaments from opposite ends of the sarcomere. In both cases the active tension is reduced. In a whole muscle contraction the passive tension caused by the stretching of the elastic elements must also be considered (Figure 1.19) in addition to the active tension developed by the contractile elements, which is similar to the active tension of an isolated fibre.

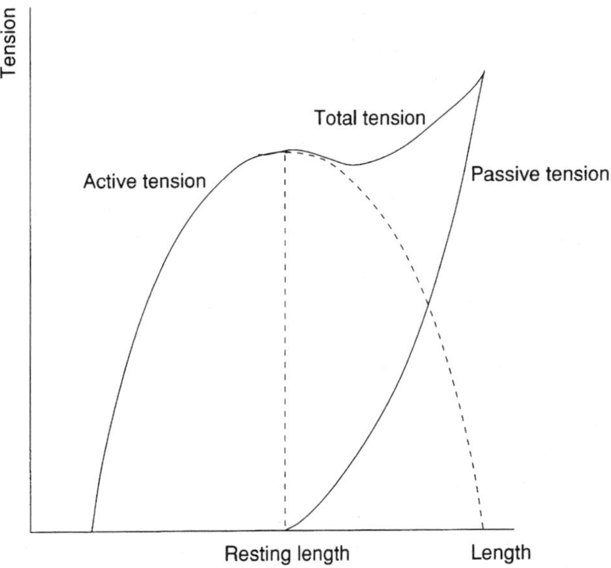

Figure 1.19 The length–tension relationship for whole muscle contraction.

The total tension (Figure 1.19) is then the sum of the active and passive tensions and depends upon the amount of connective tissue (the elastic elements) which the muscle contains. For single joint muscles, such as brachialis, the amount of stretch is not usually sufficient for the passive tension to be important. In two-joint muscles (for example, three of the four heads of the hamstrings) the extremes of the length–tension relationship may be reached, with maximal total tension being developed in the stretched muscle (as in Figure 1.19).

The force–velocity relationship

As Figure 1.20 shows, the speed at which a muscle shortens when concentrically contracting is inversely related to the external force, or load, applied. It is greatest when the applied force is zero.

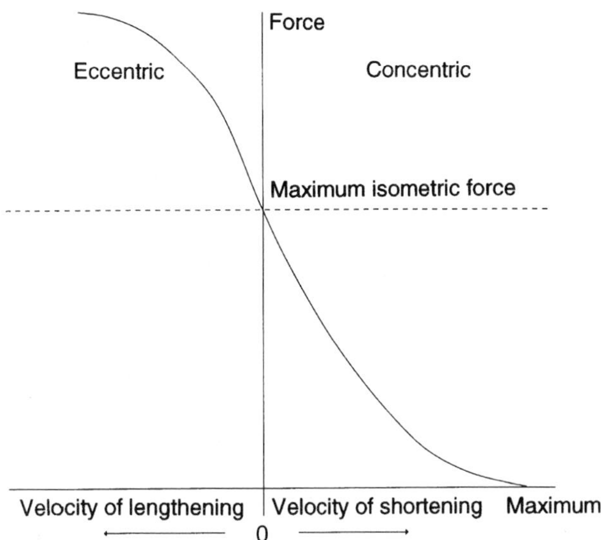

Figure 1.20 The force–velocity relationship.

When the force has increased to a level which equals the maximum force that the muscle can exert, the speed of shortening becomes zero and the muscle is contracting isometrically. The reduction of contraction speed with applied force is accompanied by an increase in the latency period and a shortening of the contraction time. Further increase of the force results in an increase in muscle length as it contracts eccentrically and then the speed of lengthening increases with the force applied (Nordin and Frankel, 1989).

The tension–time relationship

The tension developed within a muscle is proportional to the contraction time. Tension increases with the contraction time up to the peak tension (Figure 1.21).

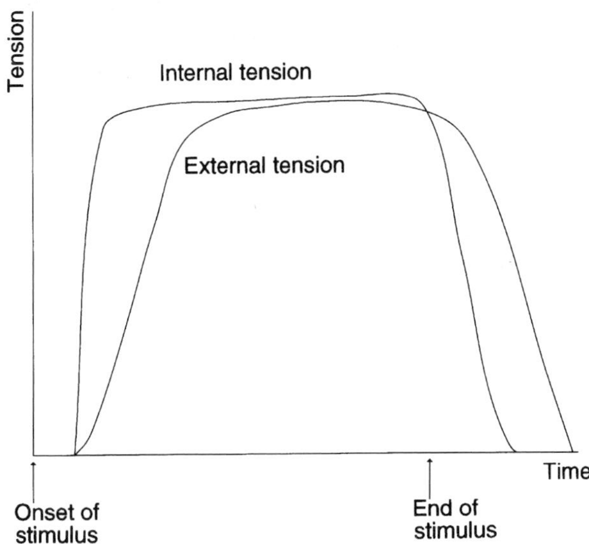

Figure 1.21 The tension–time relationship.

Slower contraction enhances tension production as time is allowed for the internal tension produced by the contractile elements (which can peak inside 10 ms) to be transmitted as external tension to the tendon through the series elastic elements, which have to be stretched (which may require 300 ms). The tension within the tendon, and that transmitted to its attachments, reaches the maximum developed in the contractile elements only if the duration of active contraction is sufficient (Nordin and Frankel, 1989). This only happens during prolonged tetanus.

1.5.5 MUSCLE FORCE COMPONENTS AND THE ANGLE OF PULL

In the general case, the overall force exerted by a muscle on a bone can be resolved into three force components, as shown in Figure 1.22 (see Chapter 2 for a detailed discussion of vector quantities such as forces).

These are:

- a component of magnitude F_g tending to rotate (or spin) the bone about its longitudinal (vertical) axis;

- a turning or rotating component, magnitude F_t. In Figure 1.23 this is shown in one plane of movement only; in reality, this muscle force component may be capable of causing movement in both the sagittal and frontal planes, for example flexor carpi ulnaris;
- a component of magnitude F_s along the longitudinal axis of the bone, which is normally stabilizing, as in Figure 1.23(c),(e). It may sometimes be dislocating, as in Figure 1.23(f). Joint stabilization is a secondary function of muscle force, which can be performed without metabolic cost by passive structures, principally ligaments.

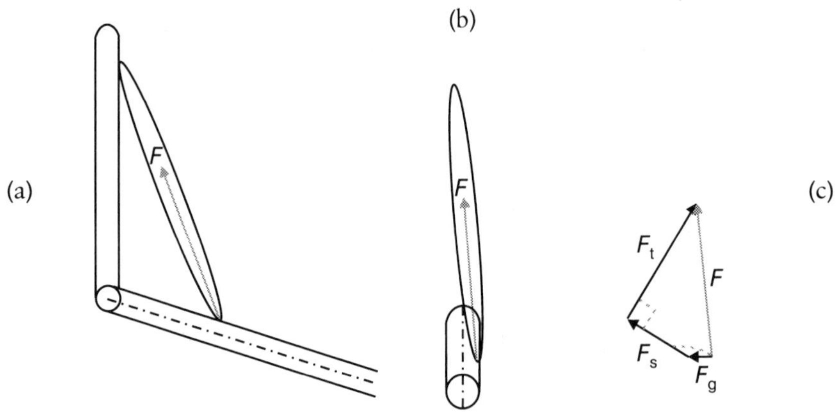

Figure 1.22 Three-dimensional muscle force components: (a) side view; (b) front view; (c) force components.

The relative importance of the last two components of muscle force is determined by the angle of pull (θ), and this is illustrated for the brachialis acting at the elbow in Figure 1.23 and assuming F_g to be zero. The optimum angle of pull is 90° when $F_t = F$ (Figure 1.23(e)) and all the muscle force is contributing to turning or rotating the bone. The angle of pull is defined as the angle between the muscle's line of pull along the tendon and the mechanical axis of the bone. It is usually small at the start of the movement (Figure 1.23(d)), increasing as the bone moves away from its reference position (Figure 1.23(e),(f)).

For collinear muscles, the ratio of the distance of the muscle's origin from a joint to the distance of its insertion from that same joint (*a*/*b* in Figure 1.23(b)) is known as the partition ratio, *p*. It is used to define two kinematically different types of muscle (MacConaill and Basmajian, 1977).

Spurt muscles are muscles for which $p > 1$, i.e. the origin is further from the joint than is the insertion. The muscle force mainly acts to turn or rotate the moving bone. Examples are biceps brachii and brachialis for forearm flexion. Such muscles are often prime movers.

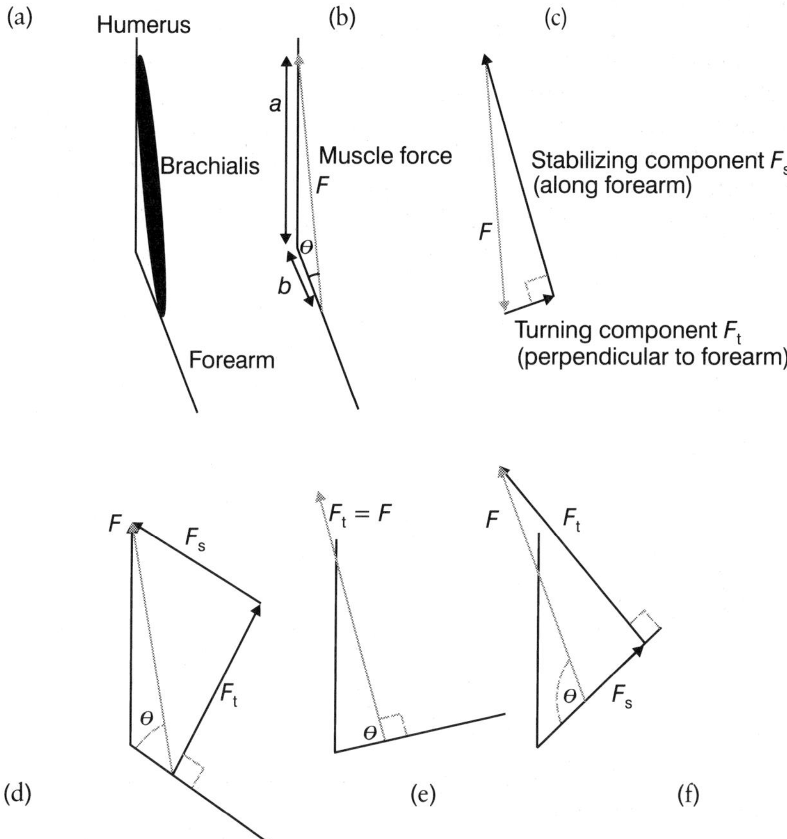

Figure 1.23 Two-dimensional muscle force components: (a) brachialis acting on forearm and humerus; (b) angle of pull; (c) stabilizing and turning components of muscle force; (d)–(f) effect of angle of pull on force components.

Shunt muscles in contrast have $p < 1$ as the origin is nearer the joint than is the insertion. Even for rotations well away from the reference position, the angle of pull is always small. The force is therefore directed mostly along the bone so that these muscles act mainly to provide a stabilizing rather than a rotating force. An example is brachioradialis for forearm flexion. They may also provide the centripetal force (largely directed along the bone towards the joint) for fast movements.

Two-joint muscles are usually spurt muscles at one joint, such as the long head of biceps brachii acting at the elbow, and shunt for the other, as for the long head of biceps brachii acting at the shoulder.

Within the human musculoskeletal system, anatomical pulleys serve to change the direction in which a force acts by applying it at a different angle and perhaps achieving an altered line of movement. One example occurs for the insertion tendon of peroneus longus, which runs down the lateral aspect of the calf and passes around the lateral malleolus of the fibula to a notch in the cuboid bone of the foot. It then turns under the foot to insert into the medial cuneiform bone and the first metatarsal bone. The pulley action of the lateral malleolus and the cuboid accomplishes two changes of direction. The result is that contraction of this muscle plantar flexes the foot about the ankle joint (among other actions). Without the pulleys, the muscle would insert anterior to the ankle on the dorsal surface of the foot and would be a dorsiflexor. An anatomical pulley may also provide a greater angle of pull, thus increasing the turning component of the muscle force. The patella achieves this effect for quadriceps femoris, improving the effectiveness of this muscle as an extensor of the knee joint.

1.6 Summary

In this chapter attention has been focused on the anatomical principles that relate to movement in sport and exercise. This included consideration of the planes and axes of movement and the principal movements in those planes. The functions of the skeleton, the types of bone, the processes of bone growth and fracture and typical surface features of bone were covered. Attention was then given to the tissue structures involved in the joints of the body, the process of joint formation, joint stability and mobility and the identification of the features and classes of synovial joints. Finally, the features and structure of skeletal muscles were considered along with the ways in which muscles are structurally and functionally classified, the types and mechanics of muscular contraction, how tension is produced in muscle and how the total force exerted by a muscle can be resolved into components depending on the angle of pull.

1.7 Exercises

Many of the following exercises can be attempted simply after reading the text. For others, a skeleton, or a good picture of one, is helpful while others, concerning movements, will require you to observe yourself or a partner. The latter observations will be easier if your experimental subject is dressed only in swimwear. If you are your own subject, a mirror will be required.

1. Make a copy of Figure 1.9, mark on it and label the cardinal planes and axes of movement.
2. List the main functions of the skeleton.
3. List all the bones labelled in Figure 1.9 and specify the class to which each belongs.
4. With your experimental subject, perform the following activities for each of these synovial joints – the shoulder, the elbow, the radioul-

nar joints (the forearm), the wrist, the thumb carpometacarpal joint, the metacarpophalangeal and interphalangeal joints of the thumb and fingers, the hip, the knee, the ankle, the subtalar joint (rearfoot), and the metatarsophalangeal and interphalangeal joints of the toes.
a) Identify its class and the number of axes of rotation (non-axial, uniaxial, biaxial or triaxial).
b) Name and demonstrate all the movements at that joint.
c) Estimate from observation the range (in degrees) of each movement.
d) Seek to identify the location of the axis of rotation for each movement and find a superficial anatomical landmark or landmarks (usually visible or palpable bony landmarks) that could be used to define this location – for example, you may find the flexion–extension axis of the shoulder to lie 5 cm inferior to the acromion process of the scapula.

5. Repeat sections b) and c) of exercise 4 for the cervical, thoracic and lumbar regions of the spine.
6. Have your experimental subject demonstrate the various movements of:
a) the shoulder joint; observe carefully the accompanying movements of the shoulder girdle (scapula and clavicle) throughout the whole range of each movement
b) the pelvis; observe the associated movements at the lumbosacral joint and the two hip joints.
7. List the structural classes of skeletal muscle and give an example of each type (try to find examples other than those in section 1.4).
8. Name the types of muscular contraction, demonstrating each in a weight-training activity, such as a biceps curl, and palpate (feel) the relevant musculature to ascertain which muscles are contracting.
9. With reference to Figure 1.15 and using your experimental subject, locate on your subject and palpate all the superficial muscles in Figure 1.15. By movements against a light resistance only, seek to identify each muscle's prime mover roles.
10. Read the appropriate chapters in the further reading (section 1.9) to gain more familiarity with the main joints of the body and their associated ligamentous, muscular and other soft tissue structures.

1.8 References

Alexander, R. McN. (1992) *Exploring Biomechanics: Animals in Motion*, Scientific American Library, New York.

Basmajian, J. V. (1976) *Primary Anatomy*, Williams & Wilkins, Baltimore, MD.

Bloomfield, J., Ackland, T. R. and Elliott, B. C. (1994) *Applied Anatomy and Biomechanics in Sport*, Blackwell Scientific, Melbourne, Victoria.

Crouch, J. E. (1989) *Functional Human Anatomy*, Lea & Febiger, Philadelphia, PA.

Enoka, R. M. (1994) *Neuromechanical Basis of Kinesiology*, Human Kinetics, Champaign, IL.

Hatze, H. (1981) *Myocybernetic Control Models of Skeletal Muscle: Characteristics and Applications*, University of South Africa Press, Pretoria.
Hay, J. G. and Reid, J. G. (1988) *Anatomy, Mechanics, and Human Motion*, Prentice-Hall, Englewood Cliffs, NJ.
Luttgens, K., Deutsch, H. and Hamilton, N. (1992) *Kinesiology: Scientific Basis of Human Motion*, W. B. Saunders, Philadelphia, PA.
Marieb, E. N. (1991) *Human Anatomy and Physiology*, Benjamin/Cummings, Redwood City, CA.
MacConaill, M. A. and Basmajian, J. V. (1977) *Muscles and Movements*, S. Krieger, Huntington, NY.
Nigg, B. M. and Herzog, W. (1994) *Biomechanics of the Musculoskeletal System*, John Wiley, Chichester.
Nordin, M. and Frankel, V. H. (1989) *Basic Biomechanics of the Musculoskeletal System*, Lea & Febiger, Philadelphia, PA.
Rasch, P. J. and Burke, R. K. (1978) *Kinesiology and Applied Anatomy*, Lea & Febiger, Philadelphia, PA.

1.9 Further reading

Many anatomy or kinesiology texts will supply useful additional information. Particularly recommended are:

Marieb, E. N. (1991) *Human Anatomy and Physiology*, Benjamin/Cummings, Redwood City, CA, chapters 6–10: a readable text with glorious colour illustrations.

Rasch, P. J. and Burke, R. K. (1978) *Kinesiology and Applied Anatomy*, Lea & Febiger, Philadelphia, PA: chapters 9–17 cover specific joints and their kinesiology, and provide summary tables of muscle action, a feature sadly omitted from more recent editions. Alas, this book is out of print but it is still available in many University libraries.

Movement (kinematic) considerations 2

This chapter is designed to provide an understanding of the kinematics of movement in sport and exercise. Kinematics is the branch of mechanics that deals with the geometry of movement without reference to the forces that cause the movement. After reading this chapter you should be able to:

- define linear, rotational and general motion;
- understand the physical model appropriate to each type of motion, and appreciate the uses and limitations of each model;
- identify the differences between vectors and scalars, and perform vector addition, subtraction and resolution graphically;
- understand the importance of linear kinematics in sport and exercise;
- appreciate the importance of being able to obtain velocities and accelerations from measured displacements and velocities and displacements from measured accelerations;
- interpret graphical presentations of one linear kinematic variable (displacement, velocity or acceleration) in terms of the other two;
- calculate the maximum vertical displacement, flight time, range and optimum projection angle of a simple projectile for specified values of the three projection parameters;
- understand simple rotational kinematics, including vector multiplication and the calculation of the velocities and accelerations due to rotation.

2.1 Fundamentals of kinematics

2.1.1 INTRODUCTION

'Mathematics is indispensable if kinesiology is to be more than a set of statements learnt by rote, if it is to be understood in such a way that it can both explain what we know about movements and muscles and also form the basis of further fruitful enquiry' (MacConaill and Basmajian, 1969).

Much of sports biomechanics is concerned with the study and evaluation of sports techniques – how sporting skills are performed. Obviously a physical and anatomical description and consideration of such techniques is vital. However, in order to be able to compare the observed technique with others it is necessary to quantify important characteristics of the technique. This requires mathematical analysis. The increased development of computer control of the main data collection equipment in sports biomechanics has increased the importance of mathematics. Fortunately, it has also lessened the need for repetitive and tedious calculations. For these reasons, an understanding of the mathematical concepts that underpin biomechanics is essential. These include aspects of calculus and vector algebra. This chapter introduces these concepts and, where necessary, basic mathematical equations, but does not include extensive mathematical development of the topics covered.

2.1.2 TYPES OF MOTION AND APPROPRIATE MODELS

Motion in sport can be linear, rotational or, more generally, a combination of both. Each type of motion is associated with a physical model of the object or sports performer being studied. The behaviour of this physical model can be represented by mathematical equations, which constitute a mathematical model of the object or performer. These mathematical models can then be used to investigate how the object or performer moves.

Linear motion and the centre of mass (point) model

Linear motion, or translation, is motion in which all parts of the body travel the same distance in the same time in the same direction. If two points on the body are joined by a straight line then, in successive positions, this straight line will remain parallel to its initial orientation. Such motion may be rectilinear, as in the case of a hang glider in level flight (Figure 2.1(a)), or curvilinear (Figure 2.1(b)).

These correspond respectively to one-dimensional and two-dimensional linear motion. It is important to note that, although in curvilinear motion the object follows a curved path, no rotation of the object is involved (Figure 2.1(b)). The overall linear motion of the object can be specified by the motion of a single point on the object. This point (which has the same mass as the object) is known as the **centre of mass.** The mass of the object is evenly distributed about this point. The motion of the centre of mass totally specifies the linear motion of the object, even if the object's overall motion is more general.

Figure 2.1 Translational motion: (a) rectilinear; (b) curvilinear.

Rotation and the rigid body model

Rotation, or angular motion, is motion in which all parts of the body travel through the same angle in the same time in the same direction about the axis of rotation. The movement of a body segment about a joint, as in Figure 2.2, is motion of this type.

An object that retains its geometrical shape, for example a cricket bat, can be studied in this manner and is known as a **rigid body**. Human body segments (such as the thigh and forearm) are often considered to be close approximations to rigid bodies. The complete human body is not a rigid body, but there are sports techniques (as in diving and gymnastics) where a body position is held temporarily (for example a tuck or pike). During that time interval the performer will behave as if he or she was a rigid body and the model is then termed a **quasi-rigid body.**

General motion and the multisegment model

Most human movements involve combinations of rotation and translation, for example the cross-country skier shown in Figure 2.3.

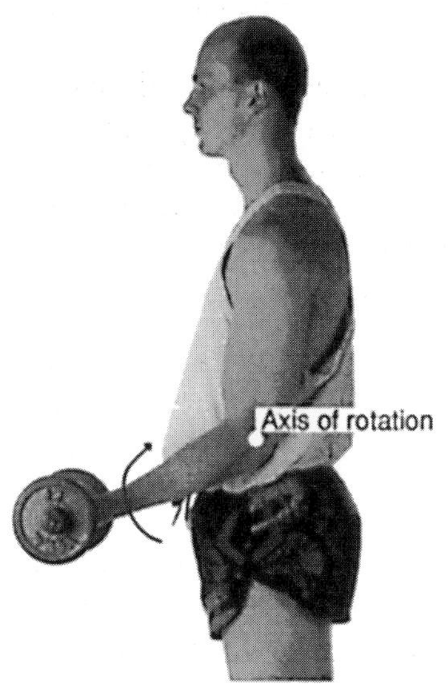

Figure 2.2 Angular motion.

Figure 2.3 General motion (adapted from Field and Walker, 1987).

Such complex motions can be represented by a link or **multisegment model** of the performer, where each of the body segments is treated as a rigid body. The rigid bodies are connected at the joints between body segments.

2.1.3 INTRODUCTION TO VECTORS AND SCALARS

Scalar quantities have a magnitude (which may be positive or negative in most cases) but no directional quality. Vectors have both a magnitude and a direction, and the behaviour of vectors cannot be considered simply in terms of their magnitude. Mass, volume, temperature and energy are scalar quantities. Many of the kinematic variables that are important for the biomechanical understanding and evaluation of movement in sport and exercise are vector quantities. These include linear and rotational position, displacement, velocity and acceleration. Many of the kinetic variables considered in Chapter 3, such as momentum, force and torque, are also vectors. The magnitude (the scalar part of the vector) of a kinematic variable usually has a name that is different from that of the vector.

Vector quantities, which are shown in bold type, can be represented graphically by a straight line (arrow) in the direction of the vector, having a length proportional to the magnitude of the vector (Figure 2.4(a)).

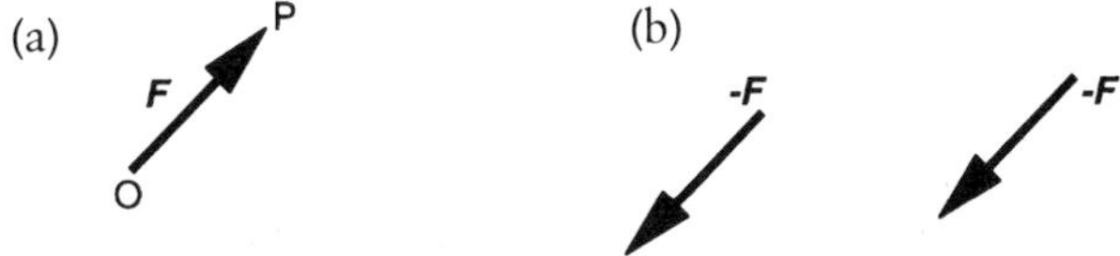

Figure 2.4 Vectors: (a) representation; (b) movement.

The vector can be designated ***F*** (magnitude *F*) or ***OP*** (magnitude *OP*). Vectors can often be moved in space parallel to their original position, as in Figure 2.4(b) (although some caution is necessary for force vectors, see Chapter 3). This allows easy graphical addition and subtraction (Figure 2.7(a)–(f)). Note that the vector in Figure 2.4(a) is equal in magnitude but opposite in direction to that in Figure 2.4(b); i.e. if the direction of a vector is changed by 180°, the sign of the vector changes.

Position, displacement and distance

The location of an object in space can be specified by its **position vector,** which may be linear, rotational or a combination of both. The position of the point P in Figure 2.5(a) can be specified in terms of its position vector (***r***).

The position of the body segment in Figure 2.5(c) can be specified in terms of the position vector of its centre of mass (***r***) and the angular position ($\boldsymbol{\theta}$) of its longitudinal axis with respect to a horizontal line drawn from its centre of mass towards the right hand side.

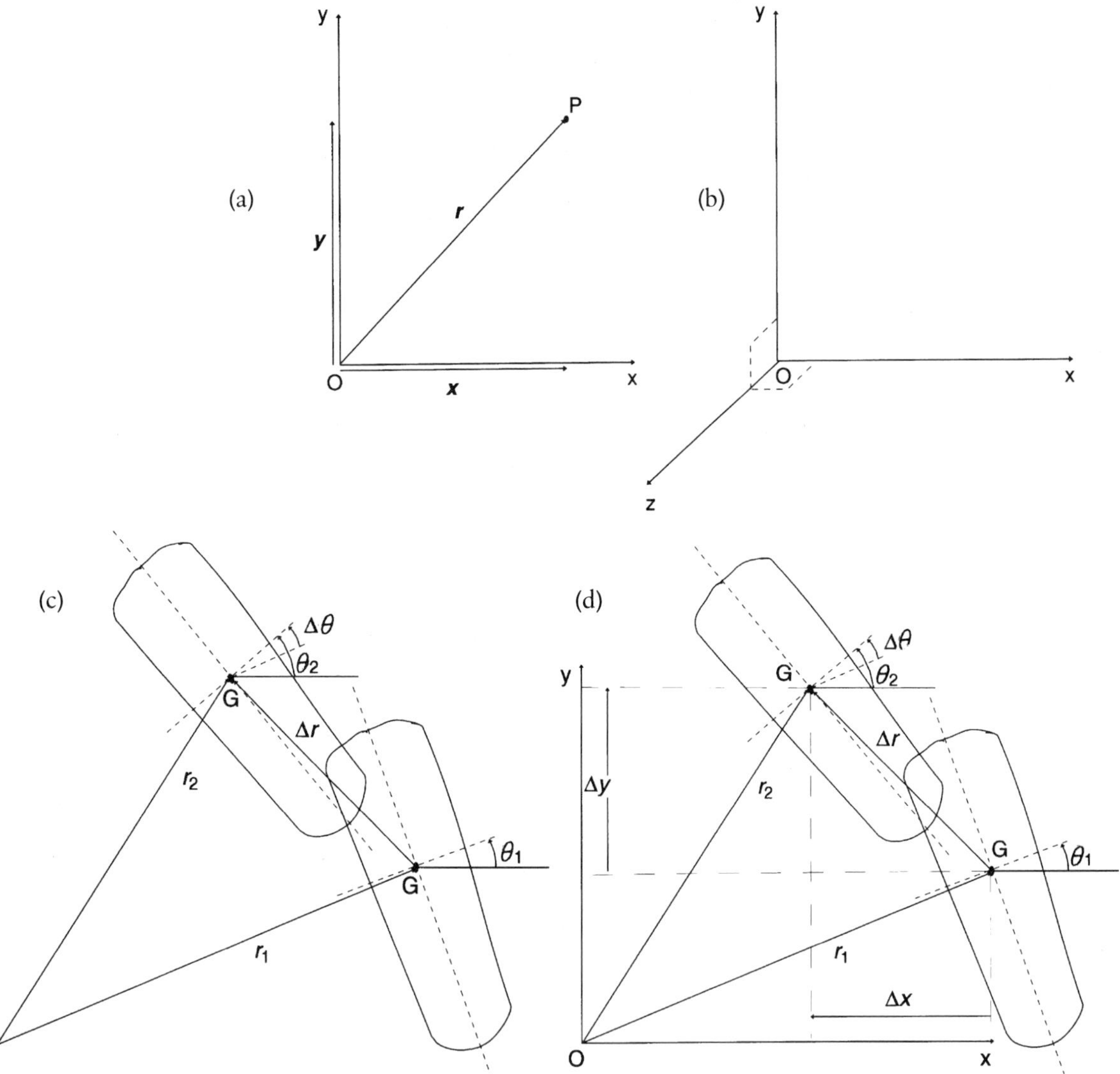

Figure 2.5 (a) Position vector of a point and its two-dimensional Cartesian coordinates; (b) three-dimensional Cartesian coordinate system; (c) position and displacement vectors of a body segment; (d) components of position and displacement vectors.

When an object changes its linear position, the change in the position vector, regardless of the path taken, is known as the **linear displacement** (or simply **displacement**), Δr. In practical terms, it is convenient to represent linear position and displacement in terms of some frame of reference or coordinate system. The most commonly used is the Cartesian coordinate system of Figure 2.5(a),(b), which show the two-dimensional and three-dimensional cases respectively. Then the position (and displacement) of a point, such as the centre of mass of a jumper, can be specified in terms of perpendicular components (see below) along the coordinate axes, as in Figure 2.5(a). For rotational motion, the change in angular position is known as **angular displacement**, (**Δθ**). The linear and angular positions and displacements of a rigid body segment are illustrated in Figure 2.5(c). The components of the linear displacements of this body segment for a two-dimensional coordinate system are shown in Figure 2.5(d). Linear displacements towards the positive axis direction are positive and those towards the negative axis direction are negative. In the two-dimensional case (Figure 2.5(a),(d)) displacements to the right or vertically upwards are usually considered to be positive and those to the left or vertically downwards to be negative. In Figure 2.5(d), the x displacement is negative and the y displacement is positive. Angular displacements are considered to be positive if they are counterclockwise when viewed from the positive direction of the axis of rotation (which would be the z axis in Figure 2.5(d)) towards the origin (O) of the coordinate system. The angular displacement of the rigid body in Figure 2.5(d) is, therefore, positive.

The linear distance travelled is a scalar quantity and is the length between the start and finish positions measured along the path taken. Measuring simply from the start to the finish of a 400 m sprint, there is no net change in the position vector as the start and finish coincide. There is, therefore, no net displacement of the sprinter, but the distance travelled is 400 m.

The SI unit for linear position and displacement is the metre (m). For angular position and displacement the SI unit is the radian (rad). However, degrees (°) are still widely used. A radian is equal to 180/π degrees, where π is approximately equal to 3.1417, so that one radian is approximately 57.3°.

Velocity and speed

Velocity is the rate of change of position (or displacement) with respect to time and **speed** is the rate of change of distance with respect to time. Velocity is a vector and speed a scalar quantity. The rate of change of angular displacement with respect to time is known as **angular velocity**. The SI unit for velocity and speed is $m{\cdot}s^{-1}$, and for angular velocity $rad{\cdot}s^{-1}$ (although °/s are still used). The abbreviation $m{\cdot}s^{-1}$ means the same as m/s, to which it is preferred in most modern biomechanical usage.

It is sometimes useful to define an average speed as the distance covered divided by the time taken. For a 400 m runner covering the distance in the 1995 UK men's record time of 44.47 s, the average speed would be 400/44.47 = 8.99 $m{\cdot}s^{-1}$.

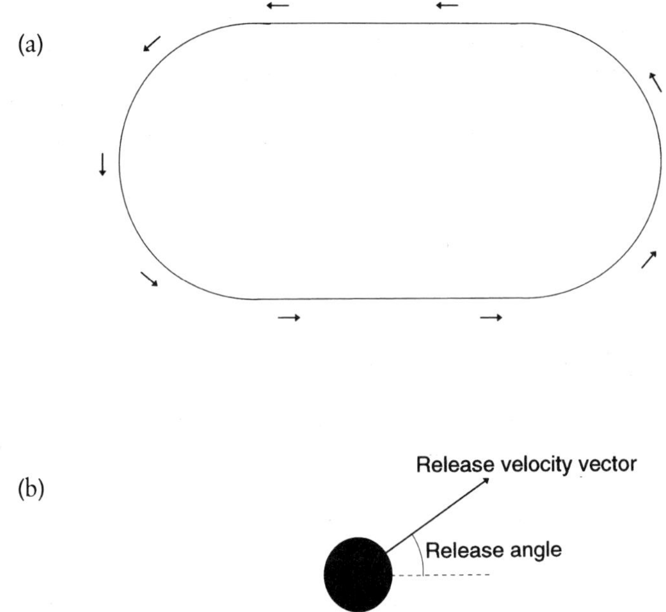

Figure 2.6 Velocity vectors: (a) 400 m running; (b) shot at release.

The mean velocity in this example would be zero, showing that the usefulness of a mean velocity is limited when velocity changes with time. A change in velocity would happen in this example even if the runner ran the race at a constant speed, because the direction of the velocity vector would change around the two bends and between the two straights (Figure 2.6(a)). In such cases, instantaneous velocities, that is velocities at specific times, are far more useful. These are represented by the vector arrows in Figure 2.6(a). Other examples of instantaneous velocities are that at take-off in a vaulting exercise (the take-off velocity) and at the instant of release of a shot (the release velocity, as in Figure 2.6(b)). The magnitudes of such instantaneous velocities are, of course, called instantaneous speeds, as in the take-off speed of a gymnast vaulting and the release speed of a shot. The direction of the velocity vector in these cases can be given by the take-off and release angles (Figure 2.6(b)) respectively. The conventions for positive and negative linear and angular velocities are similar to those for displacement (see above).

Acceleration

The name **acceleration** is used for the vector, the rate of change of velocity with respect to time, and the scalar, the rate of change of speed with respect to time. The rate of change of angular velocity with respect to time is known as **angular acceleration**. The conventions for positive and negative linear and angular accelerations are similar to those for angular displacement (see above). A positive acceleration vector indicates that an object has a positive velocity the magnitude of which is increasing or a negative velocity the magnitude of which is decreasing. The opposite applies to a negative acceleration vector. The SI unit for acceleration is $m \cdot s^{-2}$ (the same as m/s^2) and for angular acceleration $rad \cdot s^{-2}$ (although $°/s^2$ are still used).

Vector addition and subtraction

When two or more vector quantities are added together the process is called vector composition. Most vector quantities, including force, can be treated in this way. The single vector resulting from vector composition is known as the **resultant vector** or simply the **resultant**. In Figure 2.7 the vectors added are shown in black and the resultants in grey and the graphical solutions for vector addition are shown between vertical lines.

The composition of two or more vectors all having the same direction results in a single vector. This has the same direction as the original vectors and a magnitude equal to the sum of the magnitudes of the vectors being added (Figure 2.7(a)). If vectors directed in exactly opposite directions are added, the resultant has the direction of the longer vector and a magnitude that is equal to the difference in the magnitudes of the two original vectors (Figure 2.7(b)).

When the vectors to be added lie in the same plane but not in the same or opposite directions, the resultant can be found using the vector triangle approach. The tail of the second vector is placed on the tip of the first vector. The resultant is then drawn from the tail of the first vector to the tip of the second (Figure 2.7(c)). An alternative approach is to use a vector parallelogram (Figure 2.7(d)). The vector triangle is more useful as it easily generalizes to the vector polygon (Figure 2.7(e)).

Although graphical addition of vectors is very easy for two-dimensional problems, the same is not true for three-dimensional ones. Component addition of vectors, using trigonometry, can be more easily generalized to the three-dimensional case. This technique is introduced for a two-dimensional example in section 2.7.

The subtraction of one vector from another can be tackled graphically simply by treating the problem as one of addition (Figure 2.7(f)), i.e. $\boldsymbol{J} = \boldsymbol{F} - \boldsymbol{G}$ is the same as $\boldsymbol{J} = \boldsymbol{F} + (-\boldsymbol{G})$. Vector subtraction is also often used to find the relative motion between two objects. The principle of relative motion is very important in human movement. It is used to deal with the relationship that exists between a moving object (A) and a second moving object, a moving frame of reference (coordinate system) or a moving fluid (B). The principle will be illustrated here using the example of velocities (which is its most useful biomechanical form) but it applies in an identical manner to displacements and accelerations. The rule is that the velocity of A relative to B ($\boldsymbol{v}_{AB}$) is equal to the velocity of A ($\boldsymbol{v}_A$) minus the velocity of B ($\boldsymbol{v}_B$), as in Figure 2.8(a).

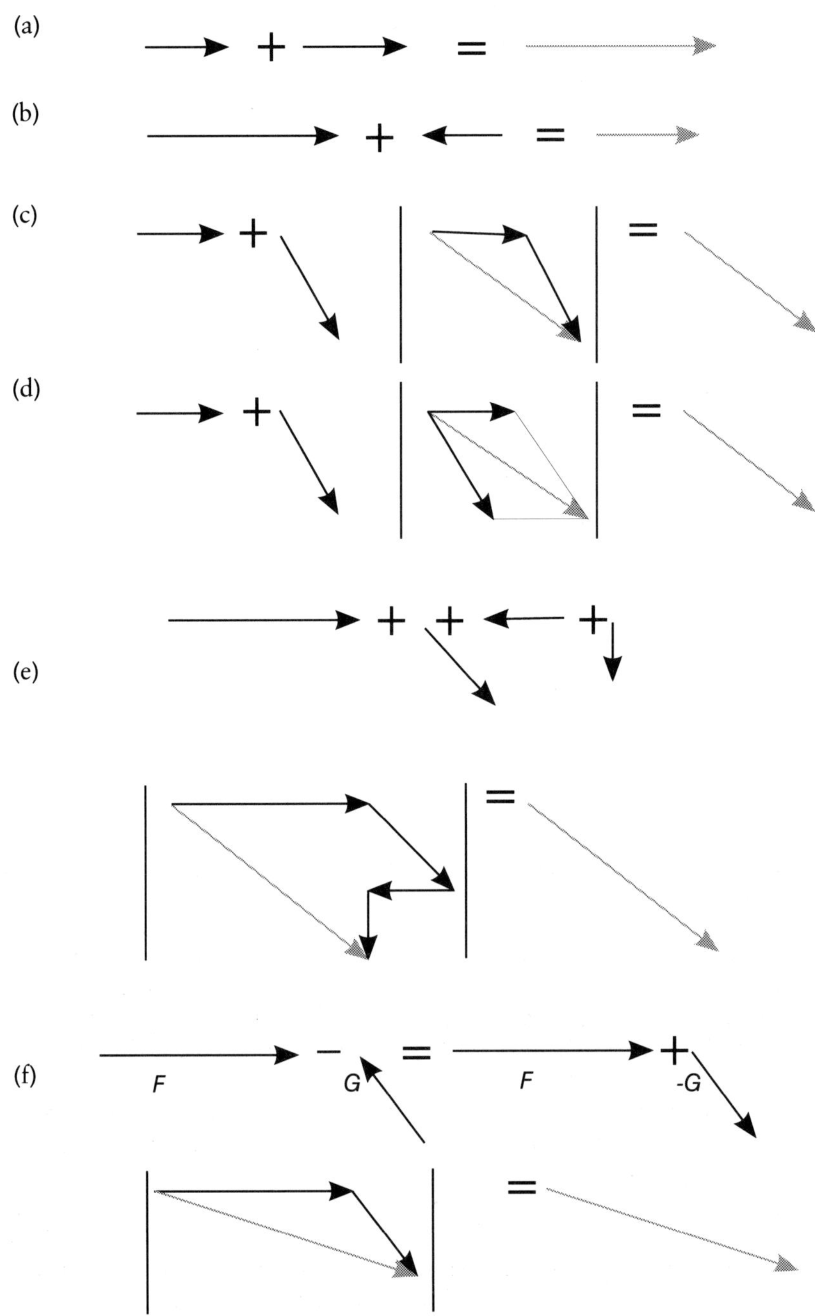

Figure 2.7 Vector composition (addition): (a) with same direction; (b) in opposite directions; (c) using vector triangle; (d) using vector parallelogram; (e) using vector polygon; (f) vector subtraction.

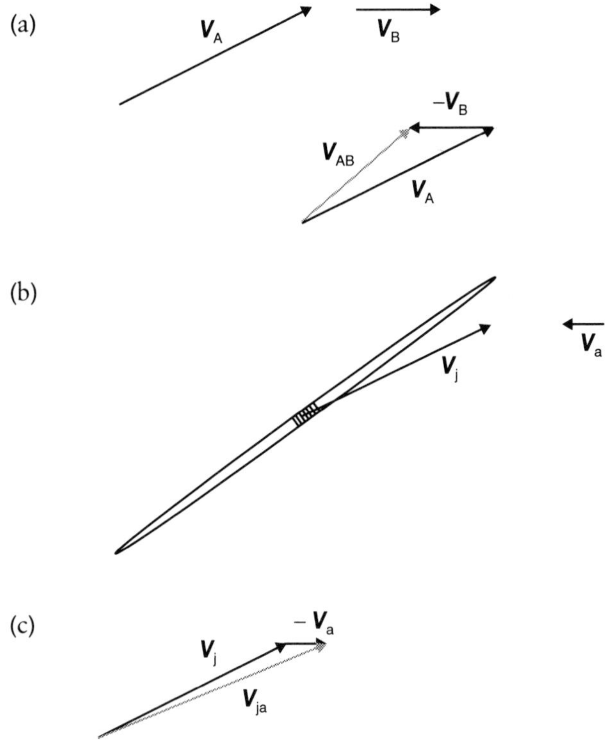

Figure 2.8 Relative motion: (a) general; (b) javelin and head wind; (c) relative velocity by vector subtraction.

In javelin throwing, the aerodynamic forces acting on the javelin depend (among other things) on the velocity of the javelin relative to the air. This is not the same as its velocity relative to the earth if there is a wind blowing, as for the example of Figure 2.8(c), where the throw is made into a head wind.

The velocity of the javelin relative to the air ($\boldsymbol{v}_{ja}$) equals the velocity of the javelin relative to the ground ($\boldsymbol{v}_j$) minus the velocity of the air relative to the ground ($\boldsymbol{v}_a$), i.e.

$$\boldsymbol{v}_{ja} = \boldsymbol{v}_j - \boldsymbol{v}_a \qquad (2.1)$$

The velocity of the javelin relative to air is found using the rules of vector subtraction (see above). This may be done graphically, as in Figure 2.8(b), and the direction and magnitude of the relative velocity can be found by use of a protractor and rule. These values can be found more accurately by simple trigonometry, in a similar way to the example in section 2.7.

Vector resolution

Determining the perpendicular components of a vector quantity is often useful in biomechanics, as in Figure 2.5(c) for displacements. Other examples include the normal and frictional components of ground reaction force, and the horizontal and vertical components of a projectile's velocity vector. It is therefore sometimes necessary to resolve a single vector into two or three perpendicular components – a process known as vector resolution. This is essentially the reverse of vector composition and can be achieved using a vector parallelogram or vector triangle. This is illustrated by the examples of Figure 2.9(a),(b), where the grey arrows are the original vector and the black ones are its components. The vector parallelogram is more usual as the components then have a common origin (Figure 2.9(a)).

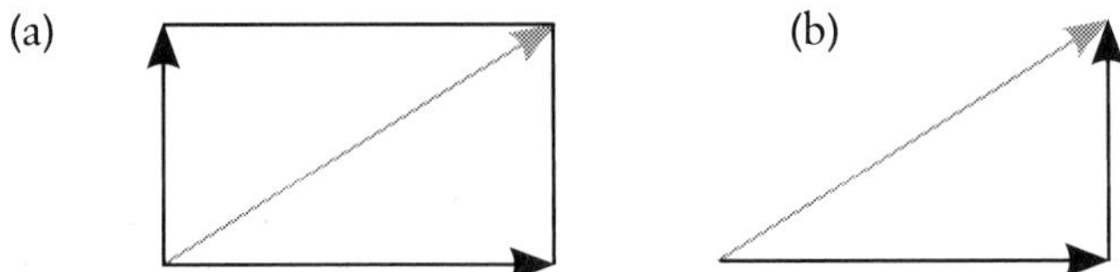

Figure 2.9 Vector resolution: (a) using vector parallelogram; (b) using vector triangle.

2.2 Further exploration of linear kinematics

2.2.1 OBTAINING VELOCITIES AND ACCELERATIONS FROM DISPLACEMENTS

A sprint coach might wish to obtain details of the speed of one of their sprinters, running a 100 m race. Specifically, the coach might wish to know the magnitude of the maximum horizontal velocity achieved by the sprinter, the distance covered before this was achieved and whether the sprinter slowed down before the tape. This problem will be treated throughout as a rectilinear movement, ignoring any vertical or lateral displacements of the sprinter. Hence, the term 'speed' will be used instead of 'magnitude of the horizontal velocity'.

From the distance run (100 m) and the time taken, say 10 s, the coach could calculate the average or mean speed of the sprinter for the race. The average speed ($\bar{v}$; the bar above the v indicates an average value) equals the distance covered divided by the time taken: $\bar{v}$ = 100 m/10 s = 10 m·s^{-1}. Although this information was very easily obtained, it is not very useful. This can be seen from the graphical presentation of Figure 2.10(a).

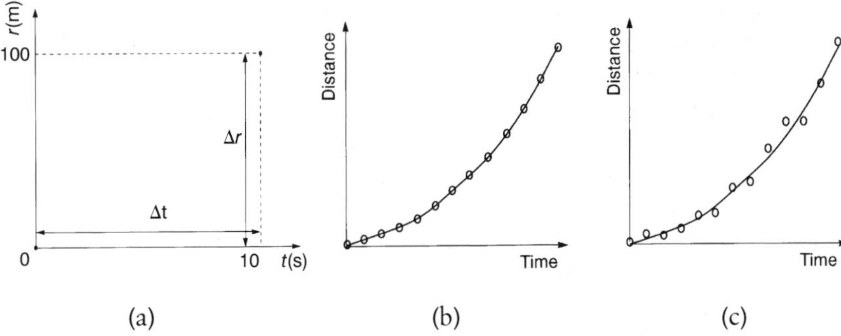

Figure 2.10 Distance–time curves for sprinter: (a) preliminary data; (b) idealized data; (c) data with errors.

Having only two experimental data points (t = 0, r = 0) (t = 10 s, r = 100 m) it is clearly not possible to plot a graph of the distance travelled by the runner against the time taken. This distance–time curve would be the first step towards obtaining the information required. It is obviously not a straight line joining the two points, as this would imply a constant running speed throughout the race.

The coach might seek to overcome this lack of data by having 10 people record the times taken by the sprinter to reach 10, 20, 30 metres (and so on) using stop watches. This might lead to the data of columns 1 and 2 of Table 2.1.

The coach could then calculate the runner's mean speed during each 10 m interval. This can be demonstrated by constructing the difference columns (3 and 4) in Table 2.1. In these, Δr refers to the change in the distance covered (r) in time interval Δt. Δ (upper case Greek letter delta) refers to a 'finite change', or 'finite difference', in the appropriate variable. The word 'finite', in this context, might be loosely considered to mean 'easily measured'. Then, by using $\bar{v} = \Delta r/\Delta t$, the coach could find the mean speed for each interval. From column 5 of Table 2.1, it appears that the runner reaches a peak speed of 12.5 m·s^{-1} after between 30 and 40 metres and maintains this speed until 90 metres or beyond. By increasing the number of data samples, which reduces the time interval Δt between measurements (the **sampling interval**), the coach has gained information. However, there still remains an uncertainty about the distance covered before peak speed is reached and when the speed begins to fall.

To produce the speed–time data that the coach needs, it is necessary to compute instantaneous speeds, that is the speeds at which the athlete is running at specific times. This can be done by reducing the sampling interval, Δt, still further so that the mean speed over that interval is an accurate approximation to the instantaneous speed. This could be done if the coach had access to a video camera and recorder (or a cine camera) so that the whole race could be recorded. Then, by frame by frame analysis, the position of the runner (or more precisely the runner's centre of mass) could be measured.

Table 2.1 Distance–time data

r (m)	t (s)	Δr (m)	Δt (s)	$\bar{v} = \Delta r/\Delta t$ $(\text{m}\cdot\text{s}^{-1})$
0	0			
		10	2.2	4.55
10	2.2			
		10	1.2	8.33
20	3.4			
		10	0.9	11.1
30	4.3			
		10	0.8	12.5
40	5.1			
		10	0.8	12.5
50	5.9			
		10	0.8	12.5
60	6.7			
		10	0.8	12.5
70	7.5			
		10	0.8	12.5
80	8.3			
		10	0.8	12.5
90	9.1			
		10	0.9	11.1
100	10.0			

As UK standard video cameras record 50 fields (25 frames) per second, Δt has now been reduced to one-25th or one-50th of a second, depending on the playback system used. This is sufficiently small for these purposes and allows a consideration of how instantaneous speeds or velocities can be obtained. The process of obtaining velocities (and accelerations) from displacements belongs to the branch of mathematics called calculus, and the process is referred to as differentiation. It can be performed graphically, numerically or analytically. Only the first of these will be considered in detail.

Assuming $\Delta t = 1/25 = 0.04$ s, there are now 251 data points from which to draw the distance–time graph rather than the 11 points in Table 2.1. An expanded portion of the distance–time graph is shown for idealized data in Figure 2.10(b), where all data points lie on the curve. More realistic experimental data are shown in Figure 2.10(c), where measurement errors exist in each of the distance values but where the time intervals are known exactly. For these data (Figure 2.10(c)), a smooth curve has been drawn by eye to minimize the measurement errors and it is assumed that this curve passes through the 'true' (error-free) data.

Speed (v) is defined as the rate of change of distance (r) with respect to time (t). This can be expressed mathematically as:

$$v = \frac{\mathrm{d}r}{\mathrm{d}t} = \lim_{\delta t \to 0} \frac{\delta r}{\delta t} \tag{2.2}$$

In Equation 2.2, dr/dt represents the instantaneous speed, i.e. the instantaneous rate of change of r with respect to t. This is defined as the limiting value, as the very small increment of time, δt, approaches zero, of the very small increment in distance, δr, that occurred during that very small time, divided by the very small increment in time. The symbol δ, lower-case Greek letter delta, represents a very small change in a quantity. Instantaneous speeds are seen to be very similar to average speeds taken over extremely small time intervals.

On a continuous curve of distance (r) against time (t), Equation 2.2 represents the gradient of the tangent to the curve at any specific time. Instantaneous speeds can, therefore, be found graphically by drawing the tangents to the curves at the required times and calculating the slopes (gradients) of the tangents. This process, known as graphical differentiation, is demonstrated in Figure 2.11.

$$\text{Speed at } (t = 4\text{s}) = \text{gradient of AB} = \frac{\text{BC}}{\text{AC}} = \frac{r_\text{B} - r_\text{A}}{t_\text{B} - t_\text{A}}$$

$$\text{Therefore } v_{(t = 4\text{s})} = \frac{35\text{m} - 20\text{m}}{4.7\text{s} - 2.05\text{s}} = \frac{15\text{m}}{2.65\text{s}} = 5.66 \text{ m}\cdot\text{s}^{-1}$$

If the data are obtained as a continuous record of distance against time, as in Figure 2.11, then graphical differentiation can be performed directly, as above. Much biomechanical data, however, are of the form discussed earlier in this section, i.e. a series of discrete data points at either equal (Figure 2.10(b),(c)) or unequal (Table 2.1) time intervals. Such discrete (digital) data are obtained from cine film or video, for example. These data have to be converted to a continuous record by drawing a smooth curve through the data points (as in Figure 2.10(c)) before they can be graphically differentiated, as in Figure 2.11.

The accuracy of graphical differentiation is dependent on the smoothness of the data, especially if they are initially discrete in nature, and the accuracy of drawing tangents to the curve. With care, good results can be obtained. From a distance–time graph speeds can be obtained at any requisite number of times. The speed–time data can then be plotted and a smooth curve constructed. This, in turn, can be graphically differentiated, by drawing tangents, and accelerations (a) obtained at various times, using $a = \mathrm{d}v/\mathrm{d}t$. The process is identical to that of obtaining speed from a distance–time curve. Velocities and accelerations can also be obtained by numerical differentiation (as in Table 2.1 and Exercise 5). Numerical differentiation can work well for data that are relatively free from error. However, in practice, biomechanical data will be contaminated with errors from various sources. The finite difference method of numerical differentiation used above cannot be directly used with noisy data. The data must be smoothed. This can be done graphically, and then new smoothed data can be read from the graph and numerically differentiated. The success of this depends on how closely the smoothed data resemble the 'true' data.

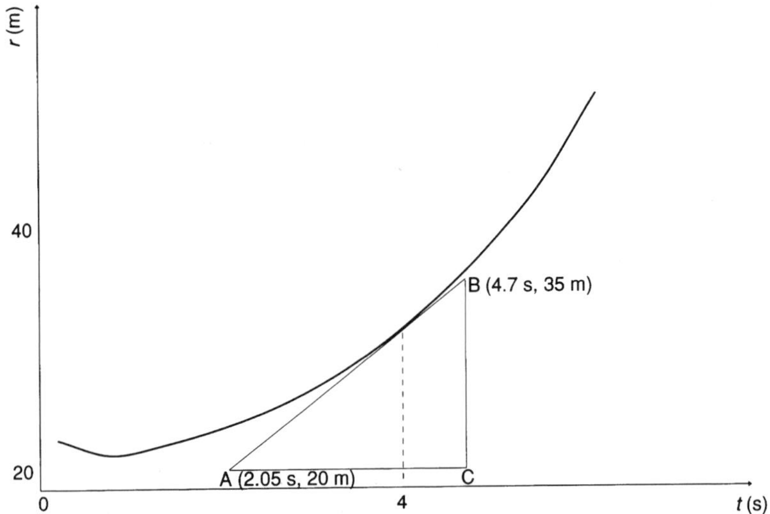

Figure 2.11 Graphical differentiation.

2.2.2 INTERPRETATION OF GRAPHS OF KINEMATIC DATA

In sports biomechanics it is common not only to present distance (or displacement) information graphically but to obtain speeds (or velocities) and accelerations from those graphs, as in the previous section. It might be interesting, for example, to ascertain, from a position or displacement–time graph when the velocity is zero or a maximum, or when the acceleration is zero. Such information can be obtained directly from a displacement–time graph without recourse to formal graphical differentiation. Familiarity with such interpretations of graphical data is essential for a sports biomechanist. These important relationships will be established for the graph of Figure 2.12(a).

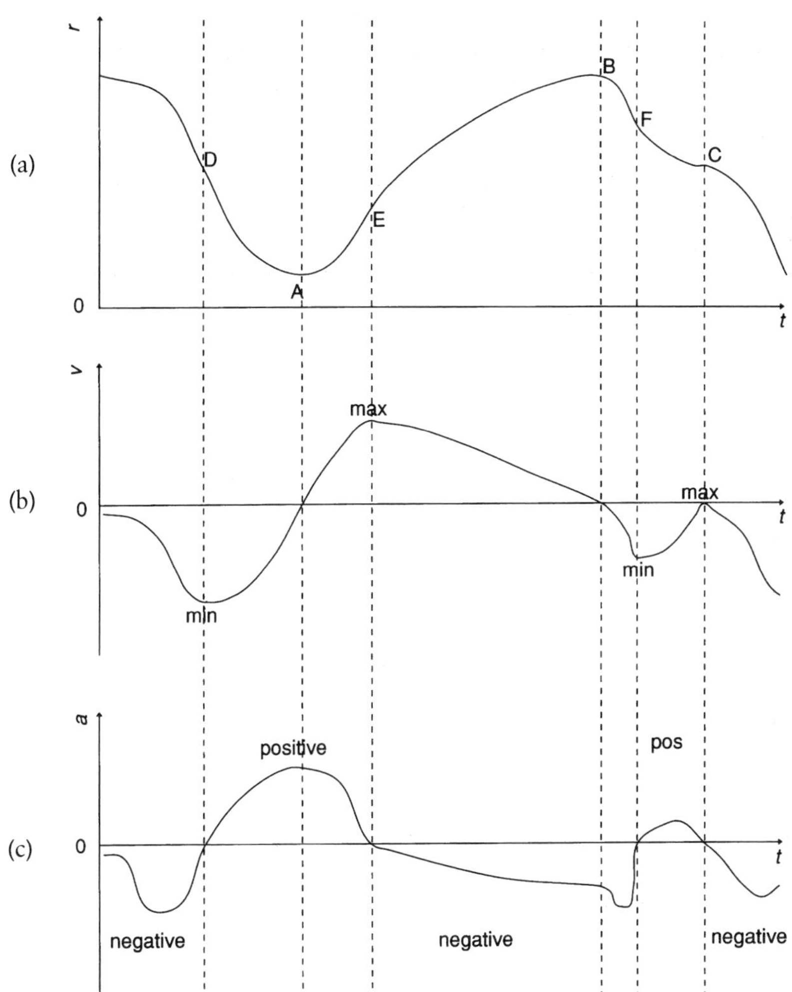

Figure 2.12 Graphs of: (a) displacement as a function of time, and associated functions of (b) velocity and (c) acceleration.

At points A, B and C in Figure 2.12(a), the gradient of the tangent to the curve is zero as the tangent is horizontal. The rate of change of displacement (*r*) with respect to time (*t*) at these points is therefore zero, that is the velocity is zero. Where this condition is satisfied, then the point is termed a **stationary point**, which may be of three types, as evidenced by A, B and C in Figure 2.12(a). Points A and B are known as

turning points. These are stationary points where the gradient of the curve (the velocity) changes sign (positive to negative or vice versa). Points C, D, E and F are known as **points of inflexion** as they are points where the curve changes its direction of curvature. This may be from concave to convex (D and F) or from convex to concave (C and E). These are considered further below.

Local minima

A point such as point A in Figure 2.12(a) is known as a local minimum, for which two conditions are necessary.

- The value of r must be less than its value at any nearby point (it is not necessarily the overall or global minimum value, hence the use of the term 'local').
- The slope of the curve, the velocity, must change from a negative value (r decreasing) to a positive value (r increasing) and hence the acceleration is positive.

For a displacement–time curve, the above can be summarized as follows: a local minimum displacement (A) corresponds to a condition of zero velocity (changing from a negative to a positive value) and a positive acceleration.

Local maxima

A point such as point B in Figure 2.12(a) is known as a local maximum, for which two conditions are necessary.

- The value of r must be greater than its value at any nearby point.
- The slope of the curve, the velocity, must change from a positive to a negative value and hence the acceleration is negative.

For a displacement–time curve, a local maximum displacement (B) again corresponds to a condition of zero velocity (changing from positive to negative) but to a negative acceleration.

Stationary points of inflexion

Point C is a point of inflexion where the gradient of the tangent happens to be zero. It therefore fulfils the conditions of a stationary point (it is called a stationary point of inflexion). It does not fulfil the extra condition required for a turning point, that the slope changes sign. The slope of the curve is negative on both sides of C.

Points of inflexion

To interpret these, consider the concave and convex portions of the curve and the point where they meet. If the curve is concave upwards (Figure 2.13(a)), the slope of the curve (and therefore the velocity) is increasing. The acceleration must therefore be positive for this portion of the curve. For a displacement–time curve, a portion with an upward concavity therefore reflects an increasing velocity, which, by definition, indicates a positive acceleration. If the curve is convex upwards (Figure 2.13(b)), the slope of the curve (and therefore the velocity) is decreasing. The acceleration must therefore be negative for this part of the curve.

For a displacement–time curve, a portion with an upward convexity therefore reflects a decreasing velocity, which, by definition, indicates a negative acceleration. The points of inflexion (C, D, E) on Figure 2.12(a) therefore indicate instances of zero acceleration.

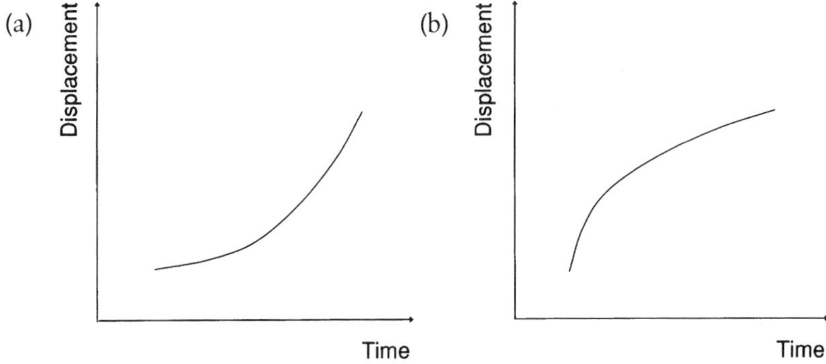

Figure 2.13 Curvature: (a) concave upwards; (b) convex upwards.

Interpretation

It is now possible to sketch approximate velocity and acceleration curves (Figure 2.12(b),(c)) from the displacement–time curve in Figure 2.12(a) by interpreting its maxima, minima and points of inflexion.

2.2.3 OBTAINING VELOCITIES AND DISPLACEMENTS FROM ACCELERATIONS

Consider an international volleyball coach who wishes to assess the vertical jumping capabilities of their squad members. Assume that this

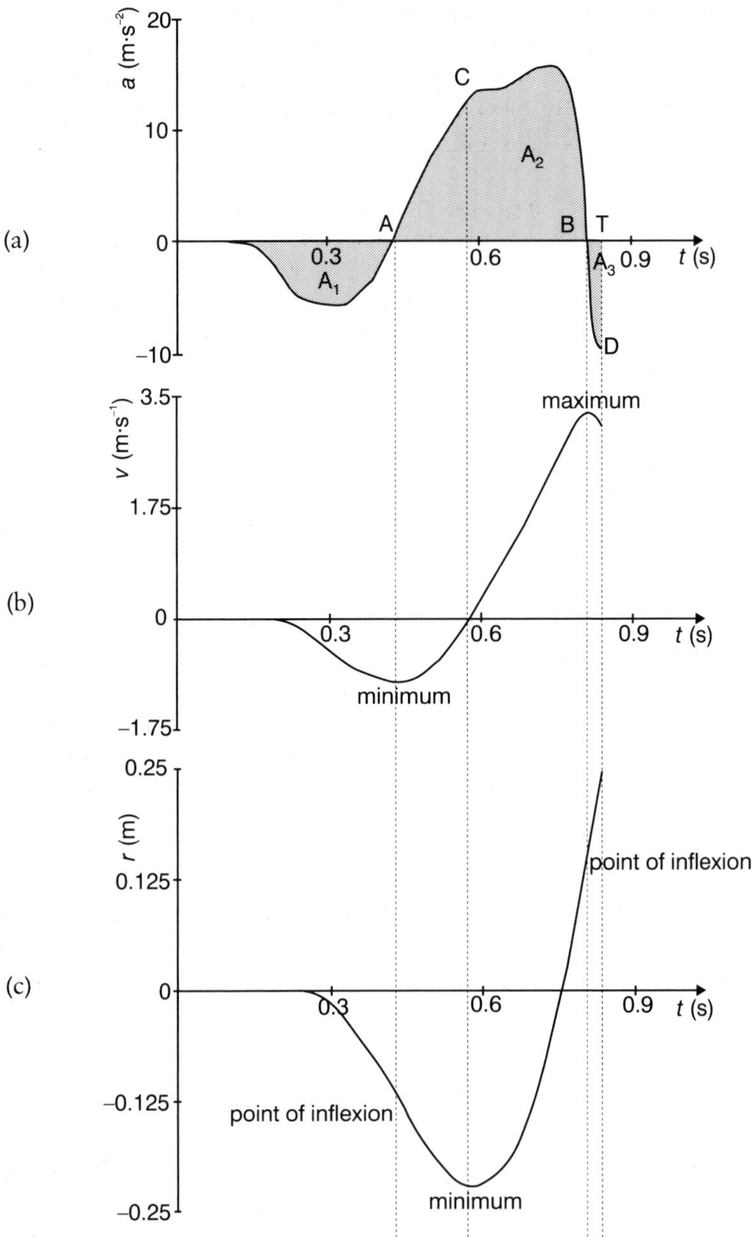

Figure 2.14 Standing vertical jump: (a) acceleration as a function of time, and associated functions of (b) speed and (c) distance travelled.

coach is a member of a university sports science department and has access to a force platform, a device that records the variation with time of the contact force between a person and the surroundings (Chapter 6). The coach uses the platform to record the forces (as force–time curves) exerted by the players performing standing vertical jumps. Each player's force–time curve easily converts to an acceleration–time curve (Figure 2.14(a)) if the player's mass is known, as acceleration equals force divided by mass.

The coach may wish to obtain, from this acceleration–time curve, values for the magnitude of the vertical velocity at take-off (referred to here as take-off speed) and the maximum height reached by the player's centre of mass. As in the previous section, this will be treated as a rectilinear problem. Here the magnitudes of the vertical components of the vector quantities are the only ones of interest.

The process of obtaining velocities (and displacements) from accelerations also belongs to the branch of mathematics called calculus, and the process is referred to as integration. As for differentiation, it can be performed graphically, numerically or analytically. Again, only the first of these will be considered in detail. Integration can basically be considered as the inverse of differentiation. Therefore, if acceleration data are available as a function of time (Figure 2.14(a)), the function can be integrated to obtain velocity (or speed) as a function of time (Figure 2.14(b)). If the speed–time data are integrated, the result is distance travelled as a function of time (Figure 2.14(c)).

The expression $a = \mathrm{d}v/\mathrm{d}t$ can be rearranged by multiplying by $\mathrm{d}t$. To indicate integration, the integration sign $\int$ is introduced. This gives the following relationship where, normally, the acceleration (a) varies with time (t).

$$\int a \, \mathrm{d}t = \int \mathrm{d}v \qquad (2.3)$$

The left side of Equation 2.3 can be easily evaluated graphically or numerically. Normally, the acceleration is performed between two specified instants of time (designated t_1 and t_2) at which the speeds are v_1 and v_2 respectively. Then:

$$\int_{t_1}^{t_2} a \, \mathrm{d}t = v_2 - v_1 \qquad (2.4)$$

This is called the **definite integral**, since it has a definite value. From Equation 2.4 it can be seen that the definite integral of the acceleration with respect to time (between t_1 and t_2) is equal to the change in speed ($v_2 - v_1$) between these time limits. The definite integral can be evaluated by graphically assessing the area under the curve between specified time limits. This is known as **graphical integration** and the technique of 'counting squares' (on graph paper) is possibly the most accurate, if the most time-consuming, method.

The volleyball coach can now calculate the take-off speed for each player from Equation 2.4 but substituting (see Figure 2.14(a)) $t_1 = 0$ and $t_2 = T$. Then, from Equation 2.4:

$$\int_0^T a\,\mathrm{d}t = v_T - v_0 \qquad (2.5)$$

In Equation 2.5, the left side is the area bounded by the recorded acceleration line ($a = a(t)$) and the lines $t = 0$, $t = T$ and $a = 0$, this is the grey-shaded area in Figure 2.14(a). Areas A_1 and A_3 are negative, as the acceleration is negative, whereas A_2 is positive as the acceleration is positive. The net area is $A_2 - (A_1 + A_3)$. This equals v_T, the take-off speed, as $v_0 = 0$ because the player is initially stationary. The maximum height (h) of the mass centre can then be calculated from $h = v_T^2/2g$, where g is the gravitational acceleration. From Figure 2.14(a) the speed at any time can also be found providing that the speed at one instant is known (as it is in this case where the value of v at $t = 0$, v_0, is zero). Thus:

$$v_1 - v_0 = \int_0^{t_1} a\,\mathrm{d}t \qquad (2.6)$$

The left side of Equation 2.6 is often written as Δv_{01} or simply Δv, the finite change of speed during a finite time interval Δt. From Equation 2.6 it is evident that the acceleration–time curve can be integrated, by taking small strips, to obtain speed as a function of time. The graph of speed as a function of time can then be graphically integrated to obtain distance as a function of time. From Equation 2.6 it should be evident that speed changes can always be computed from an acceleration–time graph even if the value of the speed at any given time is not known.

In the previous section it was seen how, from turning points and points of inflexion, it was possible to identify points where $v = 0$ and $a = 0$ from a displacement–time graph. It is instructive to seek to identify what information about velocity and displacement can be found by inspection of an acceleration–time graph such as Figure 2.14(a).

Firstly, a point of zero acceleration corresponds to a turning point on the velocity–time graph if the acceleration also changes sign as it passes through zero. The velocity is a minimum if the acceleration changes from a negative to a positive value, as for point A in Figure 2.14(a). The velocity is a maximum if the acceleration changes from a positive to a negative value, as for point B in Figure 2.14(a). Furthermore, points A and B also correspond to points of inflexion on the displacement–time graph. A is a point of inflexion with the curvature changing from convex upwards to concave upwards. B is a point of inflexion with the curvature changing from concave upwards to convex upwards.

Finding minima and maxima on the displacement–time curve is more difficult. These correspond to a zero velocity and such points can only accurately be found by performing the graphical integration. However, if the positive area bounded by the curve from A to C is equal in magnitude to the negative area bounded by the curve from 0 to A, it follows that the integral from 0 to C is zero. As $v_0 = 0$, this also gives $v_c = 0$. As the acceleration is positive here, this corresponds to a local minimum for the displacement–time curve. No local maximum exists for displacement for this example, but obviously displacement increases to take-off, $t = T$ (and beyond). The velocity–time and displacement–time curves can now be sketched as in Figure 2.14(b),(c).

2.3 Projectile motion

2.3.1 PROJECTION PARAMETERS

Bodies launched into the air that are subject only to the forces of gravity and air resistance are termed projectiles. Projectile motion occurs frequently in sport and exercise activities. Often the projectile involved is an inanimate object, such as a javelin or golf ball. In some activities the sports performer becomes the projectile, as in the long jump, high jump, diving and gymnastics. An understanding of the mechanical factors that govern the flight path or trajectory of a projectile is therefore important in sports biomechanics. The following discussion assumes that the effects of aerodynamic forces (air resistance and lift effects) on projectile motion are negligible. This is a reasonable first assumption for some, but certainly not all, projectile motions in sport. The effects of aerodynamic forces will be covered in Chapter 4.

There are three parameters, in addition to gravitational acceleration, g, that determine the trajectory of a simple projectile, such as a ball, shot or hammer. These are the projection speed, angle and height (Figure 2.15).

Projection angle

Projection angle (θ) is defined as the angle between the projectile's velocity vector and the horizontal at the instant of release or take-off. Hence the terms **release angle** or **take-off angle** are often used. The size of the projection angle depends on the purpose of the activity. For example, activities requiring maximum horizontal range (shot put, long jump, ski jump) tend to use smaller angles than those in which maximum height is an objective (high jump, volley ball spiker). In the absence of aerodynamic forces, all projectiles will follow a flight path that has a parabolic shape which depends upon the magnitude of the projection angle (Figure 2.16).

Figure 2.15 Projection parameters (adapted from Payne, 1985).

Projection speed

Projection speed (v_0) is defined as the magnitude of the projectile's velocity vector at the instant of release (Figure 2.15) or take-off. When the projection angle and height are held constant, the projection speed will determine the magnitude of a projectile's maximum vertical displacement (its apex) and its range (maximum horizontal displacement). The greater the projection speed, the greater the apex and range. It is common practice to resolve a projectile's velocity into its horizontal and vertical components and then to analyse these independently. Horizontally a projectile is not subject to any external forces (ignoring air resistance) and will therefore maintain constant horizontal speed during its airborne phase (as in a long jump or racing dive). The range (R) travelled by a projectile is the product of its horizontal projection speed ($v_{x0} = v \cos\theta$) and its time of flight (t_{max}). That is:

$$R = v_{x0}\, t_{max} \quad (2.7)$$

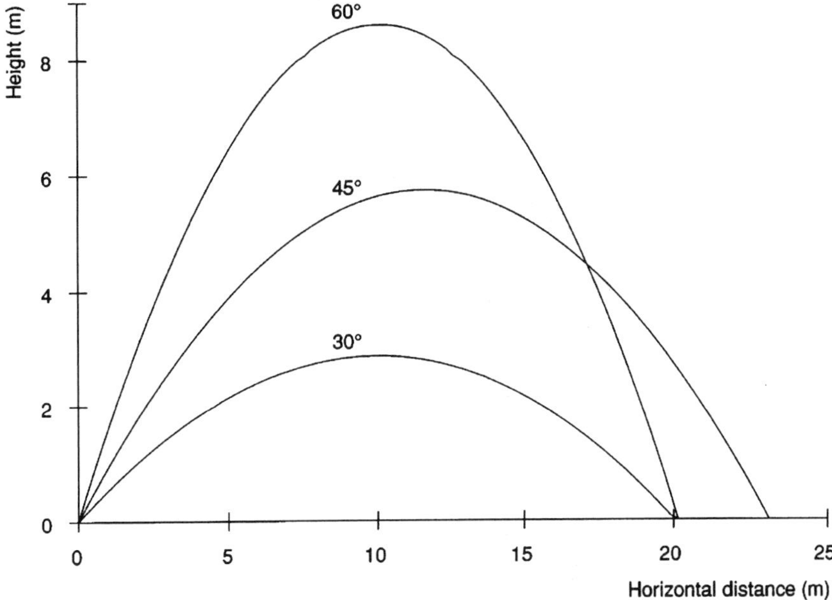

Figure 2.16 Effect of projection angle on shape of parabolic trajectory for a projection speed of 15 m·s^{-1} and zero projection height.

To calculate a projectile's time of flight it is necessary to consider the magnitude of the vertical component of its projection velocity (v_y). Vertically a projectile is subject to a constant acceleration due to gravity. The magnitude of the maximum vertical displacement (y_{max}), flight time (t_{max}) and range (R) achieved by a projectile can easily be determined from v_{y0} if it takes off and lands at the same level ($y_0 = 0$). This occurs, for example, in a football kick. In this case, the results are as follows:

$$y_{max} = v_{y0}^2/2g = v_0^2 \sin^2\theta/2g \tag{2.8}$$

$$t_{max} = 2v_{y0}/g = 2v_0 \sin\theta/g \tag{2.9}$$

$$R = 2v_0^2 \sin\theta \cos\theta/g = v_0^2 \sin 2\theta/g \tag{2.10}$$

Equation 2.10 shows that the projection speed is by far the most important of the projection parameters in determining the range achieved, because the range is proportional to the square of the release speed. Doubling the release speed would increase the range fourfold.

Projection height

Equations 2.8–2.10 have to be modified if the projectile lands at a level higher or lower than that at which it was released. This is the case with most sports projectiles, for example in a shot put, a basketball shot or a long jump. For a given projection speed and angle, the greater the relative projection height (y_0), the longer the flight time and the greater the range and maximum height. The maximum height is the same as in the last case but with the height of release added, as in Equation 2.11.

$$y_{max} = y_0 + v_0^2 \sin^2\theta/2g \quad (2.11)$$
$$t_{max} = v_0 \sin\theta/g + (v_0^2 \sin^2\theta + 2gy_0)^{1/2}/g \quad (2.12)$$
$$R = v_0^2 \sin2\theta/2g + v_0^2 \cos\theta\,(\sin^2\theta + 2gy_0/v_0^2)^{1/2}/g \quad (2.13)$$

The equations for the time of flight and range appear to be much more complicated than those for zero release height. The first of the two terms in each of Equations 2.12 and 2.13 relates, respectively, to the time and horizontal distance to the apex of the trajectory. The value of these terms is exactly half of the total values in Equations 2.9 and 2.10. The second terms relate to the time and horizontal distance covered from the apex to landing. By setting the release height (y_0) to zero (and noting that $\cos\theta \sin\theta = (\cos2\theta)/2$), you will find that the second terms in Equations 2.12 and 2.13 become equal to the first terms and that the equations are then identical to 2.9 and 2.10.

2.3.2 OPTIMUM PROJECTION CONDITIONS

In many sports events the objective is to maximize either the range or the height of the apex achieved by the projectile. As seen above, any increase in projection speed or projection height is always accompanied by an increase in the range and height achieved by a projectile. If the objective of the sport is to maximize height or range, it is important to ascertain the best (optimum) angle to achieve this.

Obviously maximum height is achieved when all of the available projection speed is directed vertically, when $\theta = 90°$. The optimum angle for maximum range can be found for zero relative projection height from Equation 2.10 as that which maximizes $\sin2\theta$, that is 45°. For the more general case of non-zero y_0, the optimum projection angle can be found from:

$$\cos2\theta = gy_0/(v_0^2 + gy_0) \quad (2.14)$$

For a good shot putter, for example, this would give a value around 42°.

Although optimum projection angles for given values of v_0 and y_0 can easily be determined mathematically, they do not always correspond to those recorded from the best performers in sporting events. This is even true for the shot put, where the object's flight is the closest to a parabola of all sports objects. The reason for this is that the calculation of an optimum projection angle assumes, implicitly, that the projection speed and projection angle are independent of one another. For a shot putter, the release speed and angle are, however, not independent, because of the arrangement and mechanics of the muscles used to generate the release speed of the shot. A greater release speed, and hence range, can be achieved at an angle (about 35°) that is less than the optimum projectile angle. If the shot putter seeks to increase the release angle to a value closer to the optimum projectile angle, the release speed decreases and so does the range.

A similar deviation from the optimum projection angle is noticed when the activity involves the projection of an athlete's body. The angle at which the body is projected at take-off can again have a large effect on the take-off speed. In the long jump, for example, take-off angles used by elite long jumpers are around 20°. Hay (1986) reported that to obtain the theoretically optimum take-off angle of around 42°, long jumpers would have to decrease their normal horizontal speed by around 50%. This would clearly result in a drastically reduced range as the range depends largely on the square of the take-off speed.

In many sporting events, such as the javelin and discus throws, badminton, sky-diving and ski jumping, the aerodynamic characteristics of the projectile can significantly influence the trajectory of the projectile. It may travel a greater or lesser distance than it would have done if projected in a vacuum. Under such circumstances, the calculations of optimal projection parameters need to be modified considerably to take account of the aerodynamic forces acting on the projectile.

2.4 Rotational kinematics

2.4.1 INTRODUCTION

In the previous section, linear motion was considered, in which the mathematical model is that of a point. Any biomechanical system can be represented as a particle situated at its centre of mass. The translation of this point is defined independently of any rotational motion taking place around it. The centre of mass generally serves as the best (and sometimes the only) point about which rotations should be considered to occur. All human motion however involves rotation (angular motion), for example the movement of a body segment about its proximal joint.

The equations of rotational motion are far more complex than those of linear motion. Those for a rigid body can be directly applied to a cricket bat or body segments. They can also be applied to a non-rigid body, which, instantaneously, is behaving as though it was rigid, such as a diver holding a fully extended body position or a gymnast holding a tuck. Such systems are classified as quasi-rigid bodies. Applications of the laws of rotational motion to non-rigid bodies, such as the complicated kinematic chains of segments which are the reality in most human movement, have to be made with considerable care. The theory of the rotation of rigid bodies in the general case is extremely complicated and many problems in this category have not yet been solved.

The kinematic vectors in rotational motion are defined in a similar manner to those for linear motion (see above), and each has a scalar magnitude. Angular displacement is the change in the orientation of a line segment. In planar (two-dimensional) motion, this will be the angle between the initial and final orientations regardless of the path taken. Angular velocity and acceleration were also defined earlier in this chapter and are, respectively, the rates of change with respect to time of angular displacement and angular velocity.

2.4.2 ANGULAR MOTION VECTORS

Angular quantities have a vector character, the direction of the vector arrow being found from the right hand rule, Figure 2.17.

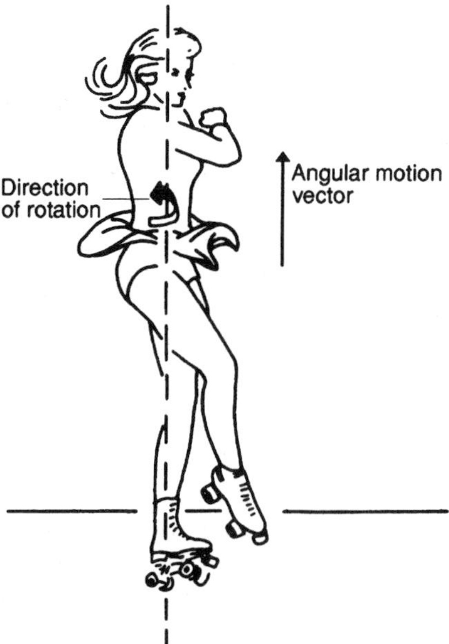

Figure 2.17 Right hand rule (adapted from Kreighbaum and Barthels, 1990).

With the right hand orientated as in Figure 2.18, the curled fingers follow the direction of rotation and the thumb points in the direction of the angular motion vector.

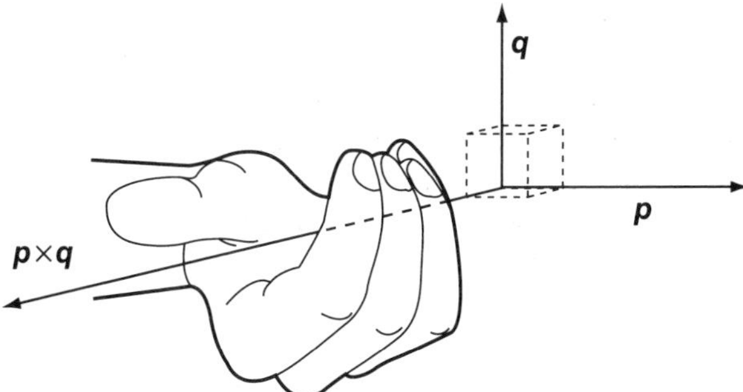

Figure 2.18 The vector cross-product.

In the case shown here, the angular motion vector and the axis of rotation coincide. This is often, but not always, the case.

Most angular motion vectors obey the rules of resolution and composition, but this is not true for angular displacements. To understand rotational motion, it is very useful to consider one of the ways in which vectors can be multiplied. The rules of vector multiplication do not follow the rules for multiplying scalars.

The vector (or cross) product of two vectors is useful in rotational motion, as, for example, it enables angular motion vectors to be related to translational motion vectors (see below). It will be stated here in its simplest case for two vectors at right angles as in Figure 2.18. The vector product of two vectors p and q inclined to one another at right angles is defined as a vector ($p \times q$) of magnitude equal to the product of the magnitudes of the two given vectors. Its direction is perpendicular to both vectors p and q in the direction in which the thumb points if the curled fingers of the right hand point from p to q through the right angle between them.

2.4.3 VELOCITIES AND ACCELERATIONS CAUSED BY ROTATION

Consider the quasi-rigid body of a gymnast, Figure 2.19, rotating in a giant circle about an axis fixed in the bar at O, with angular velocity $\boldsymbol{\omega}$.

As O is fixed and the gymnast's centre of mass G is fixed relative to O then the length (magnitude) of the position vector $\boldsymbol{r}$ of the centre of mass from O does not change. Therefore the velocity of G is tangential and given by $\boldsymbol{v} = \boldsymbol{\omega} \times \boldsymbol{r}$. The tangential velocity of G is seen, from the cross-product rule, to be mutually perpendicular to both the position and angular velocity vectors. Furthermore, the acceleration of G has two components. The first, $\boldsymbol{\alpha} \times \boldsymbol{r}$, term is called the **tangential acceleration** and has a direction identical to that of the tangential velocity and a magnitude αr. The second component is called the **centripetal accel-**

eration. It has a magnitude $\omega^2 r$ (= ωv = v^2/r) and a direction given by $\boldsymbol{\omega} \times \boldsymbol{v}$. This direction is obtained by letting $\boldsymbol{\omega}$ rotate towards $\boldsymbol{v}$ through the right angle between them and using the right hand rule. This gives a vector perpendicular to both the angular velocity and tangential velocity vectors. It should be noted that a centripetal acceleration exists for all rotational motion. This is true even if the angular acceleration is zero, in which case the angular velocity is constant.

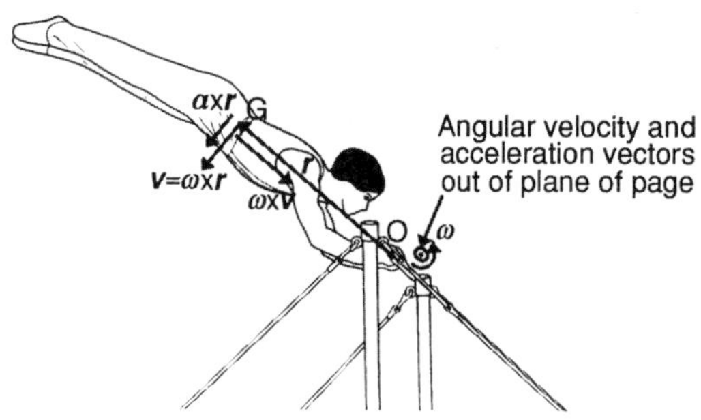

Figure 2.19 Tangential velocity and tangential and centripetal acceleration components for a gymnast performing a giant circle (adapted from Kreighbaum and Barthels, 1990).

2.4.4 ROTATIONS IN THREE-DIMENSIONAL SPACE (SPATIAL ROTATIONS)

For two-dimensional rotations of human body segments, the joint angle may be defined as the angle between two lines representing the proximal and distal segments. A similar procedure can be used to specify the relative angle between line representations of segments in three dimensions if the articulation is a simple hinge joint, but generally the process is more complex. There are many different ways of defining the orientation angles of two articulating rigid bodies (Craig, 1989) and of specifying the orientation angles of the human performer (e.g. Yeadon, 1993). Most of these conventions have certain problems, one of which is to have an angle convention that is easily understood. The one adopted throughout this book, unless otherwise stated, is that which preserves the use of the anatomical terminology of Chapter 1.

The specification of the angular orientation of the human performer as a whole is also problematic. The representation of Figure 2.20 (Yeadon, 1993) has been found to be of value in the analysis of airborne activities in gymnastics, diving and trampolining.

In this, rotation is specified in terms of: the somersault angle about a horizontal axis through the centre of mass; the twist angle about the longitudinal (vertical) axis; and the tilt angle. The last named is the angle between the longitudinal axis and the fixed plane normal to the somersault angular velocity vector.

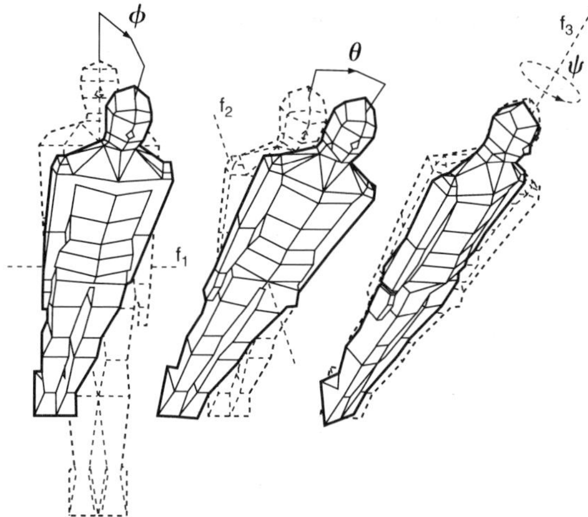

Figure 2.20 Angular orientation showing angles of somersault (ϕ), tilt (θ) and twist (ψ) (reproduced from Yeadon, 1993, with permission).

2.5 Summary

In this chapter, kinematic principles, which are important for the study of movement in sport and exercise, have been covered. This included consideration of the types of motion and the model appropriate to each, vectors and scalars, and vector addition, subtraction and resolution. The importance of differentiation and integration in sports biomechanics was addressed with reference to practical sports examples, and graphical differentiation and integration were covered. The importance of being able to interpret graphical presentations of one linear kinematic variable (displacement, velocity or acceleration) in terms of the other two was stressed. Projectile motion was considered and equations presented to calculate the maximum vertical displacement, flight time, range and optimum projection angle of a simple projectile for specified values of the three projection parameters. Deviations of the optimal angle for the sports performer from the optimal projection angle were explained. Finally, simple rotational kinematics were introduced, including vector multiplication and the calculation of the velocities and accelerations caused by rotation.

2.6 Exercises

1. Draw sketch diagrams to show the three types of motion and the model appropriate to each. List the uses and limitations of each model.
2. The following vector directions are defined in terms of angles, measured anticlockwise, to a right-facing horizontal line: $\boldsymbol{A}$ = 10 units at 45°, $\boldsymbol{B}$ = 15 units at 120°, $\boldsymbol{C}$ = 20 units at 90°. Obtain graphically, as in section 2.1.3, or trigonometrically, as in section 2.7, the following:

   a) $\boldsymbol{A} + \boldsymbol{B} + \boldsymbol{C}$

b) **A** + **B** – **C** (Hint: the vector –**C** will have a direction of –90° or 270°).

3. A vector, defined as in Exercise 2, has a magnitude of 45 units and a direction of 75°. Obtain the magnitudes of the vertical and horizontal components of this vector either graphically, as in section 2.1.3, or trigonometrically, as in section 2.7.
4. Calculate the following relative velocities, by graphical means:
a) that of a javelin relative to the air when it is thrown at 30 m·s^{-1} and at an angle of 35° relative to the ground into a wind which has a velocity (which is horizontal) of 3 m·s^{-1} relative to the ground.
b) that of the apparent wind for a yacht that is sailing from the south east with a velocity of magnitude 8 m·s^{-1} relative to the sea, in which a current is flowing (relative to the earth) with a velocity of 1.5 m·s^{-1} from the north, and the true wind (relative to the earth) is blowing at a velocity of 12 m·s^{-1} from the south west. (Hint: the apparent wind is the wind relative to the yacht).
5. Plot a distance–time graph for the noisy data of Table 2.2. Draw a smooth curve through the data. (Note r has been set to zero at t = 5.90 s. You should plot your t axis to begin at 5.90 s.) Use the smooth curve to calculate, by graphical differentiation, the speeds at times t = 5.94 s and t = 6.06 s. Also obtain the speeds from the noisy data by numerical differentiation (as in Table 2.2 for the ideal data and in section 2.2.1). Check to see how both your sets of values compare with those for the ideal data of Table 2.2.

Table 2.2 Differentiation of ideal and noisy data

		Ideal data			Noisy data		
t (s)	δt (s)	r (m)	δr (m)	$v = \delta r/\delta t$ (m·s^{-1})	r (m)	δr (m)	v (m·s^{-1})
5.90		0.00			0.02		
	0.04		0.50	12.50			
5.94		0.50			0.48		
	0.04		0.50	12.50			
5.98		1.00			0.94		
	0.04		0.50	12.50			
6.02		1.50			1.53		
	0.04		0.50	12.50			
6.06		2.00			2.01		
	0.04		0.50	12.50			
6.10		2.50			2.47		

6. Photocopy Figure 2.12(a) (not (b) and (c)!). From this, sketch the appropriate velocity and acceleration graphs. Compare your answers with Figure 12(b),(c).

7. By photocopying and tracing on to graph paper, or any other method, obtain a graph-paper copy of Figure 2.14(a). From this, obtain the magnitude of the vertical take-off velocity of the jumper by graphical integration. From the copy of Figure 2.14(a), sketch the velocity and displacement graphs. Compare your answers with Figure 2.14(b),(c).
8. A shot is released at a height of 1.89 m, with a speed of 13 $m \cdot s^{-1}$ and at an angle of 34°. Calculate the maximum height reached and the time at which this occurs, the range and the time of flight, and the optimum projection angle. Why do you think the release angle differs from the optimum projection angle?
9. Calculate the maximum height reached and the time at which this occurs, the range and the time of flight, and the optimum projection angle for a similar object projected with zero release height but with the same release speed and angle. Comment on the effects of changing the release height.
10. Calculate the magnitudes of the acceleration components and the velocity of a point on a rigid body at a radius of 0.8 m from the axis of rotation when the angular velocity and angular acceleration have magnitudes, respectively, of 10 $rad \cdot s^{-1}$ and –5 $rad \cdot s^{-2}$. Sketch the body and draw on it the velocity vector and components of the acceleration vector.

2.7 Appendix: Component addition of vectors

Vector addition (and subtraction) can also be performed on the components of the vector using the rules of simple trigonometry. For example, consider the addition of the three vectors represented in Figure 2.21(a).

Vector A is a horizontal vector with a magnitude (proportional to its length) of 1 unit. Vector B is a vertical vector of magnitude –2 units (it points vertically downwards, hence it is negative). Vector C has a magnitude of 3 units and is 120° measured anticlockwise from a right-facing horizontal line.

The components of the three vectors are summarized below (see also Figure 2.21(b)). The components (R_x, R_y) of the resultant vector $R = A + B + C$ are shown in Figure 2.21(c). The magnitude of the resultant is then obtained from the magnitudes of its two components, using Pythagoras' theorem ($R^2 = R_x^2 + R_y^2$), as $R = (0.25 + 0.36)^{1/2} = 0.78$. Its direction to the right horizontal is given by the angle, θ, whose tangent is R_y/R_x. That is $\tan\theta = -1.2$, giving $\theta = 130°$. The resultant R is rotated anticlockwise 130° from the right horizontal, as shown in Figure 2.21(d). (A second solution for $\tan\theta = -1.2$ is $\theta = -50$, which would have been the answer if R_x had been +0.5 and R_y had been –0.6).

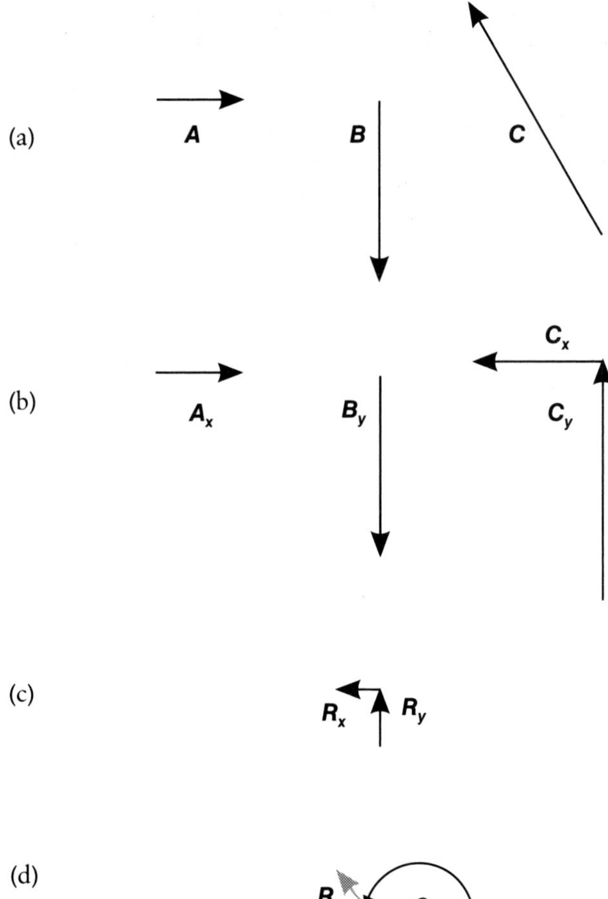

Figure 2.21 Vector addition using components: (a) vectors to be added; (b) their components; (c) components of resultant; (d) resultant.

Vector	Horizontal component (x)	Vertical component (y)
A	1	0
B	0	–2
C	3 cos120° = –3 cos60°	3 sin120° = 3 cos60°
	= –1.5	= 2.6
$A + B + C$	–0.5	0.6

2.8 References

Craig, J. J. (1989) *Introduction to Robotics: Mechanics and Control*, Addison Wesley, Wokingham.

Field, P. and Walker, T. (1987) *Cross-Country Skiing*, Crowood Press, Marlborough.

Hay, J. G. (1986) The biomechanics of the long jump, in *Exercise and Sport Sciences Reviews*, Vol. 14, (ed. K. B. Pandolf), Macmillan, New York, pp. 401–446.

Kreighbaum, E. and Barthels, K. M. (1990) *Biomechanics: a Qualitative Approach for Studying Human Movement*, Macmillan, New York.

MacConaill, M. A. and Basmajian, J. V. (1969) *Muscles and Movements: A Basis for Human Kinesiology*, S. Krieger, Huntington, NY.

Payne, H. (ed.) (1985) *Athletes in Action*, Pelham Books, London.

Yeadon, M. R. (1993) The biomechanics of twisting somersaults. *Journal of Sports Sciences*, **21**, 187–225.

2.9 Further reading

Hay, J. G. (1993) *The Biomechanics of Sports Techniques*, Prentice Hall, Englewood Cliffs, NJ, chapters 2–4.

3 Linear and angular kinetics

This chapter is designed to provide an understanding of linear (translational) and angular (rotational) kinetics. Kinetics is the branch of mechanics which covers the action of forces in producing or changing motion. After reading this chapter you should be able to:

- define force and its SI unit and identify the forces acting in sport;
- understand the laws of linear kinetics, and related concepts such as linear momentum, and classify various force systems;
- appreciate the ways in which friction and traction influence movements in sport and exercise;
- calculate, from segmental and kinematic data, the position of the centre of mass of the human performer;
- appreciate the factors that govern the impact of sports objects;
- understand the importance of the centre of percussion in sport;
- identify the ways in which rotation is acquired and controlled in sports motions;
- understand the laws of rotational kinetics and related concepts such as angular momentum.

3.1 Basics of linear kinetics

3.1.1 FORCES – DEFINITION AND TYPES

A force can be considered as the pushing or pulling action that one object exerts on another. Forces are vectors, i.e. they possess both a magnitude and a directional quality. The latter is specified by the direction and by the point of application, the point on an object where the force acts. Alternatively the total directional quality can be given by the line of action (Figure 3.1).

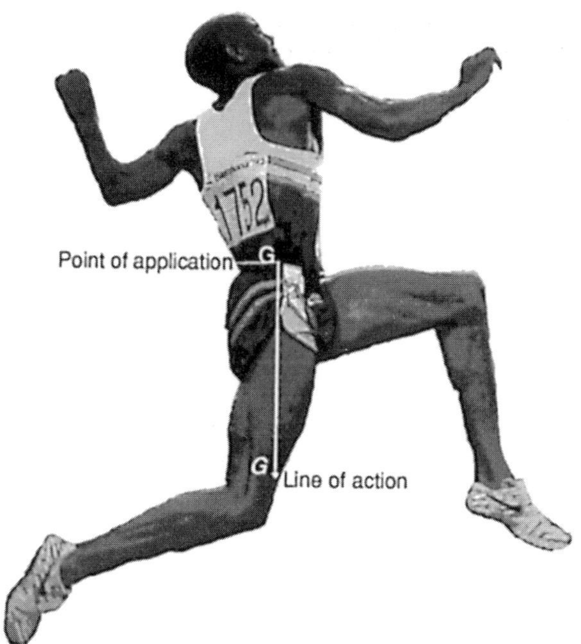

Figure 3.1 Directional quality of force.

The effects of a force are not altered by moving it along its line of action. The overall effects (though not in terms of linear motion) are changed if the force is moved parallel to the original direction but away from its line of action. A **moment** or **turning effect** is then introduced: that is an effect tending to rotate the object (see below). Care must therefore be exercised when solving systems of forces graphically and a vector approach is then often preferable. The SI unit of force is the newton (N) and the symbol for a force vector is F. One newton is that force which, when applied to a mass of one kilogram (1 kg), causes it to accelerate at 1 m·s^{-2} in the direction of the force application. A sports performer experiences forces both internal to and external to the body. Within the body forces are generated by the muscles and transmitted by tendons, bones, ligaments and cartilage, and these were considered in Chapter 1. The main external forces, the combined effect of which determine the overall motion of the body, are as follows.

Weight

Weight is a familiar force (Figure 3.1) attributable to the gravitational pull of the earth. It acts at the centre of gravity of an object towards the centre of the earth, i.e. vertically downwards. The centre of gravity (G) is an imaginary point at which the weight of an object can be considered to act. For the human performer, there is little difference between the positions of the centre of mass (see later) and the centre of gravity. The former is the term preferred in most modern sports biomechanics literature and will be used in the rest of this book. An athlete with a mass of 50 kg has a weight (*G*) at sea level (where the standard value of gravitational acceleration, *g*, is usually assumed to be 9.81 $m \cdot s^{-2}$) of approximately 490 N.

Reaction forces

Reaction forces are the forces that the ground (or other external surface) exerts on the sports performer as a reaction (Newton's third law of linear motion) to the force that the performer exerts on the ground or surface. The component tangential to the surface, friction or traction, is crucially important in sport and will be considered in detail below.

Buoyancy

Buoyancy is the force experienced by an object immersed, or partly immersed, in a fluid. It always acts vertically upwards at the centre of buoyancy (CB) as in Figure 3.2.

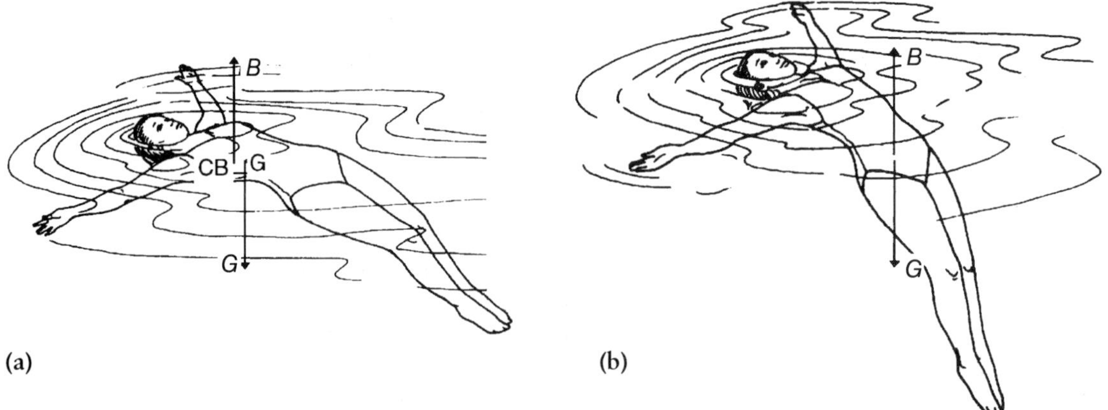

Figure 3.2 Buoyancy force: (a) forces acting; (b) forces in equilibrium (adapted from Thomas, 1989).

The magnitude of the force is expressed by Archimedes' principle 'the upthrust is equal to the weight of fluid displaced' and is given by:

$$B = V\rho g \quad (3.1)$$

where V is the volume of fluid displaced and ρ the density of the fluid. The buoyancy force is large in water ($\rho \approx$ 1000 kg·m^{-3}) and much smaller (although not entirely negligible) in air ($\rho \approx$ 1.23 kg·m^{-3}). For an object to float and not sink, the magnitudes of the buoyancy force and the weight of the object must be equal ($B = G$).

The swimmer in Figure 3.2 will only float if her average body density is less than or equal to the density of water. The degree of submersion will then depend on the ratio of the two densities. If the density of the swimmer is greater than that of the water, she will sink. It is easier to float in sea water ($\rho \approx$ 1020 kg·m^{-3}) than fresh water. For the human body, the relative proportions of tissues will determine whether sinking or floating occurs. Clauser, McConville and Young (1969) gave the following typical density values: fat 960 kg·m^{-3}, muscle 1040–1090 kg·m^{-3}, bone 1100 (cancellous) to 1800 kg·m^{-3} (compact). The amount of air in the lungs is also very important. Various researchers (for example Cureton, 1951; Malina, 1969) have established that most caucasians can float with full inhalation and that most negroes cannot float even with full inhalation because of their different body composition (this is surely a factor contributing to the paucity of world class black swimmers). Most people cannot float with full exhalation. Women float better than men because of an inherently higher proportion of body fat and champion swimmers have (not surprisingly) higher proportions of body fat than other elite athletes.

Other forces, known as fluid dynamic forces, arise whenever there is relative motion between a fluid and an object, such as the air resistance acting on a cyclist. These forces will be considered in detail in Chapter 4.

Impact forces

Impact forces occur whenever two (or more) objects collide. They are usually very large compared to other forces acting and are of short duration. The most important impact force for the sport and exercise participant is that between that person and some external object; for example a runner's foot striking a hard surface. Impacts involving sports objects, such as a ball and the ground, can affect the technique of a sports performer. For example, the spin imparted by the server to a tennis ball will affect how it rebounds, which will influence the stroke played by the receiver. Impacts of this type are termed **direct impacts** if the objects at impact are moving along the line joining their centres of mass. An example is a ball dropped vertically onto the ground. Such impacts are unusual in sport, where **oblique impacts** predominate. Examples include a ball hitting the ground at an angle of other than 90°, as in a tennis serve, and a bat or racket hitting a moving ball. These will be considered later in this chapter.

3.1.2 SYSTEMS OF FORCES AND EQUILIBRIUM

In sport, more than one external force usually acts on the performer. In such a case, the effect produced by the combination of the forces, or the force system, will depend on their magnitudes and relative directions. Figure 3.3(a) shows the biomechanical system of interest, here the runner, isolated from the surrounding world.

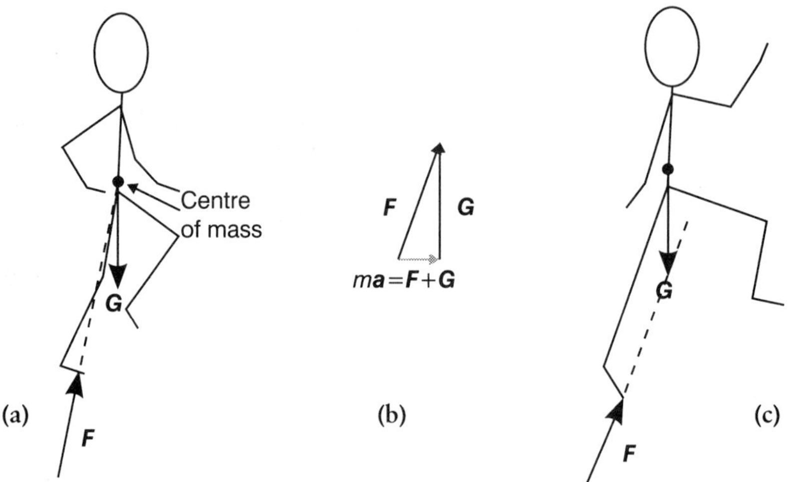

Figure 3.3 Forces on a runner: (a) free body diagram of dynamic force system; (b) resultant force; (c) free body diagram with force not through centre of mass.

The effects of those surroundings, which for the runner are weight and ground reaction force, are represented on the diagram as force vectors. Such a diagram is known as a **free body diagram** and should be used whenever carrying out a biomechanical analysis of force systems. The effects of such force systems can be considered in different ways.

Statics is a very useful and mathematically very simple and powerful branch of mechanics. It is used to study force systems in which the forces are in equilibrium, such that they have no resultant effect on the object on which they act (as in Figure 3.2(b)). Here ***B*** and **G** share the same line of action and are equal in magnitude but have opposite directions:

$$B = -G \text{ or}$$
$$B + G = 0 \qquad (3.2)$$

This approach may seem to be somewhat limited in sport, in which the net, or resultant, effect of the forces acting is usually to cause the object to accelerate, as in Figure 3.3(a). Here the resultant force can be obtained by moving ***F*** along its line of action (which passes through the centre of mass in this case), giving Figure 3.3(b).

As the resultant force in this case passes through the runner's centre of mass, the runner can be represented as a point (the centre of mass) because only changes in linear motion will occur for such a force system. The resultant of F and G will be the net force acting on the runner. By Newton's second law of linear motion (see below), the net force equals mass times acceleration ($m\,a$) or:

$$F + G = m\,a \qquad (3.3)$$

More generally, as in Figure 3.3(c), the resultant force will not act through the centre of mass and a moment of force will then tend to cause the object (the runner in this case) to rotate. The magnitude of this moment of force about a point is the product of the force and its moment arm, which is the perpendicular distance of the line of action of the force from that point. Rotation will be considered in detail in Section 3.4.

It is possible to treat dynamic systems of forces, such as those represented in Figure 3.3, using the equations of static equilibrium. This involves the introduction of an imaginary, or inertia, force to the dynamic system, which is equal in magnitude to the resultant force but opposite in direction, to produce a quasi-static force system. This allows the use of the general equations of static equilibrium for forces (F) and moments of force (M):

$$\begin{aligned}\Sigma F &= 0\\ \Sigma M &= 0\end{aligned} \qquad (3.4)$$

That is the sum (Σ) of all the forces, including the imaginary inertia forces, is zero and the sum of all the moments of force, including inertia ones, is also zero. The vector equations of static equilibrium (Equation 3.4) can be applied to all force systems that are static or have been made quasi-static through the use of inertia forces. The way in which the vector equations simplify to the scalar equations used to calculate the magnitudes of forces and moments of force will depend on the nature of the system of forces. Such force systems can be classified as follows.

Linear (collinear), planar (coplanar) and spatial force systems

Collinear systems consist of forces with the same line of action, such as the forces in a tug-of-war rope. No moment equilibrium equation ($\Sigma M = 0$) is relevant for such systems as all the forces act along the same line. Planar force systems have forces acting in one plane only and spatial force systems are three-dimensional.

Concurrent force systems

These are systems in which the lines of action of the forces pass through a common point. The collinear system is a special case. The runner in Figure 3.3(a) is an example of a planar concurrent force system and spatial ones can also be found in sport and exercise movements. As all forces pass through the centre of mass, no moment equilibrium equation is relevant.

Parallel force systems

These have the lines of action of the forces all parallel and can be planar, as in Figure 3.2(a), or spatial. The tendency of the forces to rotate the object about some point means that the equation of moment equilibrium must be considered. The simple cases of first- and third-class levers in the human musculoskeletal system are examples of planar parallel force systems and are shown in Figure 3.4(a),(b).

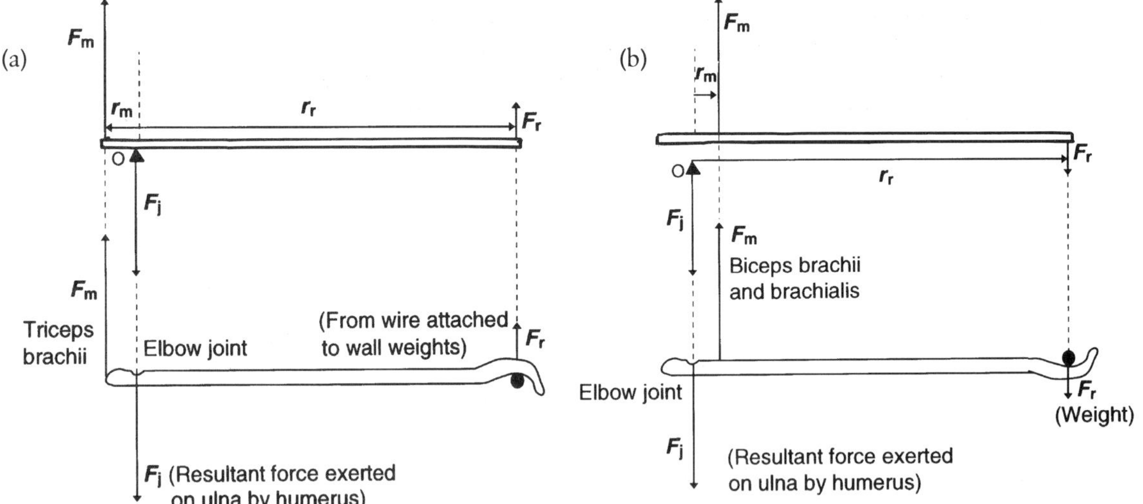

Figure 3.4 Levers as examples of parallel force systems: (a) first-class lever; (b) third-class lever.

The moment equilibrium equation in these examples reduces to the principle of levers. This states that the product of the magnitudes of the (muscle) force and its moment arm (sometimes called the force arm) equals the product of the resistance and its moment arm (the resistance arm), or $F_m r_m = F_r r_r$. The force equilibrium equation leads to $F_j = F_m + F_r$ and $F_j = F_m - F_r$ for the joint force (F_j) in the first- and third-class levers of Figures 3.4(a) and 3.4(b) respectively. It is worth mentioning here that the example of a second-class lever often quoted in sports biomechanics textbooks – that of a person rising on to the toes treating the floor as the fulcrum – is contrived. Few, if any, such levers exist in the human musculoskeletal system. This is not surprising as they represent a class of mechanical lever intended to enable a large force to be moved by a small one, as in a wheelbarrow. The human musculoskeletal system, by contrast, achieves speed and range of movement but requires relatively large muscle forces to accomplish this against resistance.

General force systems

These may be planar or spatial, have none of the above simplifications and are the ones normally found in sports biomechanics, such as when analysing the various soft tissue forces acting on a body segment. These force systems will not be covered further in this chapter.

The vector equations of statics (Equation 3.4) and the use of inertia forces can aid the analysis of the complex force systems that are commonplace in sports biomechanics. However, many biomechanists feel that the use of the equations of static equilibrium obscures the dynamic nature of force in sport, and that it is more revealing to deal with the dynamic equations of motion. This latter approach will be preferred in this book. In sport, force systems almost always change with time, as in Figure 3.5.

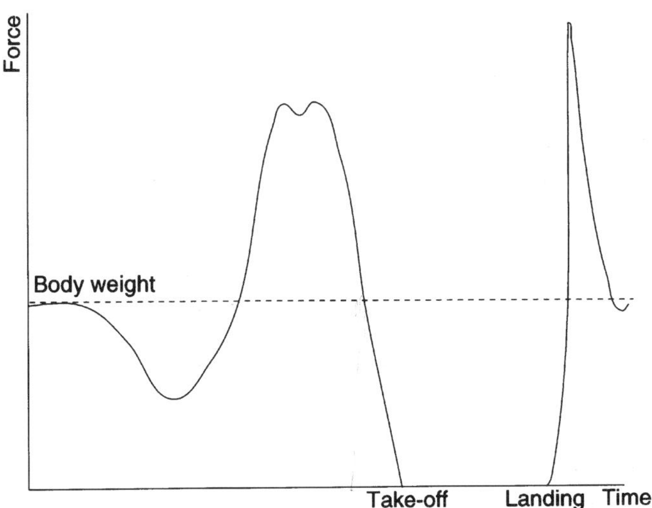

Figure 3.5 Vertical component of ground reaction force during a standing vertical jump.

This shows the vertical component of ground reaction force recorded from a force platform during a standing vertical jump. The effect of the force at any instant is reflected in an instantaneous acceleration of the performer's centre of mass. The change of the force with time determines the way in which the velocity and displacement of the centre of mass change, and it is important to remember this!

3.1.3 MOMENTUM AND NEWTON'S LAWS OF LINEAR MOTION

Inertia and mass

The inertia of an object is its reluctance to change its state of motion. Inertia is directly measured or expressed by the mass of the object, which is the quantity of matter of which the object is composed. It is more difficult to accelerate an object of large mass (such as a shot) than one of small mass (such as a dart). Mass is a scalar quantity, symbol m, and the SI unit of mass is the kilogram (kg).

Momentum

Linear momentum (symbol p) is the quantity of motion possessed by a particle or rigid body measured by the product of its mass and the velocity of its mass centre. It is a very important quantity in sports biomechanics. Being a product of a scalar and a vector, it is itself a vector whose direction is identical to that of the velocity vector. The unit of linear momentum is $kg{\cdot}m{\cdot}s^{-1}$.

First law (law of inertia)

An object will continue in a state of rest or of uniform motion in a straight line (constant velocity) unless acted upon by external forces that are not in equilibrium. Constant speed straight line skating is a close approximation to this. A skater can glide across the ice at almost constant speed as the coefficient of friction (see below) is so small. To change speed, the blades of the skates need to be turned away from the direction of motion to increase the force acting on them. In the flight phase of a long jump the horizontal velocity of the jumper remains almost constant as air resistance is small. However, the vertical velocity of the jumper changes continuously because of the jumper's weight – an external force caused by the gravitational pull of the earth.

Second law (law of momentum)

The rate of change of momentum of an object (or, for one of constant mass, its acceleration) is proportional to the force causing it and takes place in the direction in which the force acts. This is so vitally important as almost to justify, with its rotational counterpart, the name of the fundamental law of sports biomechanics. When a ball is kicked, in soccer for example, the acceleration of the ball will be proportional to the force applied to the ball by the kicker's foot and inversely proportional to the mass of the ball.

Third law (law of interaction)

For every action (force) exerted by one object on a second, there is an equal and opposite reaction (force) exerted by the second object on the first. The ground reaction force experienced by the runner of Figure 3.3 is equal and opposite to the force exerted by the runner on the ground (F). This latter force could be shown on a free body diagram of the ground.

Impulse of a force

Newton's second law of linear motion (the law of momentum) can be expressed mathematically as:

$$F = \frac{dp}{dt} = \frac{d(mv)}{dt} \qquad (3.5)$$

That is, F, the net external force acting on the body, equals the rate of change (d/dt) of momentum (p or mv). Equation 3.5 can be rewritten for an object of constant mass as:

$$F = \frac{dp}{dt} = m\frac{dv}{dt} = ma. \qquad (3.6)$$

This gives the familiar relationship that force is mass (m) times acceleration (a). Equations 3.5 and 3.6 can be rearranged, by multiplying by dt, and integrated to give:

$$\int F dt = \int d(mv) \; (= m\int dv, \text{ if } m \text{ is constant}) \qquad (3.7)$$

The left side of this equation is the impulse of the force (the SI unit is newton-seconds, N·s). This equals the change of momentum of the object ($\int d(mv)$ or $m\int dv$ if m is constant). Equation 3.7 is known as the impulse–momentum relationship and, with its equivalent form for rotation, forms a major foundation of studies of human dynamics in sport. The impulse is usually evaluated by numerical or graphical integration (as in Chapter 2). This can be performed from a force–time record, such as Figure 3.5, which is easily obtained from a force platform.

Equation 3.7 can be re-written for an object of constant mass (m) as $\bar{F}\Delta t = m\Delta v$, where $\bar{F}$ is the mean value of the force acting during a time interval Δt during which the velocity of the object changes by Δv. The change in the horizontal velocity of a sprinter from the gun firing to leaving the blocks depends on the impulse of the force exerted by the sprinter on the blocks (from the second law) and is inversely proportional to the mass of the sprinter. In turn, the impulse of the force exerted by the blocks on the sprinter is equal in magnitude but opposite in direction to that exerted, by muscular action, by the sprinter on the blocks (third law). Obviously, a large horizontal velocity off the blocks is desirable. However, a compromise is necessary as the time spent in achieving the required impulse (Δt) adds to the time spent running after leaving the blocks to give the recorded race time. The production of a large impulse of force is also important in many sports techniques of hitting, kicking and throwing in order to maximize the speed of the object involved. In javelin throwing, for example, the release speed of the javelin depends on the impulse applied to the javelin by the thrower during the delivery stride, and the impulse applied by ground reaction and gravity forces to the combined thrower–javelin system throughout the preceding phases of the throw. In catching a ball, the impulse required to stop the ball is determined by the mass (m) and velocity (Δv) of the ball. The catcher can reduce the mean force ($\bar{F}$) acting on the hands by increasing the duration of the contact (Δt) by 'giving' with the ball.

3.1.4 DETERMINATION OF THE CENTRE OF MASS OF THE HUMAN BODY

The centre of mass of an object is the unique point about which the mass of the object is evenly distributed. The effects of external force systems upon the human body can be studied by the linear (translational) kinematics of the mass centre and by rotations about the mass centre. It is often found that the movement patterns of the centre of mass vary between subjects of different skill level, hence providing a good, simple tool for evaluating technique (Page, 1978). The position of the mass centre is a function of age, sex and body build and changes with breathing, ingestion of food and disposition of body fluids. It is doubtful whether it can be pinpointed to better than 3 mm. In the fundamental or anatomical positions, the centre of gravity lies about 56–57% of a male's height from the soles of the feet, the figure for females being 55%. In this position, the centre of mass is located about 4 cm inferior to the navel roughly midway between the anterior and posterior skin surfaces. The position of the centre of mass is highly dependent on the orientation of a person's body segments. For example in a piked body position the centre of mass of a gymnast may lie outside the body.

Historically, several techniques were used to measure the position of the centre of mass of the sports performer. These included reaction board (boards and scales) and mannikin methods. The interested reader is referred to Page (1978) and Hay (1993) for descriptions of these techniques. They are now rarely, if ever, used in sports biomechanics, having been superseded by the segmentation method. In this method, the following information is required to calculate the position of the whole body centre of mass: the masses of the individual body segments and the locations of the centres of mass of those segments in the position to be analysed (e.g. the left calf centre of mass, x, in Figure 3.6). The latter requirement is usually met by a combination of the pre-established location of each segment's centre of mass with respect to the endpoints of the segment, and the positions of those endpoints on a photographic or video image. The endpoints (joint centres and terminal points) are estimated in the analysed body position from the camera viewing direction (e.g. Figure 3.6).

Figure 3.6 Centre of mass determination (adapted form Payne, 1985).

The ways of obtaining body segment data, including masses and locations of mass centres, will be considered in Chapter 5. The positions of the segmental endpoints are obtained from some visual record, such as cine film, still photographs or a video recording of the movement. For the centre of mass to represent the system of segmental masses, the moment of mass (similar to the moment of force) of the centre of mass must be identical to the sum of the moments of body segment masses about any given axis. The calculation can be expressed mathematically as:

$$mr = \Sigma m_i r_i$$

or

$$r = \Sigma(m_i/m)r_i \tag{3.8}$$

where m_i is the mass of segment number i; m is the mass of the whole body (the sum of all of the individual segment masses Σm_i); m_i/m is the fractional mass ratio of segment number i; and $\mathbf{r}$ and $\mathbf{r}_i$ are the position vectors respectively of the mass centre of the whole body and segment number i. In practice, the position vectors are specified in terms of their two-dimensional (x,y) or three-dimensional (x,y,z) coordinates. Table 3.1, at the end of the chapter, shows the calculation process for a two-dimensional case with given segmental mass fractions and position of centre of mass data.

3.2 Friction

3.2.1 TYPES OF FRICTION

The ground, or other, contact force acting on an athlete (Figure 3.7(a)) can be resolved into two components, one normal to (F_n) and one tangential to (F_t) the contact surface (Figure 3.7(b)).

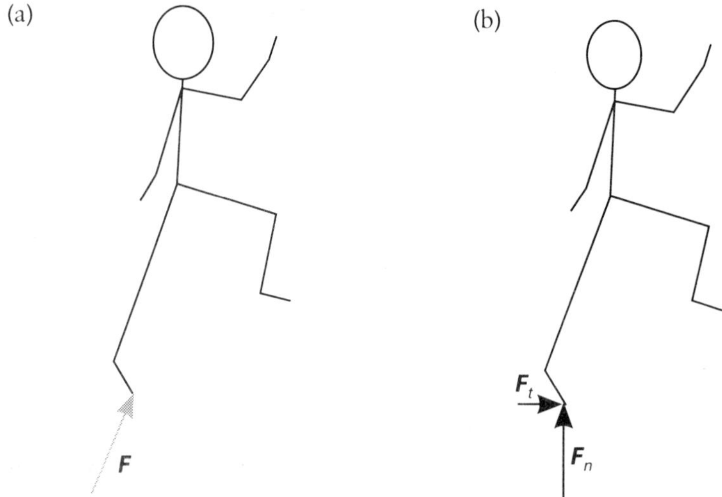

Figure 3.7 (a) Ground reaction force and (b) its components.

The former component is the normal force and the latter is the friction (or traction) force. Traction is the term used when the force is generated by interlocking of the contacting objects, such as spikes penetrating a Tartan track, known as **form locking.** In friction, the force is generated by **force locking.** Without friction or traction, movement in sport would be very difficult.

If an object, such as a training shoe (Figure 3.8(a)), is placed on a material such as Tartan, it is possible to investigate how the friction force changes.

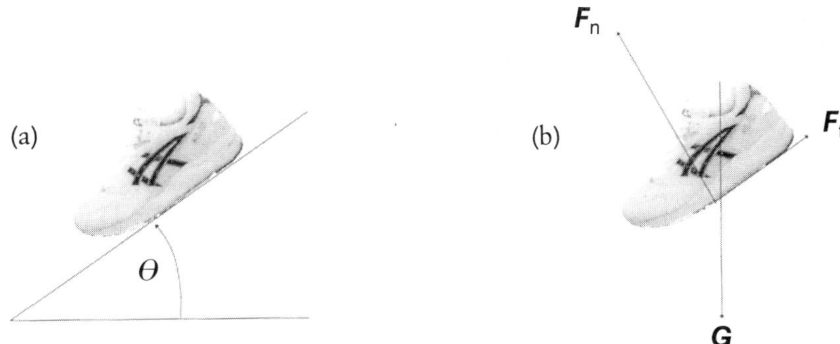

Figure 3.8 (a) Training shoe on an inclined plane and (b) its free body diagram.

The forces acting on the plane are shown in the free body diagram (Figure 3.8(b)). As the shoe is not moving, these forces are in equilibrium. Resolving the weight of the shoe (**G**) along and normal to the plane, the magnitudes of the components are equal respectively to F_t and F_n:

$$F_t = G \sin\theta \; ; \; F_n = G \cos\theta \qquad (3.9)$$

and, by dividing F_t by F_n:

$$F_t / F_n = \tan\theta \qquad (3.10)$$

If the angle of inclination of the plane (θ) is increased, the friction force will eventually be unable to resist the component of the shoe's weight down the slope and the shoe will begin to slide. The ratio of F_t/F_n (= $\tan\theta$) at which this occurs is called the coefficient of (limiting) static friction (μ_s). The maximum sliding friction force that can be transmitted between two bodies is:

$$F_{t\,max} = \mu_s \, F_n \qquad (3.11)$$

This is Newton's law of friction and refers to static friction, just before there is any relative movement between the two surfaces. It also relates to conditions where only the friction force prevents relative movement (referred to as **force locking**). For such conditions Equation 3.11 indicates that the maximum friction force depends only on the magnitude of the normal force pressing the surfaces together and the coefficient of static friction (μ_s). This coefficient depends only on the materials and nature (such as roughness) of the contacting surfaces and is, to a large extent, independent of the area of contact.

In certain sports, the tangential (usually horizontal) force is transmitted by interlocking surfaces (traction) rather than by friction, such as when spikes or studs penetrate or substantially deform a surface. It is this **form locking** that then generates the tangential force, which is usually greater than that obtainable from static friction. For such force generation, a traction coefficient can be defined similarly to the friction coefficient in Equation 3.11.

Once the two surfaces are moving relative to one another, as when skis slide over snow, the friction force between them decreases and a coefficient of kinetic friction is defined such that:

$$F_t = \mu_k F_n \tag{3.12}$$

noting $\mu_k < \mu_s$. This coefficient is relatively constant up to about 10 $m{\cdot}s^{-1}$. Kinetic friction always opposes relative sliding motion between two surfaces.

Friction not only affects translational motion, it also influences rotation, such as when swinging around a high bar or pivoting on the spot. At present there is no agreed definition of, nor agreed method of measuring, rotational friction coefficients in sport. Frictional resistance also occurs when one object tends to rotate or roll along another, as for a hockey ball rolling across an Astroturf pitch. In such cases it is possible to define a 'coefficient of rolling friction' using the relationship of Equation 3.12. The resistance to rolling is considerably less than the resistance to sliding and can be established by allowing a ball to roll down a slope from a fixed height (1 m is often used) and then measuring the horizontal distance that it rolls on the surface of interest. In general, for sports balls rolling on sports surfaces, the coefficient of rolling friction is of the order of 0.1 (see Bell, Baker and Canaway, 1985).

3.2.2 REDUCING FRICTION

From Equations 3.15 and 3.16, it can be seen that to reduce friction (of any type) or traction between two surfaces it is necessary to reduce the normal force or the coefficient of friction. In sport, the latter can be done by changing the materials of contact, and the former by technique. Such a technique is known as unweighting, where the performer imparts a downward acceleration to his or her centre of mass (Figure 3.9(a)), thus reducing the normal ground contact force to below body weight (Figure 3.9(b)). This technique is used, for example, in some turning techniques in skiing and is often used to facilitate rotational movements.

High coefficients of friction are detrimental when speed is wanted and friction opposes this. In skiing straight runs, kinetic friction is minimized by treating the base of the skis with wax. This can reduce the coefficient of friction to below 0.1. At the high speeds associated with skiing, frictional melting occurs, which further reduces the coefficient of friction to as low as 0.02 at speeds above 5 $m{\cdot}s^{-1}$. The friction coefficient is also affected by the nature of the snow–ice surface. In ice hockey, figure and speed skating, the sharpened blades minimize the friction coefficient in the direction parallel to the blade length. The high pressures involved cause localized melting of the ice, which, along with the smooth blade surface, reduces friction. Within the human body, where friction causes wear, synovial membranes of one form or another excrete synovial fluid to lubricate the structures involved, resulting in frictional coefficients as low as 0.001. This occurs in synovial joints and in the synovial sheaths (such as that of the biceps brachii long head) and synovial sacs and bursae (for example at the tendon of quadriceps femoris near the patella) which protect tendons.

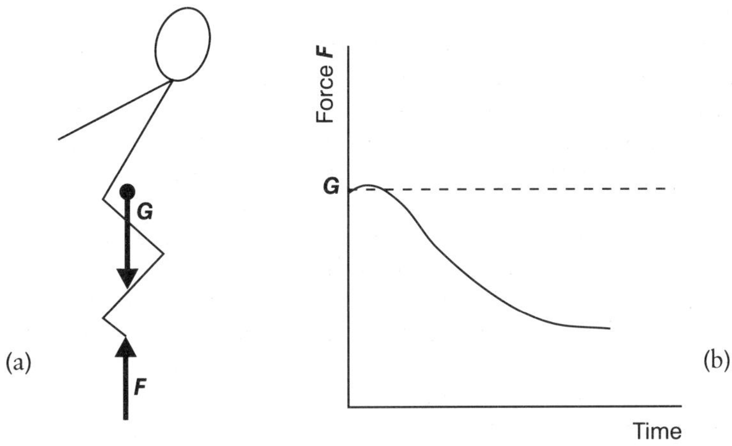

Figure 3.9 Unweighting: (a) free body diagram; (b) force platform record.

3.2.3 INCREASING FRICTION

A large coefficient of friction or traction is often needed to permit quick changes of velocity (large accelerations). To increase friction, it is necessary to increase either the normal force or the friction coefficient. The normal force can be increased by weighting, the opposite process from unweighting. Other examples of this include the use of inverted aerofoils on racing cars and the technique used by skilled rock and mountain climbers of leaning away from the rock face. To increase the coefficient of friction, a change of materials, conditions or locking mechanism is necessary. The last of these can lead to a larger traction coefficient through the use of spikes (and to a lesser extent studs) in, for instance, javelin throwing and running, where large velocity changes occur.

3.2.4 STARTING, STOPPING AND TURNING

As seen above, when large accelerations are needed, sliding should be minimized. The use of spikes or studs makes rotation more difficult. This requires a compromise in general games, where stopping and starting and rapid direction changes are combined with turning manoeuvres necessitating sliding of the shoe on the ground. When starting, the coefficient of friction or traction limits performance. A larger value on synthetic tracks compared with cinders (about 0.85 compared with 0.65, Stucke, Baudzus and Baumann, 1984) allows the runner a greater forward inclination of the trunk and a more horizontally directed leg drive and horizontal impulse. The use of blocks has similar benefits with form locking replacing force locking. With lower values of the coefficient of friction, the runner must accommodate by using shorter strides. Although spikes can increase traction, energy is necessary to pull them out of the surface and it is questionable whether they confer any benefit when running on dry, dust-free, synthetic tracks. When stopping, a sliding phase can be beneficial unless a firm anchoring of the foot is required as in the delivery stride of javelin throwing. Sliding is possible on cinder tracks or grass and this potential is used by the grass and clay court tennis player. On synthetic surfaces larger friction coefficients limit sliding and the athlete has to accommodate this by unweighting, by flexion of the knee. When turning, a high coefficient of friction requires substantial unweighting to permit a reduction in the moment needed to generate the turn. Without this, the energy demands of turning increase with the rotational friction. It has been suggested that the coefficient of friction should exceed 0.4 for safe walking on normal floors, 1.1 for running and 1.2 for all events (cited by Stucke, Baudzus and Baumann, 1984). In turning and stopping techniques in skiing, force locking is replaced by form locking (using the edges of the skis) and this substantially increases the force on the skis.

3.2.5 EFFECTS OF CONTACT MATERIALS

Greatest friction coefficients occur between two absolutely smooth, dry surfaces owing to microscopic force locking at atomic or molecular level. Many dry, clean, smooth metal surfaces *in vacuo* adhere when they meet and in most cases attempts to slide one past the other produce complete seizure. This perfect smoothness is made use of with certain rubbers in rock climbing shoes and racing tyres for dry surfaces. In the former case, the soft, smooth rubber adheres to the surface of the rock, and in the latter, localized melting of the rubber occurs at the road surface. However in wet (or very dirty) conditions the loss of friction is very marked, as the adhesion between the rubber and surface is prevented. The coefficient of friction reduces for a smooth tyre from around 5.0 on a dry surface to around 0.1 on a wet one. A compromise is afforded by treaded tyres with dry and wet surface friction coefficients of around 1.0 and 0.4 respectively. The treads allow water to be removed from the contact area between the tyre and road surface. Likewise, most sports shoes have treaded or cleated soles, and club and racket grips are rarely perfectly smooth. In some case, the cleats on the sole of a sports shoe will also provide some form locking with certain surfaces (see above).

3.2.6 PULLEY FRICTION

Passing a rope around the surface of a pulley makes it easier to resist a force of large magnitude at one end by a much smaller force at the other, because of the friction between the rope and the pulley. This principle is used for example in abseiling techniques in rock climbing and mountaineering. It also explains the need for synovial membranes (to prevent high friction forces) when tendons pass over bony prominences.

3.3 The impact of sports objects

3.3.1 THE NATURE OF IMPACT

Many sports involve the impact of one object with another in ways that affect the performance or technique of the participant. It is therefore important to have a sound understanding of the factors that affect the results of such impacts. An impact involves the collision of two objects over a short time, during which the two objects exert relatively large forces on each other. For example, a golf ball driven at 70 $m{\cdot}s^{-1}$ experiences a maximum impact force in the region of 10 000 N during a clubhead contact of around 0.5 ms. The forces involved in an impact are inversely related to the duration of the impact. When stopping a moving object the contact should be prolonged, reducing the impact force, as when boxers ‘ride a punch’. The behaviour of two objects after an impact depends on a number of factors including the nature of the impact, the relative momentums of the objects before impact and energy losses during impact.

During the initial part of an impact, both of the objects will deform to some extent, although the relative magnitudes of the two deformations may differ considerably, and the greater the deformation the longer the impact lasts. The energy stored in the deformation will be wholly regained in the latter part of impact only if the deformation is perfectly elastic: the impact of two snooker balls is a reasonable approximation to this. In a totally plastic deformation, such as a shot embedding itself in turf, no energy is regained and the kinetic energy of the objects before impact is downgraded to heat energy. Most impacts involve a combination of elastic and plastic deformation.

3.3.2 DIRECT IMPACT AND THE COEFFICIENT OF RESTITUTION

Direct impacts between two objects occur when their velocity vectors are parallel (one may be zero) and the forces of impact between the colliding objects act through their centres of mass. Linear momentum is conserved during such impacts. As noted above, energy is lost during an impact unless it is perfectly elastic. This is expressed by an empirical law:

$$v_{b2} - v_{a2} = e(v_{b1} - v_{a1}) \tag{3.13}$$

where the subscript a and b refer to the two objects; 1 and 2 refer to the velocities of the objects before and after impact respectively; $v_{b2} - v_{a2}$ is the rebound or separation velocity; $v_{b1} - v_{a1}$ is the approach velocity; e is the coefficient of restitution, which is 0 for plastic and 1 for perfectly elastic impacts. When the scalar (magnitude only) form of Equation 3.13 is used, a negative sign is necessary on the right hand side to account for the change in direction of the relative velocity vector during the impact. The velocities of the two bodies after impact can be expressed in terms of their masses and velocities before impact and the coefficient of restitution. Daish (1972) provides some interesting developments of these relationships.

The coefficient of restitution depends greatly on the materials and construction of the colliding objects. For sports balls, the rules often specify a range of acceptable values of e. For example, a tennis ball dropped from a height of 2.54 m on to a concrete surface is required to rebound to a height between 1.35 and 1.47 m, which needs a coefficient of restitution in the range 0.64–0.75. The coefficient of restitution decreases with impact speed and varies with temperature. Air-filled balls have a value of e that increases with temperature and depends on inflation pressure where relevant. This temperature effect accounts for squash balls becoming faster during play. The rise in temperature causes an increase of the air pressure inside the ball making it more difficult to deform, and the coefficient of restitution increases. The relationship between ball-bounce (rebound) resilience (e^2) and the 'speed' of sports surfaces was considered by Bell, Baker and Canaway (1985).

3.3.3 OBLIQUE IMPACT

This is far more common in sport than is direct impact. Examples include ball–surface impacts in the basketball bounce pass and lay-up shot, and bat–ball collisions in cricket shots. In the basketball examples, one object is stationary, whereas in the cricket shots both objects are moving. For simplicity, only the former will be considered in this section, although the extension to the latter can be done through use of the principle of relative motion (see Daish, 1972). Some of the following section assumes that the bouncing ball and surface are rigid, except as accounted for by the energy loss expressed through the coefficient of restitution. This rigidity assumption will not be true, especially for air-filled balls. The mathematics of this problem are dealt with in Daish (1972) and will not be repeated here. The results are, however, worthy of consideration. The problem is summarized in Figure 3.10.

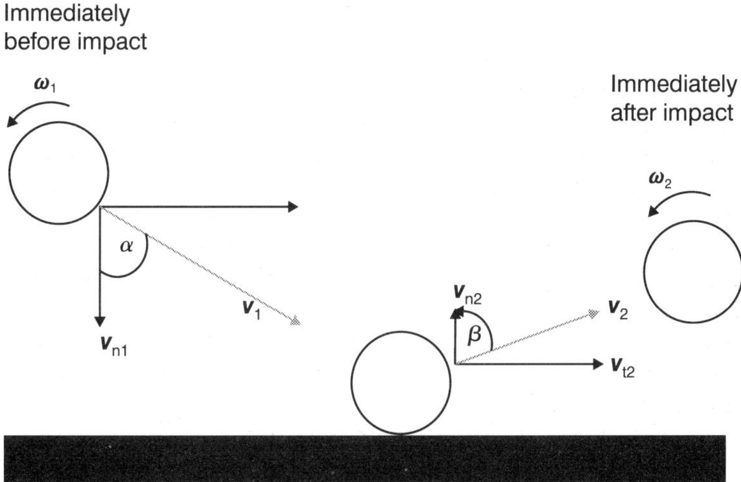

Figure 3.10 Oblique impact between a ball and a stationary surface.

The ball approaches the surface with a linear velocity v_1 and angular velocity (backspin) $\boldsymbol{\omega}_1$ and rebounds with velocity v_2 and angular velocity $\boldsymbol{\omega}_2$. Figure 3.10 also shows the approach and rebound angles α and β and the velocity components normal and tangential to the surface. As for direct impact, the problem is solved by applying conservation of momentum principles and the principle of energy loss, using e. The latter will mean that the normal rebound speed is always less than or equal to the normal approach speed ($v_{n2} = e\, v_{n1}$). In addition, the rotational form of Newton's second law of motion (see section 3.4) has to be considered.

The behaviour of the ball following impact depends on whether the ball is still sliding when it rebounds or whether there is sufficient friction between the surface and ball to stop sliding. The ball will then be instantaneously rolling on the surface at rebound. In this context, two critical values can be defined, a critical approach angle (α_c) and a critical magnitude of the approach angular velocity (ω_c). Using the nomenclature of Figure 3.10, with the coefficients of friction and restitution as μ and e, and for a solid ball of radius r:

$$\tan\alpha_c = 3.5\mu\,(1 + e) \qquad (3.14)$$

$$\omega_c = v_1\,[3.5\mu\,(1 + e)\cos\alpha - \sin\alpha] \qquad (3.15)$$

The following summarizes the solution, which is also represented, for certain cases as discussed below, in Figure 3.11.

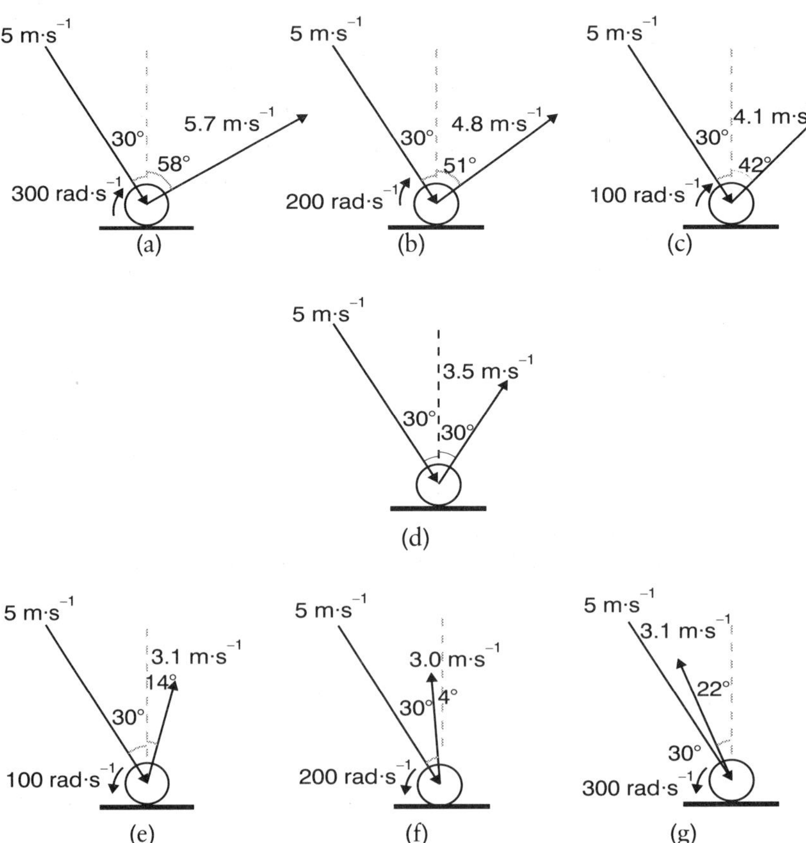

Figure 3.11 The predicted effects of varying degrees of spin on rebound: (a)–(c) decreasing topspin; (d) no spin; (e)–(g) increasing backspin (adapted from Daish, 1972).

Sliding throughout contact

This will occur if the angular velocity of the ball before impact exceeds the critical value ($\omega_1 > \omega_c$). The ball will lose tangential velocity and acquire some topspin. If the approach angle is less than the critical value

($\alpha < \alpha_c$), then ω_c is positive (backspin) and no topspinning ball (ω_1 negative and thus $\omega_1 < \omega_c$) will slide throughout contact. For initially non-spinning balls, the ball will always slide throughout contact if the approach angle exceeds the critical angle ($\alpha > \alpha_c$).

Rolling at rebound

Rolling at rebound will occur if the angular velocity of the ball before impact does not exceed the critical value ($\omega_1 \leqslant \omega_c$). All the examples of Figure 3.11 (for which ω_c was estimated to be > 350 rad·s^{-1}) satisfied this condition. For initially non-spinning balls, this will always occur if the approach angle does not exceed the critical angle ($\alpha \leqslant \alpha_c$). The ball will lose tangential velocity and acquire some topspin. A ball with backspin will also lose tangential velocity and acquire some topspin (i.e. it will lose backspin). If the product of the ball's radius and angular velocity before impact exceeds 2.5 times its tangential approach speed ($r\omega_1 > 2.5v_{t1}$), the ball will rebound with some residual backspin and with negative values of the tangential rebound speed and rebound angle (v_{t2} and β). That is it will rebound backwards (Figure 3.11(f),(g)), a phenomenon familiar to many players of ball sports such as table tennis. The behaviour of topspun balls depends on whether the product of their radius and angular velocity before impact (ignoring the negative sign for topspin) is greater than the tangential approach speed ($r\omega_1 > v_{t1}$). If this is the case, as in Figure 3.11(a)–(c), then the bottom of the ball is moving backwards relative to the surface at impact. The ball then rebounds with less topspin, but with increased tangential velocity. This effect is familiar to tennis players. Otherwise, increased topspin but decreased tangential velocity result.

3.3.4 INSTANTANEOUS CENTRES OF ROTATION AND THE CENTRE OF PERCUSSION

Consider a rigid body that is simultaneously undergoing translation and rotation about its mass centre, as in Figure 3.12(a).

The whole body shares the same translational velocity (v) (Figure 3.12(b)) but the tangential velocity due to rotation depends on the displacement (r) from the centre of mass G ($v_t = \boldsymbol{\omega} \times \boldsymbol{r}$) (Figure 3.12(c)). Adding $\boldsymbol{v}$ and $\boldsymbol{v}_t$ gives the net velocity ($\boldsymbol{v} + \boldsymbol{v}_t$) (Figure 3.12(d)) which at a point P (which need not lie within the body) has a zero value. This point of zero velocity is known as the **instantaneous centre of rotation** (or velocity pole). Its position often changes with time.

Consider now a similar rigid body to that of Figure 3.12 acted upon by an impact force F (Figure 3.13(a)).

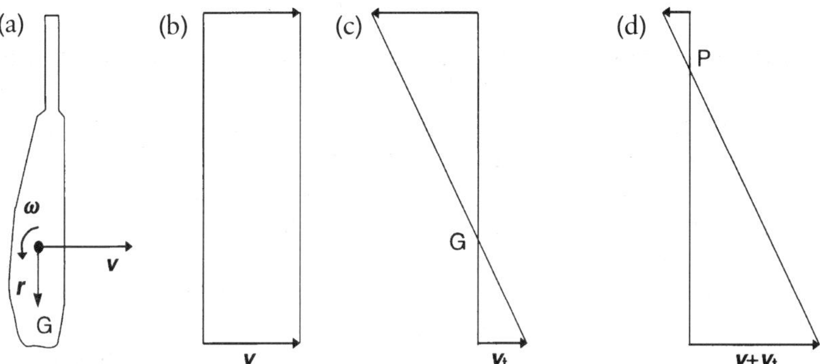

Figure 3.12 Instantaneous centre of rotation: (a) translating and rotating rigid body; (b) translational velocity profile; (c) tangential velocity profile; (d) net velocity profile and velocity pole (P).

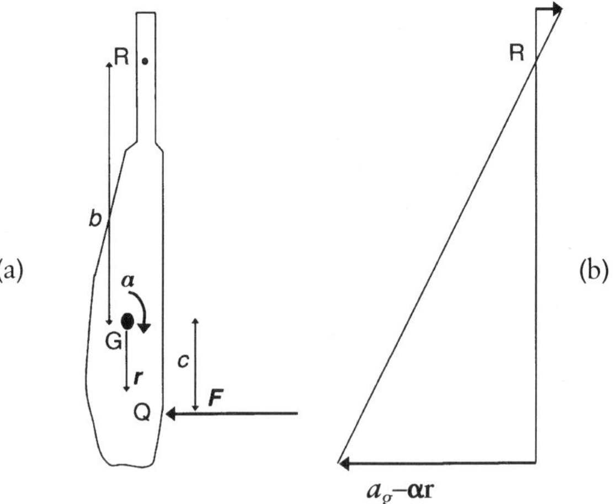

Figure 3.13 Acceleration pole and centre of percussion: (a) impact force on rigid body; (b) net acceleration profile and centre of percussion (Q).

There will, in general, be a point R which will experience no net acceleration. By Newton's second law of linear motion, the magnitude of the acceleration of the centre of mass of the rigid body, mass m, is $a_g = F/m$. From Newton's second law of rotation (see below), the magnitude (Fc) of the moment of force F is equal to the product of the moment of inertia of the body about its centre of gravity (I_g) and its angular acceleration (α). That is $Fc = I_g\,\alpha$. This gives a tangential acceleration (αr) that increases linearly with distance (r) from G. The net acceleration profile then appears as in Figure 3.13(b). At point R ($r = b$), the two accelerations are equal and opposite and hence cancel to give a zero net acceleration. The position of R is given by equating F/m to Fc/I_g. That is:

$$bc = I_g/m. \tag{3.16}$$

Point R is known as the **acceleration pole.** If R is a fixed centre of rotation then Q is known as the **centre of percussion**, defined as that point at which a force may be applied without causing an acceleration at another specific point, the centre of rotation. Equation 3.16 shows that Q and R can be reversed. The centre of percussion is important in sports where objects, such as balls, are struck with other objects such as bats and rackets. If the impact occurs at the centre of percussion, no force is transmitted to the hands. For an object such as a cricket bat, the centre of percussion will lie some way below the centre of gravity, whereas for a golf club, with the mass concentrated in the club head, the centres of percussion and gravity more nearly coincide. Variation in grip position will alter the position of the centre of percussion. If the grip position is a long way from the centre of gravity and the centre of percussion is close to the centre of gravity, then the position of the centre of percussion will be less sensitive to changes in grip position. This is achieved by moving the centre of gravity towards the centre of percussion. Examples include golf clubs with light shafts and heavy club heads and cricket bats where there is a build up of the mass of the bat around the centre of percussion. Much tennis racket design has been evolving towards positioning the centre of percussion nearer to the likely impact spot. The benefits of such a design feature include less fatigue and a reduction in injury. Page (1978) reported that playing oneself in (in cricket) involved, among other things, accommodating the stroke to hit the ball nearer the centre of percussion of the bat.

The application of the centre of percussion concept to a generally non-rigid body, such as the human performer, is problematic. However, some insight can be gained into certain techniques. Consider a reversal of Figure 3.13 so that Q is high in the body and R is at the ground. Let *F* be the reaction force experienced by a thrower who is applying force to an external object. If *F* is directed through the centre of percussion, there will be no resultant acceleration at R (the foot–ground interface). A second example relates Figure 3.13 to the braking effect when the foot lands in front of the body's mass centre. The horizontal component of the impact force will oppose relative motion and cause an acceleration distribution, as in Figure 3.13, with all body parts below R decelerated and only those above R accelerated. This is important in, for example, javelin throwing, where it is desirable not to slow the speed of the object to be thrown during the final foot contacts of the thrower.

3.4 Fundamentals of rotational kinetics

Almost all human motion in sport and exercise involves rotation (angular motion), for example the movement of a body segment about its proximal joint. In section 2.3, the kinematics of rotational motion was considered. In this section, the focus will be on the kinetics of such motions.

3.4.1 MOMENT OF INERTIA

In linear (translational) motion, the reluctance of an object to move, its inertia, is expressed by its mass. In angular (rotational) motion, the reluctance of the object to rotate also depends on the distribution of that mass about the axis of rotation, as in Figure 3.14, and is expressed by the **moment of inertia**. The SI unit for moment of inertia is kilogram-metres^2 (kg·m^2) and moment of inertia is a scalar quantity.

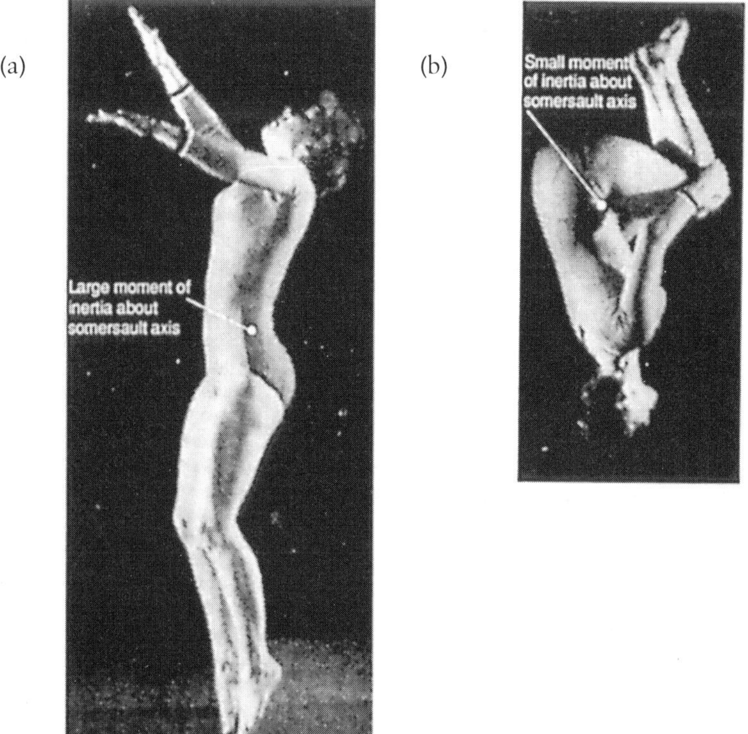

Figure 3.14 Gymnast in: (a) extended and (b) tucked positions (reproduced from Coulton, 1977, with permission).

The extended (straight) gymnast of Figure 3.14(a) has a greater moment of inertia than the tucked gymnast of Figure 3.14(b) and is therefore more 'reluctant' to rotate. This is a factor which makes it more difficult to somersault in an extended (layout) position rather than in a piked or tucked position.

Formally stated, the moment of inertia is the measure of an object's resistance to accelerated angular motion about an axis. It is equal to the sum of the products of the masses of the object's elements and the squares of the distances of those elements from the axis. Consider a small element of mass δm of a thin rectangular plate, as in Figure 3.15(a).

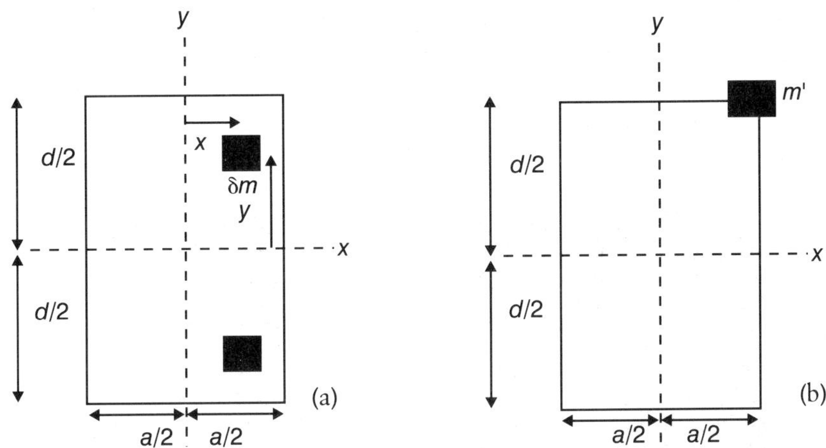

Figure 3.15 Products of inertia: (a) balanced masses; (b) unbalanced masses.

Two moments of inertia about the centre of mass are defined in the plane xy:

$$\begin{aligned} I_{yy} &= \Sigma x^2\,\delta m \quad (= m\,a^2/12) \\ I_{xx} &= \Sigma y^2\,\delta m \quad (= m\,d^2/12) \end{aligned} \tag{3.17}$$

These are the moments of inertia about the x and y axes respectively. Moments of inertia can be expressed in terms of the radius of gyration, k, such that the moment of inertia is the product of the mass and the square of the radius of gyration: $I = m\,k^2$. For the moments of inertia in Equation 3.17, this gives $k_y = a/12^{1/2}$ and $k_x = d/12^{1/2}$.

It is also possible to define:

$$I_{xy} = \Sigma xy\,\delta m \tag{3.18}$$

It should be evident that every elemental mass δm having a positive value of y in Figure 3.15(a) is balanced, for the same value of x, by an elemental mass δm having a negative y. Therefore $I_{xy} = 0$ in this case. I_{xy} is known as a **product of inertia** and those axes through a body for which the products of inertia are zero are known as the **principal axes of inertia**. There are three such axes through any point on a three-dimensional rigid body and they are mutually perpendicular. The moments of inertia about these axes are the **principal moments of inertia** (such as I_{xx} and I_{yy} for the body of Figure 3.15(a)). One of the principal moments of inertia is the maximum moment of inertia for any axis through that point, another principal moment of inertia is a minimum and the third is necessarily an intermediate value. For the sports performer in the reference or anatomical position, the three principal axes of inertia through the centre of mass correspond with the three cardinal axes. The moment of inertia about the vertical axis is much smaller (approximately one-tenth) of the moment of inertia about the other two axes and that about the frontal axis is slightly less than that about the sagittal axis. It is worth noting that rotations about the intermediate principal axis are unstable, whereas those about the principal axes with the greatest and least moments of inertia are stable. This is another factor making layout somersaults difficult, unless the body is realigned to make the moment of inertia about the frontal (somersault) axis greater than that about the sagittal axis.

Consider now the addition of a small mass (m') to the object of Figure 3.15(a), as in Figure 3.15(b). Then:

$$I_{xy} = m'ad/4 \tag{3.19}$$

and the x and y axes are no longer principal axes of inertia. With respect to the sports performer, movements of the limbs (other than in symmetry) away from the standing position result in a misalignment between the body's cardinal axes and the principal axes of inertia. This has important consequences for the generation of aerial twist in a twisting somersault.

The three principal axes of inertia are the obvious axes about which to express moments of inertia. The centre of mass is usually chosen as the point through which the axes pass, as the moments of inertia about this point are less than those about any other point. The parallel axis theorem (Equation 3.20) allows the calculation of the moments of inertia about a parallel axis through another point. In Figure 3.16, the moment of inertia of the forearm about the shoulder axis (I_s) is the sum of the moment of inertia of the forearm about its centre of mass (I_g) and the product of the mass (m) of the forearm and the square of the distance of its centre of mass from the shoulder axis (r). That is:

$$I_s = I_g + mr^2 \tag{3.20}$$

This theorem can also be used to calculate the moments of inertia and the radii of gyration for body segments about their proximal and distal ends from the values about the centre of mass (see Table 3.2 at the end of this chapter).

Equation 3.20 can further be adapted to compute the moment of inertia of several segments (i = 1, 2, etc.) about a common axis:

$$I = \Sigma I_i = \Sigma(I_{gi} + m_i r_i^2) \tag{3.21}$$

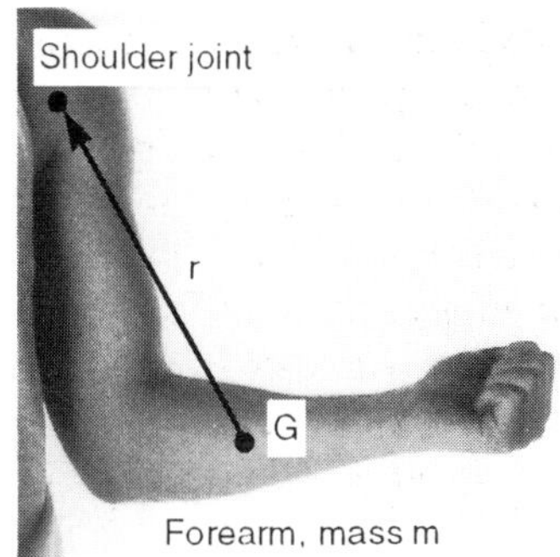

Figure 3.16 Moment of inertia of the forearm about the shoulder.

where I_{gi}, m_i and r_i are, respectively, the moment of inertia about the segment centre of mass, the mass, and the perpendicular distance between the chosen axis and the parallel axis through the centre of mass of segment number *i*. Equation 3.21 can be used to calculate a moment of inertia for the human body as a whole. This is useful when the performer is behaving as a quasi-rigid body, as in the extended or tucked positions of Figure 3.21. It is not adequate to explain rotation of a multisegmental system in which some segments move relative to others, as is the case in the great majority of human movement in sport.

3.4.2 LAWS OF ROTATIONAL MOTION

These are analogous to Newton's three laws of linear motion. Some applications of these laws to sports movements are also considered in sections 3.4.5 and 3.4.6.

Principle of conservation of angular momentum (law of inertia)

A rotating body will continue to turn about its axis of rotation with constant angular momentum unless an external moment of force (or a torque) acts on it. The magnitude of the moment of force about an axis of rotation is the product of the force and its moment arm, which is the perpendicular distance of the line of action of the force from the axis of rotation.

Law of momentum

The rate of change (d/d*t*) of angular momentum (*L*) of a body is proportional to the moment (*M*) causing it and has the same direction as the moment. This is expressed mathematically by Equation 3.22, which is true for rotation about any axis fixed in space:

$$M = \frac{dL}{dt} \tag{3.22}$$

Rearranging this equation, by multiplying by dt, and integrating:

$$\int M dt = \int dL = \Delta L \tag{3.23}$$

That is, the impulse of the moment equals the change of angular momentum (ΔL). If the moment impulse is zero, then Equation 3.23 reduces to:

$$\Delta L = 0 \text{ or } L = \text{constant} \tag{3.24}$$

which is a mathematical statement of the first law.

Equation 3.22 can be modified by writing $\boldsymbol{L} = I\boldsymbol{\omega}$ where $\boldsymbol{\omega}$ is the angular velocity if, and only if, the axis of rotation is a principal axis of inertia of a rigid body or quasi-rigid body. Then, and only then:

$$M = \frac{d(I\omega)}{dt} \tag{3.25}$$

Further if, and only if, I is constant (for example for an individual body segment):

$$M = I\boldsymbol{\alpha}, \tag{3.26}$$

where $\boldsymbol{\alpha}$ is the angular acceleration. The restrictions on the use of Equations 3.25 and 3.26 compared with the universality of Equation 3.22 should be carefully noted. The angular motion equations for spatial (three-dimensional) rotations are far more complex than the ones above and will not be considered here.

The law of reaction

For every moment that is exerted by one object on another, there is an equal and opposite moment exerted by the second object on the first. Thus in Figure 3.17(a), the forward and upward swing of the long jumper's legs (thigh flexion) evokes a reaction causing the forward and downward motion of the trunk (trunk flexion).

As the two moments here are both within the jumper's body, there is no change in her angular momentum. From Equation 3.23, the two moments involved produce equal but opposite impulses. In Figure 3.17(b), the ground provides the reaction to the action moment generated in the racket arm and upper body, although it is not apparent because of the extremely high moment of inertia of the earth. If the tennis player played the same shot with his feet off the ground there would be a moment on the lower body causing it to rotate counter to the movement of the upper body, lessening the angular momentum in the arm–racket system.

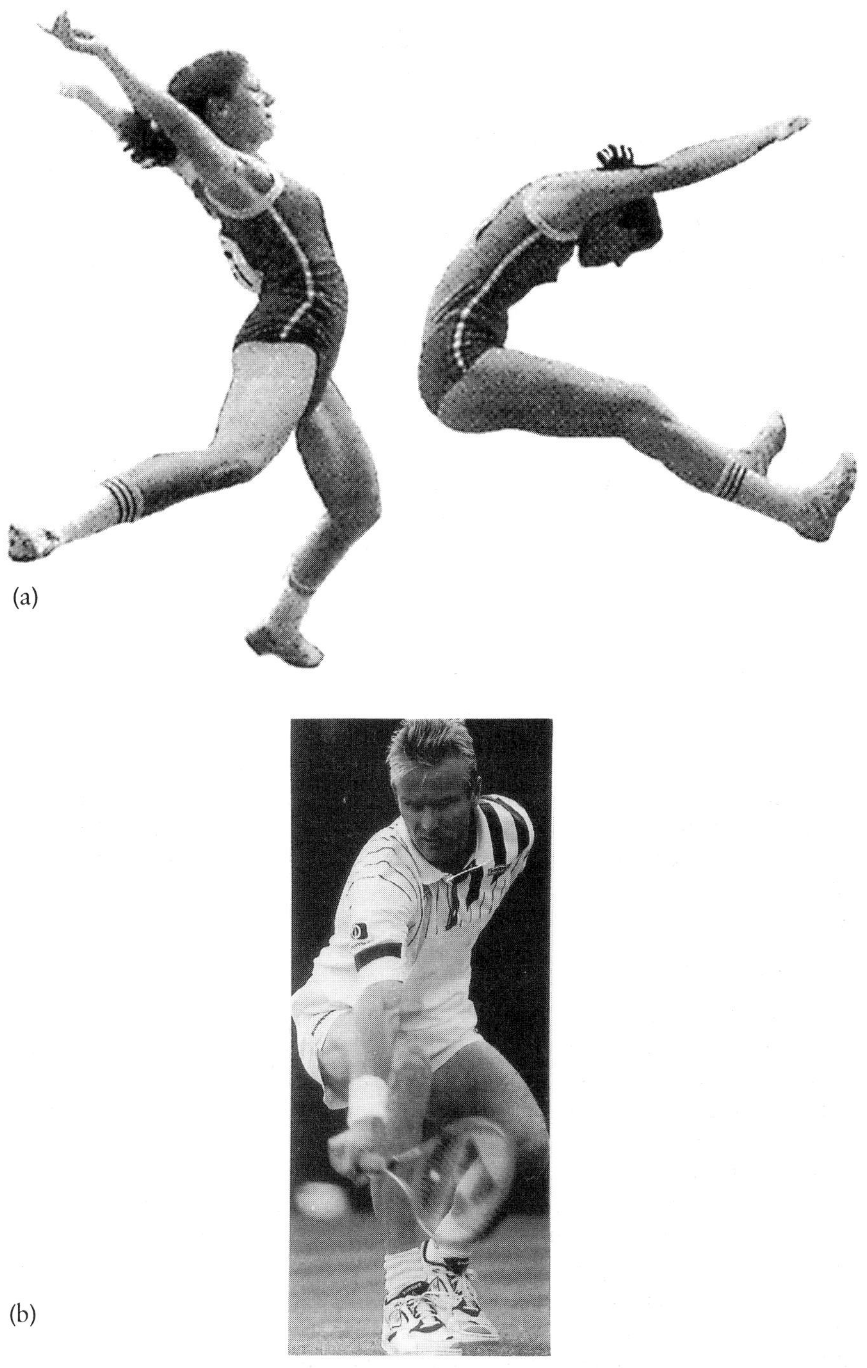

(a)

(b)

Figure 3.17 Action and reaction: (a) airborne; (b) with ground contact ((a) adapted from Payne, 1985).

Minimizing inertia

The law of momentum allows the derivation of the principle of minimizing inertia. From Equation 3.25, the increase in angular velocity (and thus the reduction in the time taken to move through a specified angle) will be greater if the moment of inertia of the whole chain of body segments about the axis of rotation is minimized. Hence, for example, the flexing of the knee of the recovery leg in running (especially in sprinting) to minimize the duration of the recovery phase.

3.4.3 ANGULAR MOMENTUM OF A RIGID BODY

The angular momentum of a particle (such as δm in Figure 3.18(a)) about a point can be expressed as the product of its linear momentum and the perpendicular distance of the linear momentum vector from the point.

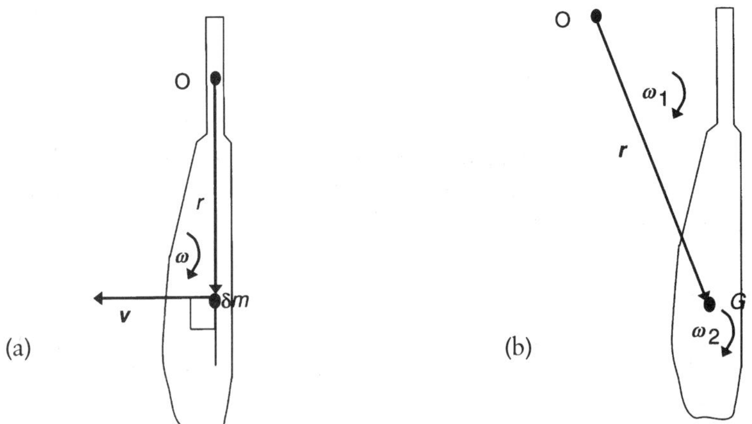

Figure 3.18 Angular momentum: (a) single rigid body; (b) part of a system of rigid bodies.

Such moments can be taken for all elemental masses, δm, of any rigid body rotating about either an axis fixed in space or a principal axis of inertia, as for the bat in Figure 3.18(a). It can then be shown that, for planar motion, the angular momentum ($\boldsymbol{L}$) is the product of the body's moment of inertia about the axis (I) and its angular velocity ($\boldsymbol{\omega}$), that is $\boldsymbol{L} = I\boldsymbol{\omega}$. The direction of vector $\boldsymbol{L}$ (as that of $\boldsymbol{\omega}$) is given by allowing vector $\boldsymbol{r}$ to rotate towards vector $\boldsymbol{v}$ through the right angle indicated. By the right hand rule, the angular momentum vector is into the plane of the page. The SI unit of angular momentum is $kg \cdot m^2 \cdot s^{-1}$.

3.4.4 ANGULAR MOMENTUM OF A SYSTEM OF RIGID BODIES

For planar rotations of systems of rigid bodies (for example the sports performer), each rigid body can be considered to rotate about its mass centre (G), with an angular velocity $\boldsymbol{\omega}_2$. This centre of mass rotates about the mass centre of the whole system (O), with an angular velocity $\boldsymbol{\omega}_1$, as for the bat of Figure 3.18(b). The derivation will not be provided here; the result is that the magnitude of the angular momentum of the bat is:

$$L = m\omega_1 r^2 + I_g\omega_2 \qquad (3.27)$$

The first term, owing to the motion of the body's mass centre about the system's mass centre, is known as the remote angular momentum. The latter, owing to the rigid body's rotation about its own mass centre, is the local angular momentum. For an interconnected system of rigid body segments, representing the sports performer, the total angular momentum is the sum of the angular momentums of each of the segments calculated as in the above equation. The ways in which angular momentum is transferred between body segments can then be studied for sports activities such as the flight phase of the long jump.

3.4.5 GENERATION OF ANGULAR MOMENTUM

To alter the angular momentum of a sports performer requires a net external turning effect (or torque) acting on the system. Traditionally in sports biomechanics three mechanisms of inducing rotation have been identified, although they are, in fact, related.

Force couple

A force couple consists of a parallel force system of two equal and opposite forces which are not collinear, i.e. they are a certain distance apart (Figure 3.19(a)).

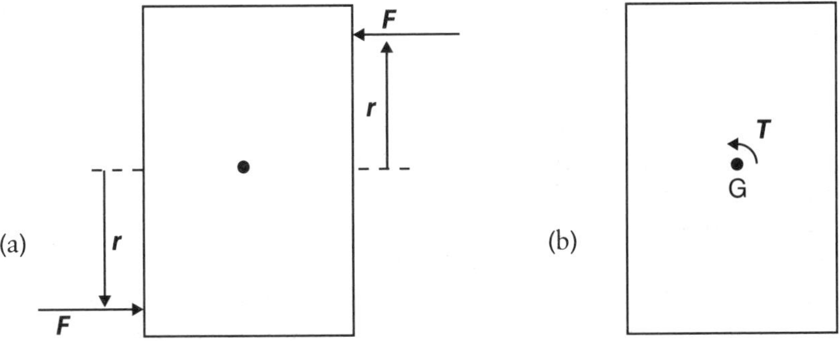

Figure 3.19 Force couple: (a) the couple; (b) the torque or moment of the couple.

The net translational effect of these two forces is zero and they cause only rotation. The net moment of the couple, the torque, is: $\mathbf{T} = \mathbf{r} \times \mathbf{F}$. The moment (or torque) vector has a direction perpendicular to and into the plane of this page. Its magnitude is the magnitude of one of the forces multiplied by the perpendicular distance between them (Fr).

The torque (moment of the couple) can be represented as in Figure 3.19(b) and has the same effect about a particular axis of rotation wherever it is applied along the body. In the absence of an external axis of rotation, the body will rotate about an axis through its centre of mass. The swimmer in Figure 3.2(a) is acted upon by a force couple of her weight and the equal, but opposite, buoyancy force.

Eccentric force

An eccentric force ('eccentric' means 'off-centre') is effectively any force (or resultant of a force system) that is not zero and that does not act through the centre of mass of an object. This constitutes the commonest way of generating rotational motion, as in Figure 3.20(a).

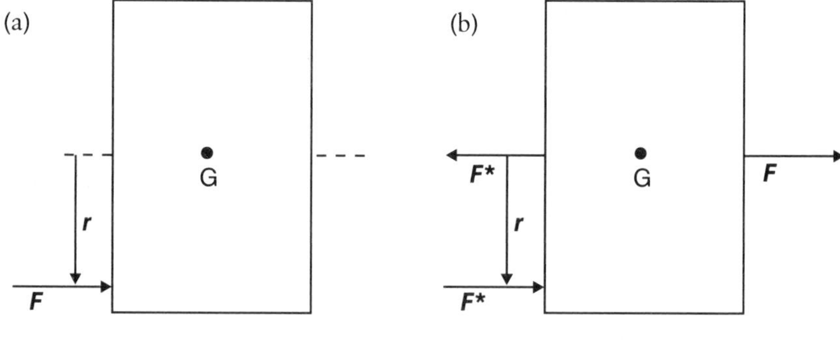

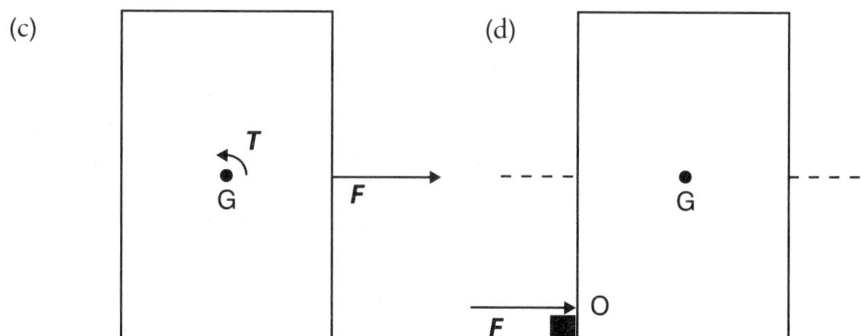

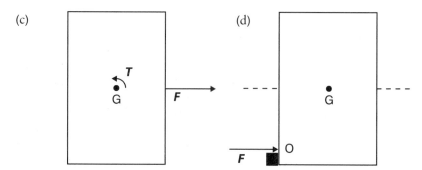

Figure 3.20 Generation of rotation: (a) an eccentric force; (b) addition of two equal and opposite collinear forces; (c) equivalence to a 'pure' force and a torque; (d) checking of linear motion.

The eccentric force here can be transformed by adding two equal and opposite forces at the centre of mass, as in Figure 3.20(b), which will have no net effect on the object. The two forces indicated in Figure 3.20(b) with an asterisk can then be considered together and constitute a force couple (anticlockwise), which can be replaced by a torque T as in Figure 3.20(c). This leaves a 'pure' force acting through the centre of mass. This force F causes only translation, $F = \mathrm{d}(mv)/\mathrm{d}t$, and the torque T causes only rotation ($T = \mathrm{d}L/\mathrm{d}t$). The magnitude of T is Fr.

This example could be held to justify the use of the term 'torque' for the turning effect of an eccentric force although, strictly, 'torque' is defined as the moment of a force couple. The two terms, torque and moment, are often used interchangeably. There is a case for abandoning the use of the term moment (of a force or couple) entirely in favour of torque, given the various other uses of the term moment in biomechanics. At present, however, 'moment' persists in this context and will be used throughout this book.

Checking of linear motion

This occurs when an already moving body is suddenly stopped at one point. An example is the foot plant of a javelin thrower in the delivery stride, although the representation of such a system as a quasi-rigid body is of limited use. This (as shown in Figure 3.20(d)) is merely a special case of an eccentric force. It is best considered in that way to avoid misunderstandings which exist in the literature, such as the misconception that O is the instantaneous centre of rotation (see above).

3.4.6 INTERPRETATION OF PLANAR SPORTS MOTIONS

Limitations of the quasi-rigid body model

For the sequence of Figure 3.21, the angular momentum acquired at take-off is constant throughout the airborne phase of the dive (tucked 1½ backward somersault). Use of the quasi-rigid body model shows that, by tucking, the diver reduces the moment of inertia about his transverse axis (a principal axis of inertia). This increases his rotational speed (angular velocity) to facilitate the completion of the 1½ somersaults. He then returns to his original position, with a high moment of inertia, to reduce rotational speed and enter the water with 'good form'. During the quasi-rigid sequences – a couple of frames after leaving the board and before entering the water, and during the tuck (0.45 s to 0.65 s approximately) – the formula $\boldsymbol{L} = I\boldsymbol{\omega}$ would apply. During the remainder of the sequence, as the body orientation changes, this equation and the quasi-rigid body model are somewhat meaningless as there is no single angular velocity for the body as a whole.

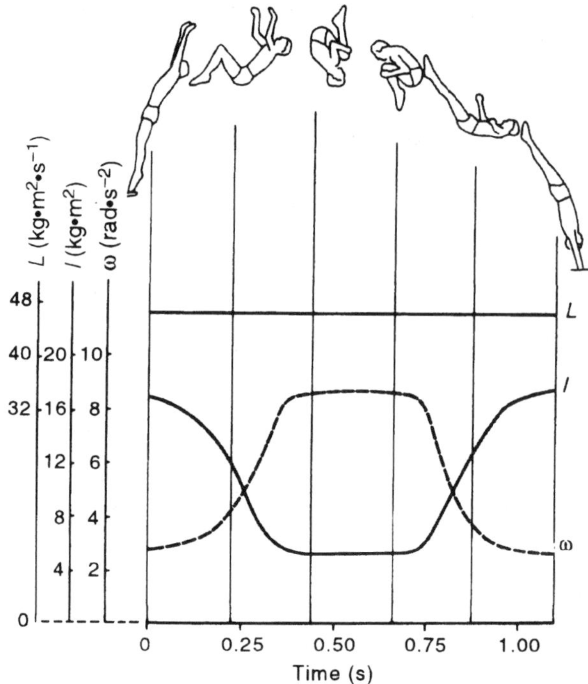

Figure 3.21 Angular momentum and limitations of the quasi-rigid body model (reproduced from Hay, 1993, with permission).

Transfer of angular momentum

The principle of transfer of angular momentum from segment to segment is often considered to be a basic principle of coordinated movement. Consider the diver performing the simple piked dive in Figure 3.22. The diver is represented by two 'gross' body segments attached in the pelvic region, about which they mutually rotate by muscular contractions evoking the torque–countertorque principle. Here take-off is followed by contraction of the trunk and hip flexors, generating a clockwise (forward) angular momentum in the trunk and an equal but opposite anticlockwise angular momentum in the legs. These add to and subtract from the take-off values to give the trend up to approximately 0.4 s. Angular momentum has been transferred from the legs (which remain more or less fixed in space) to the upper body. Thereafter the leg and trunk extensors contract to transfer angular momentum from the trunk to the legs until approximately 1.2 s. During this second phase of angular momentum transfer the orientation of the trunk in space remains fairly constant. Prior to water entry, the original angular momentum values are more or less restored. A more complex example is the long jump hitch kick technique, where the arm and leg motions transfer angular momentum from the trunk, to prevent the jumper from rotating forwards too early.

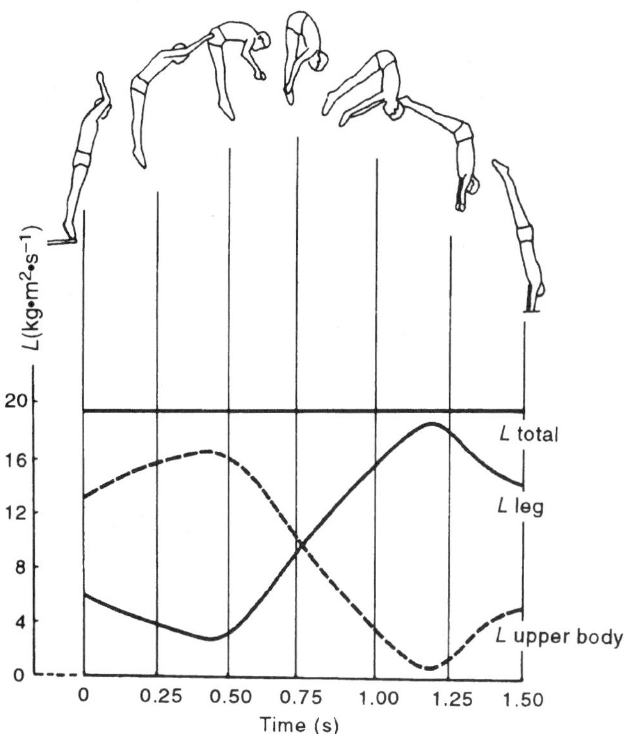

Figure 3.22 Transfer of angular momentum (reproduced from Hay, 1993, with permission).

Trading of angular momentum

This term is often used to refer to the transfer of angular momentum from one axis of rotation to another. For example, the model diver – or gymnast – (from Yeadon, 1993) in Figure 3.23 takes off with angular momentum about the somersault (horizontal) axis.

The diver then adducts his left arm (or some other asymmetrical movement) by a muscular torque, which evokes an equal but opposite counter rotation of the rest of the body to produce an angle of tilt. No external torque has been applied so the angular momentum is still constant about a horizontal axis but now has a component about the twisting axis. The diver has 'traded' some somersaulting angular momentum for twisting (longitudinal) angular momentum and hence will now both somersault and twist. It is often argued that this method of generating twisting angular momentum is preferable to contact twist, as it can be more easily removed by re-establishing the original body position before landing. This can avoid problems in gymnastics, trampolining and diving caused by landing with residual twisting angular momentum. The crucial factor in generating airborne twist is to establish a tilt angle and, approximately, the twist rate is proportional to the angle of tilt. In practice, many sports performers use both the contact and the airborne mechanisms to acquire twist (see for example Yeadon, 1993).

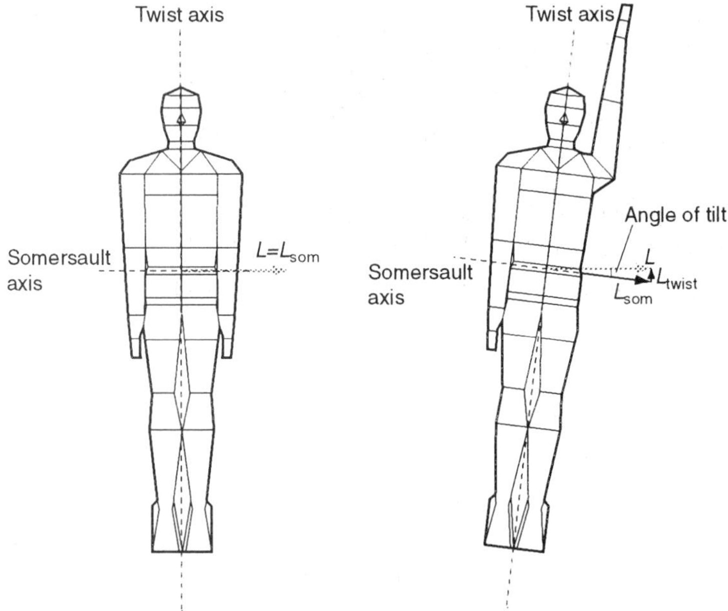

Figure 3.23 Trading of angular momentum between axes.

3.5 Spatial rotation

Rotational movements in airborne sports activities in diving, gymnastics and trampolining, for example, often involve multisegmental movements of a very three-dimensional nature. The mathematical analysis of the dynamics of such movements is beyond the scope of this book. The trading of angular momentum between the body's axes of rotation was briefly discussed in the previous section. Even the spatial dynamics of a rigid or quasi-rigid body are far from simple. For example, in planar rotation of such bodies, the angular momentum and angular velocity vectors coincide in direction. If such a body with principal moments of inertia which are not identical (as for the sports performer), rotates about an axis that does not coincide with one of the principal axes, then the angular velocity vector and the angular momentum vectors do not coincide. The movement known as **nutation** can result (see Figure 3.24). Nutation also occurs, for example, when performing an airborne pirouette with asymmetrical arm positions. The body's longitudinal axis is displaced away from its original position of coincidence with the angular momentum vector (sometimes called the axis of momentum), and will describe a cone around that vector.

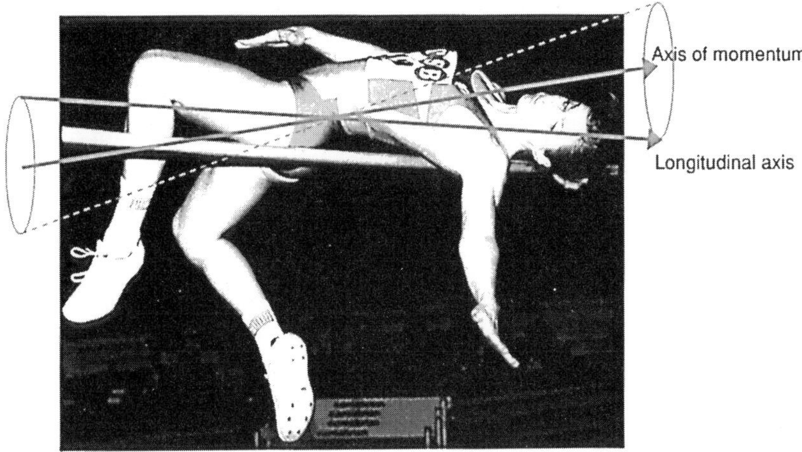

Figure 3.24 Nutation.

Furthermore, the equation of conservation of angular momentum (Equation 3.24) applies to an inertial frame of reference, such as one moving with the centre of mass of the performer but always parallel to a fixed, stationary frame of reference. The conservation of angular momentum does not generally apply to a frame of reference fixed in the performer's body and rotating with it. For a further discussion of some aspects of spatial rotation, see Yeadon (1993).

3.6 Summary

In this chapter linear kinetics were considered, which are important for an understanding of human movement in sport and exercise. This included the definition of force and its SI unit, the identification of the various external forces acting in sport, the laws of linear kinetics and related concepts such as linear momentum, and the ways in which force systems can be classified. The segmentation method for calculating the position of the whole body centre of mass of the sports performer was explained. Some important forces were considered in more detail. The ways in which friction and traction influence movements in sport and exercise were addressed, including reducing and increasing friction and traction. An appreciation was provided of the factors that govern impact, both direct and oblique, of sports objects, and the centre of percussion was introduced and related to sports objects and performers. The vitally important topic of rotational kinetics was covered, including the laws of rotational kinetics and related concepts such as angular momentum and the ways in which rotation is acquired and controlled in sports motions. The chapter concluded with a very brief introduction to spatial (three-dimensional) rotation.

3.7 Exercises

1. Define force in terms of its SI unit. List the external forces that act on the sports performer and, for each force, give an example of a sport or exercise in which that force will be very important. Draw sketches of examples from sport and exercise of each of the force system classifications in section 3.1.2.
2. Define and explain the three laws of linear kinetics and give at least two examples from sport or exercise, other than the examples in this chapter, of the application of each law.
3. Photocopy Figure 3.6, or trace on to graph paper the outline of the thrower. Use the x,y axes and the segment end points of Figure 3.6. Measure the x and y coordinates of each of the segment end points. Then use a photocopy of Table 3.1 to calculate the position of the thrower's whole body centre of mass in the units of the image. Finally, as a check on your calculation, mark the resulting centre of mass position on your figure. If it looks silly, check your calculations and repeat until the centre of mass position appears reasonable.
4. Carry out the inclined plane experiment of section 3.2.1 to calculate the coefficient of friction between the plane's surface and training shoes and other sports objects. (You only need a board of material, a shoe and a protractor.)
5. List some examples, other than those of section 3.2, of sports in which methods are used to increase and decrease friction to aid performance.
6. Using the equations for the critical angular velocity and critical approach angle in section 3.3, establish for the following examples, in all of which $e = 0.7$, whether the solid ball will slide throughout impact or roll. Where appropriate, establish any other facts you can about the ball's behaviour after impact from the information in section 3.3.
   a) Ball radius 35 mm, $\mu = 0.5$, $v = 20\ \text{m·s}^{-1}$, $\alpha = 80°$, $\omega = 0$.
   b) Ball radius 35 mm, $\mu = 0.5$, $v = 20\ \text{m·s}^{-1}$, $\alpha = 30°$, $\omega = -200\ \text{rad·s}^{-1}$.
   c) Ball radius 70 mm, $\mu = 0.6$, $v = 5\ \text{m·s}^{-1}$, $\alpha = 30°$, $\omega = 800\ \text{rad·s}^{-1}$.
7. Distinguish between moments of inertia and products of inertia. Define, and explain, the three laws of angular kinetics and give at least two examples from sport or exercise of the application of each law.
8. Consider the ways in which angular momentum can be generated (section 3.4.5). Give examples of each from sport and exercise movements.
9. Obtain a video recording of top class diving, trampolining or gymnastics. Carefully analyse some airborne movements that do not involve twisting, including the transfer of angular momentum between body segments.
10. Repeat Exercise 9 for movements involving twisting. Consider in particular the ways in which the performers generate twist in somersaulting movements.

Table 3.1 Calculation of the two-dimensional whole body centre of mass position (data from Dempster (1955), adjusted to correct for fluid loss)

Segment	Mass fraction[1]	Length ratio[1,2]	x-coordinate				y-coordinate			
			Proximal	Distal	Mass centre	Moment	Proximal	Distal	Mass centre	Moment
	$m' = m_i / m$	lr	x_p	x_d	$x_m = x_p + lr(x_d - x_p)$	$m' x_m$	y_p	y_d	$y_m = y_p + lr(y_d - y_p)$	$m' y_m$
Head and neck	0.582	0.396								
R Upper arm	0.028	0.436								
R Forearm	0.016	0.430								
R Hand	0.006	0.506								
R Thigh	0.099	0.433								
R Calf	0.046	0.433								
R Foot	0.014	0.429								
L Upper arm	0.028	0.436								
L Forearm	0.016	0.430								
L Hand	0.006	0.506								
L Thigh	0.099	0.433								
L Calf	0.046	0.433								
L Foot	0.014	0.429								
Whole body	1.000	-------	-------	-------	-------		-------	-------	-------	

Notes:
1. Using simple ratios, although many biomechanists use regression equations (see also Chapter 5).
2. Defined as the distance of the mass centre from the proximal joint (the seventh cervical vertebra for the head–neck) towards the distal point divided by the proximal to distal point distance. The distal point is the distal joint, the vertex of the head, the midpoint of the line joining the hips (for the trunk), third metacarpophalangeal joint or distal end of the second toe.

Table 3.2 Radii of gyration for body segments (values calculated from the data of Chandler *et al.*, 1975)

Segment	Radii of gyration[1,2] about the frontal axis through:		
	Segment mass centre (k_g)	Proximal joint (k_p)	Distal joint/point (k_d)
Head and neck	0.419	0.784	-------
Trunk	0.224	0.551	0.482
Upper arm	0.300	0.589	0.577
Forearm	0.288	0.505	0.653
Hand	0.320	0.606	0.587
Thigh	0.329	0.516	0.691
Calf	0.291	0.510	0.653
Foot	0.255	0.507	0.610

Notes: 1. Using simple ratios, although many biomechanists use regression equations (see also Chapter 5).
2. Defined as a fraction of the segment length and can be related to one another by the parallel axis theorem.

3.8 References

Bell, M. J., Baker, S. W. and Canaway, P.M. (1985) Playing quality of sport surfaces: a review. *Journal of the Sports Turf Research Institute*, **61**, 30–38.

Chandler, R. F., Clauser, C. E., McConville, J. T. *et al.* (1975) Investigation of inertial properties of the human body. Report DOT HS-801 430, US Department of Transportation, Washington, DC.

Clauser, C. E., McConville, J. T. and Young, J. W. (1969) Weight, volume and center of mass of segments of the human body. Report AMRL-TR-69-70, Wright-Patterson Air Force Base, Aerospace Medical Research Laboratory, Dayton, OH.

Coulton, J. (1977) *Women's Gymnastics*, EP, Wakefield.

Cureton, T. K. (1951) *Physical Fitness of Champion Athletes*, University of Illinois Press, Urbana, IL.

Daish, C. B. (1972) *The Physics of Ball Games*, EUP, Cambridge.

Dempster, W. T. (1955) Space requirements for the seated operator. WADC Technical Report 55159, Wright-Patterson Air Force Base, Wright Air Development Centre, Dayton, OH.

Hay, J. G. (1993) *The Biomechanics of Sports Techniques*, Prentice Hall, Englewood Cliffs, NJ.

Malina, R. M. (1969) Growth and physical performance of African negro and white children. *Clinical Pediatrics*, **8**, 476–483.

Page, R. L. (1978) *The Physics of Human Movement*, Wheaton, Exeter.

Payne, H. (ed.) (1985) *Athletes in Action*, Pelham Books, London.

Stucke, H., Baudzus, W. and Baumann, W. (1984) On friction characteristics of playing surfaces, in *Sports Shoes and Playing Surfaces*, (ed. E. C. Frederick), Human Kinetics, Champaign, IL, pp.87–97.

Thomas, D. G. (1989) *Swimming: Steps to Success*, Leisure Press, Champaign, IL.

Yeadon, M. R. (1993) The biomechanics of twisting somersaults. *Journal of Sports Sciences*, **11**, 187–225.

3.9 Further reading

Daish, C. B. (1972) *The Physics of Ball Games*, EUP, Cambridge: chapters 2 and 3 present a non-mathematical treatment of impact. The mathematically inclined reader will also benefit from reading chapters 10 and 15.

Page, R. L. (1978) *The Physics of Human Movement*, Wheaton, Exeter: chapter 3 deals with centre of mass measurement.

Stucke, H., Baudzus, W. and Baumann, W. (1984) On friction characteristics of playing surfaces, in *Sports Shoes and Playing Surfaces*, (ed. E. C. Frederick), Human Kinetics, Champaign, IL: pp.87–97.

4 Fluid mechanics and energetics

This chapter is intended to provide an understanding of the forces involved in moving through a fluid and the energetics of both linear and rotational motion. These topics are of considerable importance for human movement in sport and exercise. After reading this chapter you should be able to:

- understand the different types of fluid flow;
- appreciate how drag forces affect motion through fluids (air and water);
- explain how differential boundary layer separation can cause a sideways force, as exemplified by cricket ball swing;
- appreciate the other mechanisms that cause lift, or sideways, forces on bodies moving through fluids;
- use the principles of thermodynamics to study the energetics of the human performer;
- understand the models of inter- and intrasegmental energy transfers and how they can be applied to movements in sport and exercise.

4.1 Moving through a fluid

4.1.1 VISCOSITY

All sports take place within a fluid environment which may be gaseous (air), liquid (underwater turns in swimming) or both (sailing). A fluid is defined as a substance which will continue to distort even when subjected to a very small shear force or shear stress, which tends to slide layers of the fluid past each other, as in Figure 4.1.

The shear stress is the shear force divided by the area over which it acts. Fluids differ from solids which distort only a fixed amount under the action of a shear stress. The distinction can be expressed as follows. **Solids** have a fixed shape and volume with their particles arranged in a fixed structure. **Fluids** flow freely; their shape is only retained if enclosed in a container and particles of the fluid alter their relative positions whenever a force acts. **Liquids** have a volume that stays the same while the shape changes. **Gases** expand to fill the whole volume available by changing density. This ability of fluids to distort continuously under the action of shear forces is vital to sports motions, as it permits movement.

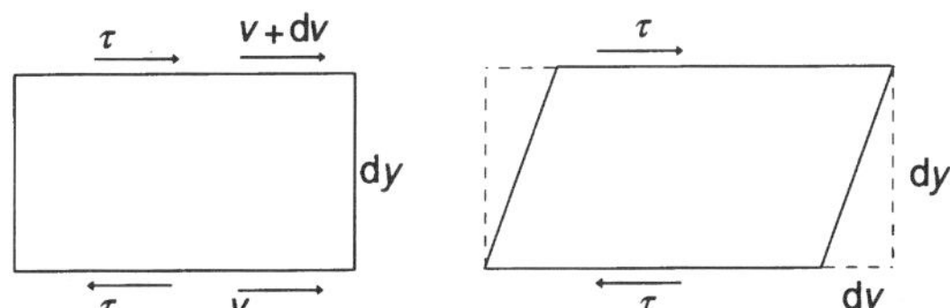

Figure 4.1 Shear stress and rate of shear on an element of fluid.

Although fluids flow freely, there is a resistance to this flow known as the **viscosity** of the fluid. This is a property causing shear stresses between adjacent layers of moving fluid, leading to an irreversible loss of energy (as heat) and a resistance to motion through the fluid. The relationship between shear stress (τ) and rate of shear (dv/dy) depends on the fluid. The rate of shear is the difference in velocity between adjacent layers of fluid divided by the distance between them. For most common fluids, including those in which sports take place, the shear stress is directly proportional to the rate of shear. The constant of proportionality is called the **coefficient of dynamic viscosity** (η), which has an SI unit of pascal-seconds (Pa·s). This coefficient depends in general on both the pressure and the temperature of the fluid. For moderate pressures, the effect of pressure changes is small, but temperature has a significant effect. An increase in temperature leads to a decrease in viscosity for liquids and an increase for gases. The ratio of the coefficient of dynamic viscosity to the density of the fluid is known as the **coefficient of kinematic viscosity** (ν). The SI unit for kinematic viscosity is $m^2 \cdot s^{-1}$.

4.1.2 FLUID KINEMATICS

In general, the instantaneous velocity of a small element of fluid will depend both on time and its spatial position. An element of fluid will generally follow a complex path, such as that in Figure 4.2(a).

This is known as the **path line** of the element of fluid. In steady motion, the velocity at any fixed point in the fluid is independent of time. The path lines are then reproduced at different times. They can be seen by introducing some visualization material, such as a dye for a liquid or smoke for a gas, into the fluid flow. An imaginary line that lies tangential to the direction of flow of the fluid particles at any instant is called a **streamline** (Figure 4.2(b)). Streamlines have no fluid flow across them. In steady flow the streamlines are the same at all times and coincide with the path lines.

Figure 4.2 Fluid kinematics: (a) path line; (b) streamlines.

4.1.3 BASIC EQUATIONS OF FLUID MOTION

The three basic equations of fluid motion are as follows.

Conservation of mass (continuity equation)

If the motion of a fluid is steady, the mass of fluid flowing across any two sections through the flow must be the same. Thus:

$$\rho A v = \text{constant} \tag{4.1}$$

where ρ is the density of the fluid, A is the cross-sectional area of flow, v is the mean fluid speed across the section. For incompressible fluids (water and also air at moderate speeds), the fluid density is constant, so the continuity equation (4.1) becomes:

$$A v = \text{constant} \tag{4.2}$$

This shows that reducing the flow area increases the fluid speed and *vice versa*.

Bernoulli's equation

The energy form of Bernoulli's equation relates pressure, potential and kinetic energies (per unit mass of fluid):

$$p/\rho + gh + \tfrac{1}{2}v^2 = \text{constant} \tag{4.3}$$

where h is the height above some arbitrary and convenient datum and the other symbols have been defined above. If Equation 4.3 is multiplied by ρ, the result is the pressure form of Bernoulli's equation. In this p is referred to as the **static pressure**, ρgh as the **hydrostatic pressure** and $\tfrac{1}{2}\rho v^2$ as the **dynamic pressure.** The sum of the three terms is known as the **total** or **stagnation pressure** and would occur if the fluid was brought to rest. If height changes are negligible, as is often the case in sport applications, then:

$$p/\rho + \tfrac{1}{2}v^2 = \text{constant} \tag{4.4}$$

Within a closed volume of fluid, an increased speed will result in a decrease in pressure, providing the flow is subsonic. With flow in a confined volume of incompressible fluid, a combination of Equations 4.2 and 4.4 shows that reducing the flow area (as for fluid flow past a ball or runner) results in an increase in fluid speed and a decrease in fluid pressure. This is sometimes referred to as **Bernoulli's principle**. In Figure 4.3, the region from (1) to (2) is an accelerating, decreasing pressure region and that from (2) to (3) is a region of deceleration and increasing pressure.

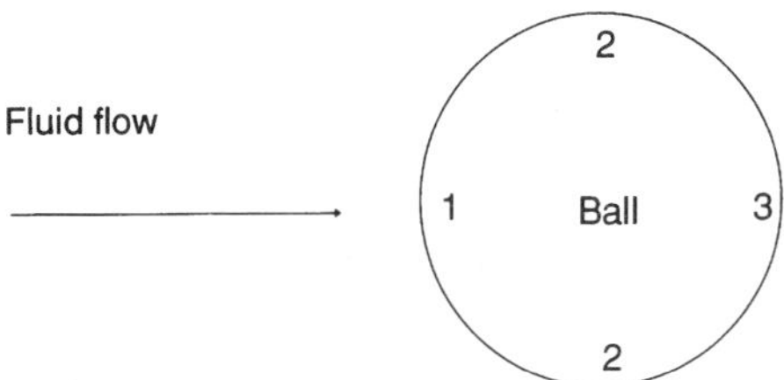

Figure 4.3 Regions of accelerating (1–2) and decelerating (2–3) fluid flow.

Momentum equation

Newton's second law of motion also applies to fluids, so that the force acting on the fluid is equal to its rate of change of momentum. Thus if an external force acts on a fluid it will change the fluid's momentum vector. Conversely, if a momentum change occurs as a fluid flows past an object, a force must have been exerted on the fluid by the object in the required direction to cause that momentum change. By Newton's third law of motion, the force that the fluid exerts on the object is equal and opposite to that exerted by the object on the fluid.

4.1.4 NON-DIMENSIONAL GROUPS

Because of the complexity of the general equations of fluid flow, many problems are approached empirically. The equipment tested, for example in the design of racing yachts, is usually a scale model of the full-size equipment. For the results of such tests to be valid, the model and the full-scale equipment must possess physical similarity. This means that the two objects must be geometrically similar in all respects and that the flow of fluid around them must be dynamically similar. This latter requirement is met only when important forces acting on the corresponding fluid particles have the same ratio to one another for both cases. The most important forces in sport applications are: inertial force (mass × acceleration), viscous (or shear) force (see above), and gravity force (mass × gravitational acceleration, g). Ratios of these forces are known as non-dimensional groups, because they have no unit, being force divided by force. Two are of considerable importance in sports biomechanics.

Reynolds number

This is important in all fluid flow in sport. It is the ratio of the inertial forces in the fluid to the viscous forces. It is calculated from the expression:

$$Re = \rho l v / \eta = l v / \nu \tag{4.5}$$

In this equation, l is some characteristic dimension of the flow, for example the diameter of a ball or the waterline length of a yacht hull.

Froude number

This is the square root of the ratio of inertial to gravity forces:

$$Fr = v/(lg)^{1/2} \tag{4.6}$$

The Froude number is important whenever a free surface, that between two fluids, occurs and waves are generated. This happens in almost all water sports.

4.1.5 TYPES OF FLUID FLOW

The type of flow which exists under given conditions depends on the Reynolds number. For low Reynolds numbers the fluid flow will be laminar. At the critical Reynolds number, the flow passes through a transition region and then becomes turbulent at a slightly higher Reynolds number.

Laminar flow

In this type of flow (Figure 4.4(a)), the fluid particles move only in the direction of the flow. The fluid can be considered to consist of discrete plates or laminae flowing over one another. The streamline pattern will remain constant if the flow is steady. Laminar flow occurs at moderate speeds past objects of small diameter (for example a table tennis ball). In laminar flow energy is exchanged only between adjacent layers of the flowing fluid.

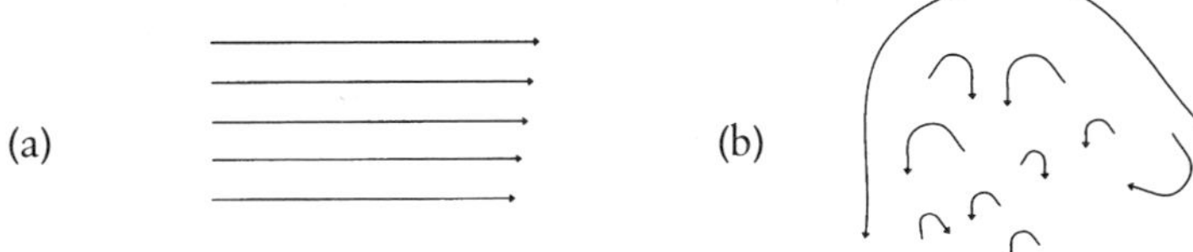

Figure 4.4 Types of fluid flow: (a) laminar flow; (b) turbulent flow.

Turbulent flow

This is the predominant type of fluid flow in sport. The particles of fluid have fluctuating velocity components in both the main flow direction and perpendicular to it (Figure 4.4(b)). It is best conceived of as a random collection of eddies or vortices. In turbulent flow energy is exchanged by the turbulent eddies on a greater scale than in laminar flow.

Transition between flow types – the critical Reynolds number

This depends very much on the object past which the fluid flows and the choice of the characteristic length and speed. Consider a fairly flat boat hull moving through stationary water. The distance along the hull from the bow is the characteristic length in the equation for the 'local' Reynolds number of the water flow past that point on the hull. The characteristic speed is that of the boat moving through the water. For this example, the critical Reynolds number is in the range 100 000–3 000 000, depending on the nature of the flow in the water away from the boat and the surface roughness of the hull. The boundary layer flow will change from laminar to turbulent at the point along the hull where the 'local' Reynolds number equals the critical value. If this does not happen, the flow will remain laminar along the whole length of the hull. For flow past a ball, the ball diameter is used as the characteristic length and the characteristic speed is the speed of the ball relative to the air. The critical Reynolds number, depending on ball roughness and flow conditions outside the boundary layer, is in the range 100 000–300 000.

The boundary layer

When relative motion occurs between a fluid and an object, as for flow of air or water past the sport performer, the fluid nearest the object is slowed down because of its viscosity. The region of fluid affected in this way is known as the boundary layer. Within this layer, the relative velocity of the fluid and object changes from zero at the surface of the object to the free stream velocity. The free stream velocity is the difference between the velocity of the object and the velocity of the fluid outside the boundary layer. The slowing down of the fluid is accentuated if the flow of the fluid is from a wider to a narrower cross-section of the object, as the fluid is then trying to flow from a low-pressure to a high-pressure region. Some of the fluid in the boundary layer may lose all its kinetic energy. The boundary layer then **separates** from the body (at the **separation points**, S, Figure 4.5), leaving a low-pressure area, known as the **wake**, behind the object.

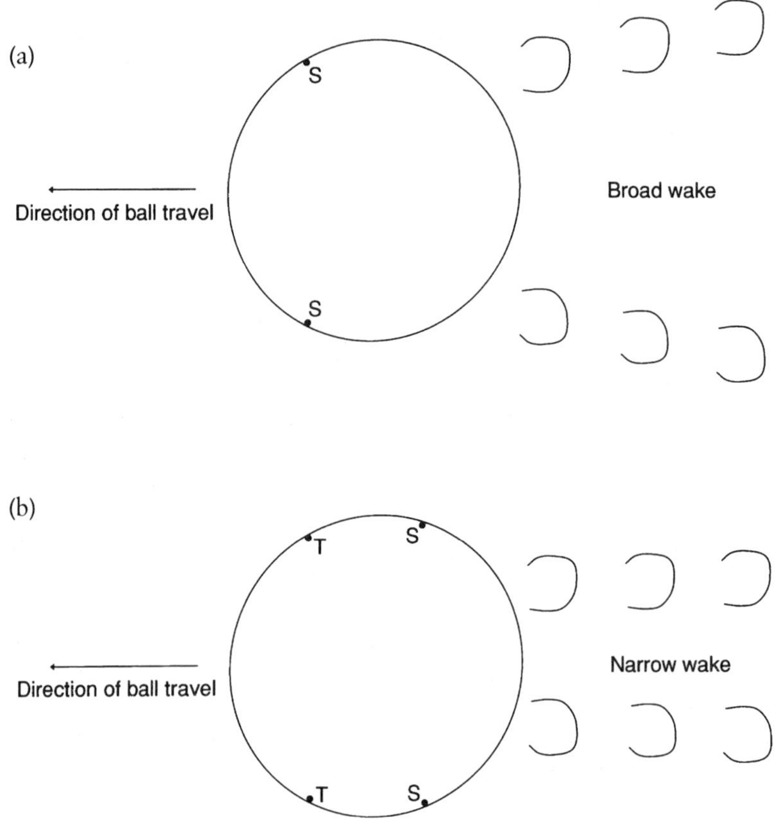

Figure 4.5 Separation points on a smooth ball for boundary layer flow which is: (a) laminar; (b) turbulent.

The object, moving from a low- to a high-pressure region, experiences a drag force, which will be discussed below. The **boundary layer separation**, which leads to the formation of the wake, occurs far less readily if the air flow in the boundary layer is turbulent (Figure 4.5(b)) than if the flow is laminar (Figure 4.5(a)). This is because kinetic energy is more evenly distributed across a turbulent boundary layer, enabling the fluid particles near the boundary to better resist the increasing pressure.

Figure 4.5 shows the difference in the separation point (S) positions and the size of the wake between laminar and turbulent boundary layers on a ball. The change from laminar to turbulent boundary layer flow will occur, for a given object and conditions, at a speed related to the critical Reynolds number. The change occurs at the **transition point** (T, Figure 4.5(b)), and the relationship between the transition and separation points is very important. If separation precedes transition a large wake is formed (Figure 4.5(a)), whereas transition to turbulent flow upstream of the separation point results in a smaller wake (Figure 4.5(b)).

4.2 Forces in a fluid

4.2.1 DRAG FORCES

If an object is symmetrical with respect to the fluid flow (such as a non-spinning soccer ball) the fluid dynamic force acts in the direction opposite to the motion of the object and is termed a **drag force**. Drag forces resist motion and therefore generally restrict sports performance. They can have beneficial, propulsive effects, as in swimming and rowing. To maintain a runner in motion at a constant speed against a drag force requires an expenditure of energy equal to the product of the drag force and the speed. If no such energy is present, as for a projectile, the object will decelerate at a rate proportional to the frontal area A (the area presented to the fluid flow) and inversely proportional to its mass m. The ratio m/A is crucial in determining how much effect air resistance has on projectile motion. A shot, with a very high ratio of m/A, is hardly affected by air resistance whereas a cricket ball (only one-16th the value of m/A of the shot) is far more affected. A table tennis ball (one-250th the value of m/A of the shot) has a greatly altered trajectory.

Pressure drag

Pressure drag (or wake drag) contributes to the fluid resistance experienced by, for example, projectiles and runners. This is the major drag force in the majority of sports and is caused by boundary layer separation leaving a low-pressure wake behind the object. The object, tending to move from a low- to a high-pressure region, experiences a drag force. The pressure drag can be reduced by minimizing the disturbance that the object causes to the fluid flow, a process known as **streamlining**. The effect of streamlining on drag is shown in Figure 4.6.

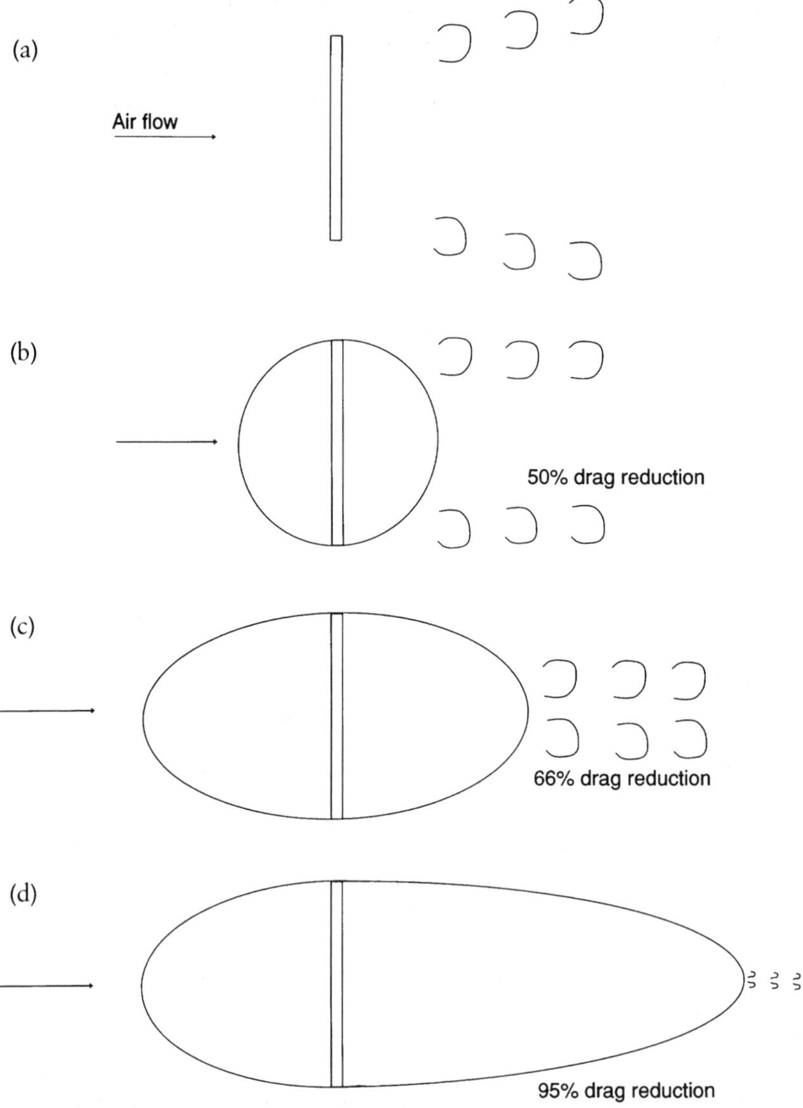

Figure 4.6 Effect of streamlining: (a) disc; (b) spherical ball; (c) oval shape; (d) simple aerofoil shape.

The oval shape (Figure 4.6(c)) has only two-thirds of the drag of a spherical ball (Figure 4.6(b)) with the same frontal area. The drag is very small on the streamlined aerofoil shape of Figure 4.6(d). Streamlining is important in motor car and motor cycle racing, discus and javelin throwing. Swimmers and skiers can reduce the pressure drag forces acting on them by adopting streamlined shapes. The adoption of a streamlined shape is of considerable advantage to downhill skiers.

To enable comparisons to be made between objects of similar geometry, it is conventional to plot graphs of drag coefficient (C_D), that is the drag force (D) divided by the product of the frontal area of the body (A) and the free stream dynamic pressure ($\frac{1}{2}\rho v^2$), against Reynolds number.

$$D = C_D \tfrac{1}{2}\rho v^2 A \tag{4.7}$$

Such a graph for a ball is shown in Figure 4.7.

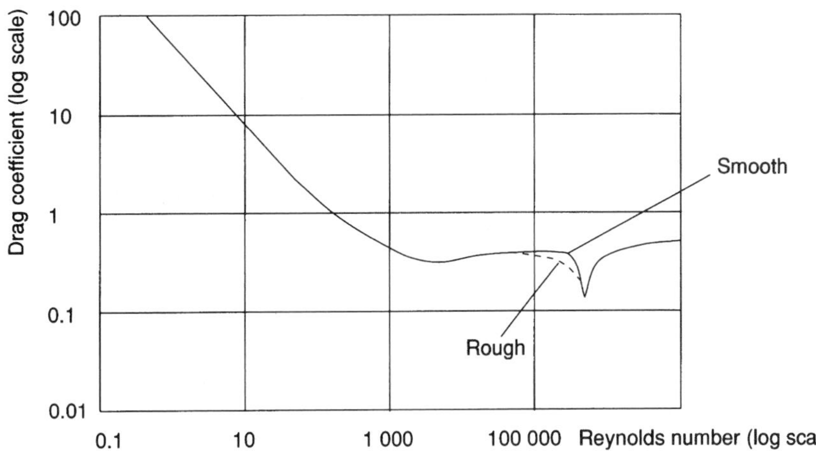

Figure 4.7 Drag coefficient as a function of Reynolds number for spherical balls.

The continuous line applies to the behaviour of any smooth ball moving through any fluid. It clearly shows the dramatic change in the drag coefficient on a smooth ball as the boundary layer flow changes from laminar to turbulent. As this transition occurs, the drag coefficient decreases by about 65% with no change of speed. This occurs at the critical Reynolds number. Promoting a turbulent boundary layer is an important mechanism in reducing pressure drag if the speed is close to the value necessary to achieve the critical Reynolds number. At such speeds, which are common in ball sports, roughening the surface promotes turbulence in the boundary layer. The decrease in drag coefficient then occurs over a wider speed range and starts at a lower speed, as shown by the dashed region of the curve in Figure 4.7. The nap of tennis balls, the dimples on a golf ball and the seam on a cricket ball are examples of roughness helping to induce boundary layer transition, hence reducing drag. Within the Reynolds number range 110 000–175 000, which corresponds to ball speeds off the tee of 45–70 $m{\cdot}s^{-1}$, the dimples on a golf ball cause the drag coefficient to decrease proportionally to speed. The drag force then increases only proportionally to speed, rather than speed squared, benefiting the hard-hitting player. Many sport balls are not uniformly rough.

Then, within a speed range somewhat below the critical Reynolds number, it is possible for roughness elements on one part of the ball to stimulate transition of the boundary layer to turbulent flow, while the boundary layer flow on the remaining, smoother portion of the ball remains laminar. The results of this are discussed below for cricket ball swing.

Skin friction drag

Skin friction drag is the force caused by friction between the molecules of fluid and a solid boundary. It is only important for streamlined bodies for which separation and hence pressure drag have been minimized. Unlike pressure drag, skin friction drag is reduced by having a laminar as opposed to a turbulent boundary layer. This occurs because the rate of shear at the solid boundary is greater for turbulent flow. Reduction of skin friction drag is an important consideration in racing cars, racing motor cycles, gliders, hulls of boats, skiers and ski-jumpers and, perhaps, swimmers. It is minimized by reducing the roughness of the surfaces in contact with the fluid.

Wave drag

This occurs only in sports in which an object moves through both water and air. As the object moves through the water, the pressure differences at the boundary cause the water level to rise and fall and waves are generated. The energy of the waves is provided by the object, which experiences a resistance to its motion. The greater the speed of the body, the larger the wave drag, which is important in most aquatic sports. Wave drag also depends on the wave patterns generated (Wellicome, 1967) and the dimensions of the object, especially its waterline length. The drag is often expressed as a function of the Froude number. Speed boats and racing yachts are designed to plane (to ride high in the water) at their highest speeds so that wave drag (and pressure drag) are then very small. In swimming the wave drag is small compared with the pressure drag, unless the swimmer's speed is above about 1.6 $m \cdot s^{-1}$ when a bow wave is formed. The design of racing yacht hulls to optimize drag and lift forces has now reached a very high level of sophistication.

Other forms of drag

Spray-making drag occurs in some water sports because of the energy involved in generating spray. It is usually negligible, except perhaps during high-speed turns in surfing and windsurfing. **Induced drag** arises from a three-dimensional object that is generating lift. It can be minimized by having a large aspect ratio (the ratio of the dimension of the object perpendicular to the flow direction to the dimension along the flow direction). Long, thin wings on gliders minimize the induced drag whereas a javelin has entirely the wrong shape for this purpose.

4.2.2 THE SWING OF A CRICKET BALL

It is the asymmetrical disposition of the seam of a cricket ball that accounts for the lateral movement of the ball known as swing. Because of the geometrical simplicity of the cricket ball seam in comparison with seams on some other sports balls, the behaviour of the cricket ball is relatively well understood and well researched. The asymmetrical disposition of the seam with respect to the relative air velocity (Figure 4.8(a)) causes the seam to trip the boundary layer to become turbulent on one side of the ball, delaying separation to S_T.

The boundary layer remains undisturbed on the other hemisphere, and separates at S_L. For this to happen, the ball speed must be sufficiently slow for the boundary layer to remain laminar on one side of the ball. This requires a ball release speed below the critical speed (about 40 $m{\cdot}s^{-1}$). The speed must be high enough (probably around 25–30 $m{\cdot}s^{-1}$), however, to cause the boundary layer flow to become turbulent (referred to as boundary layer 'tripping'). A second crucial factor is that the ball surface must remain sufficiently smooth so as not to reduce the critical speed below the bowler's release speed. This is not helped by the quarter seams on many cricket balls, which can cause boundary layer tripping, especially at the higher end of the range of swing bowling speeds (although this might facilitate reverse swing). A further problem is the progressive roughening of the ball during the game because of repeated abrasive impacts with the ground. This is minimized by the fact that, for the ball to swing, the seam must be aligned in a vertical plane, so that the ball–ground impact should generally take place at the seam. This alignment has an additional benefit as it increases the coefficient of friction between the ball and the ground, facilitating movement off the ground **(seaming)**.

A humid atmosphere might also help minimize roughening of the ball as the surface of the pitch will be more moist. A full explanation of the reported link between balls swinging and humid conditions has not been provided. A humid atmosphere is less dense than a dry one at the same temperature and pressure, thus is not 'heavy' and this reduced density diminishes rather than aids any swing force. The water vapour in a humid atmosphere might cause the stitches of the seam to swell, helping the seam to trip the boundary layer. The greater kinematic viscosity of moist air will lower the Reynolds number of the ball and enable laminar flow to be more easily maintained on the smooth side of the ball. It has been suggested that the lower free stream turbulence associated with humid days will also help to maintain laminar flow on the smooth side of the ball, but this has been refuted by Mehta (1985).

Some balls will swing, others will not, a phenomenon which is probably entirely owing to the geometry and surface roughness of the ball. Nearly all balls have a measurable misalignment of the centres of mass and volume. This will generate a couple, caused by the buoyancy force

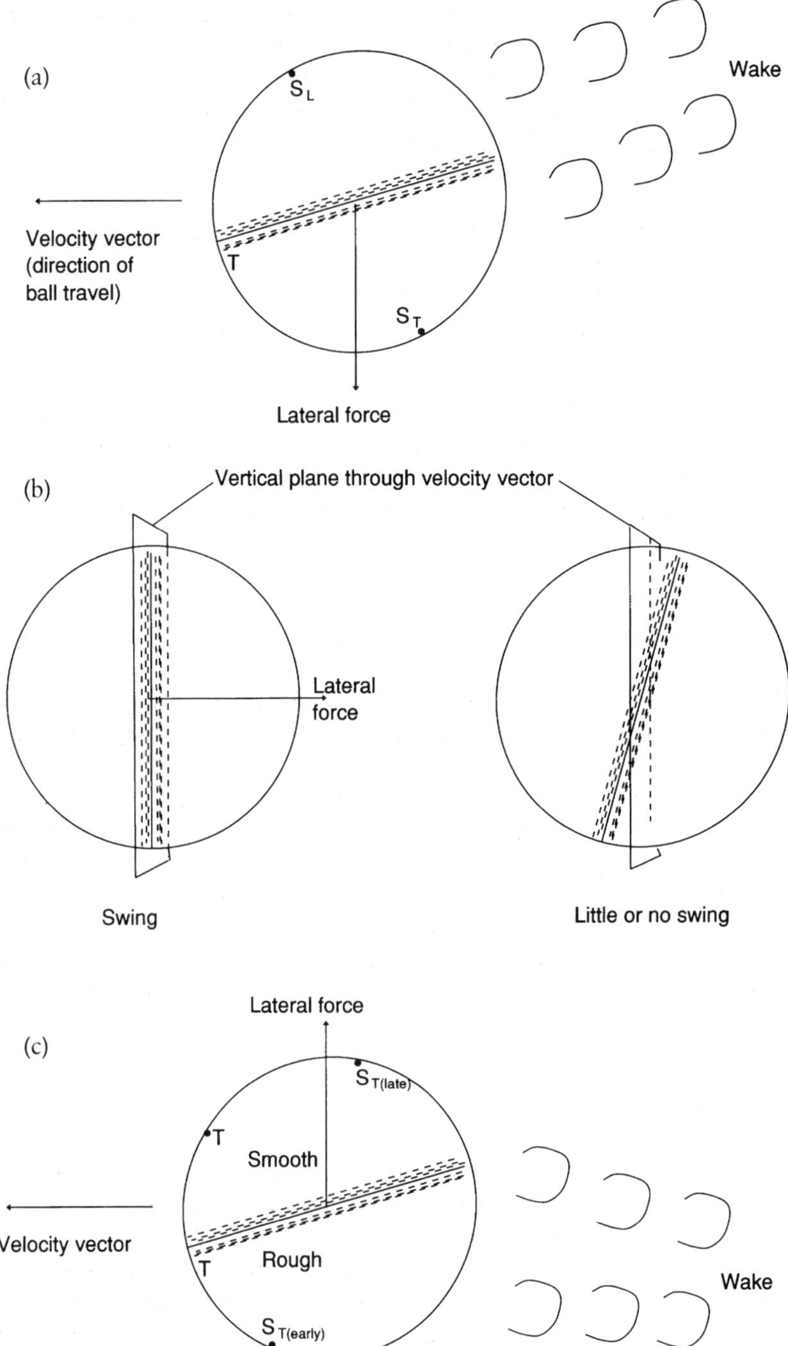

Figure 4.8 Cricket ball swing: (a) normal swing from above; (b) effect of non-vertical seam as seen from around first slip; (c) reverse swing from above.

on the ball, tending to rotate the seam of the ball. If the angle through which it rotates is large enough to move part of the seam to the other side of the vertical plane through the ball's direction of travel, the swing force will be reduced or negated (Figure 4.8(b)).

The phenomenon of **late swing** has been explained as being caused by an initially supercritical ball slowing to a subcritical speed in flight. This seems unlikely given:

- the high critical speed for a smooth ball (over 40 $m \cdot s^{-1}$) – few bowlers achieve such speeds;
- a small deceleration of less than 5 $m \cdot s^{-1}$ during flight;
- the reluctance of a turbulent boundary layer to revert to a laminar one.

The probable explanation is simpler. If the swing force was constant, the sideways movement would depend on the square of the time of ball flight. This would result in a parabolic swing path, with 75% of the deviation occurring in the second half of flight (Daish, 1972). However, the swing force depends on the angle of the seam of the ball to the direction of travel. Most good swing bowlers use an initially small seam angle, which increases in flight owing to the greater skin friction drag on the seam side of the ball. The swing force then tends to increase, causing even more of the swing to occur late in flight.

Reverse swing has been explained, for example by Bown and Mehta (1993). For this effect to occur, the ball must be released above the critical speed for the smooth side of the ball. This can obviously only be done by bowlers who can achieve such speeds. The boundary layer becomes turbulent on both hemispheres before separation. On the rough side, the turbulent boundary layer thickens more rapidly and separates earlier (at $S_{T(early)}$) than on the smooth side (at $S_{T(late)}$). This effect may be helped by the quarter seam on the rough side of the ball and, possibly, by illegal gouging of the surface and lifting of the seam. The result is the reversal of the directions of wake displacement and, therefore, swing (Figure 4.8(c)).

4.2.3 LIFT FORCES

If there is an asymmetry in the fluid flow around a body, the fluid dynamic force will act at some angle to the direction of motion and can be resolved into two component forces. These are a drag force opposite to the flow direction and a lift force perpendicular to the flow direction. Such asymmetry may be caused in three ways. For a discus and javelin, for example, it arises from an inclination of an axis of symmetry of the body to the direction of flow (Figure 4.9(a)).

Another cause is asymmetry of the body (Figure 4.9(b)), which is the case for sails and the hands of a swimmer. The Magnus effect (Figure

4.9(c)) occurs when rotation of a symmetrical body, such as a ball, produces asymmetry in the fluid flow.

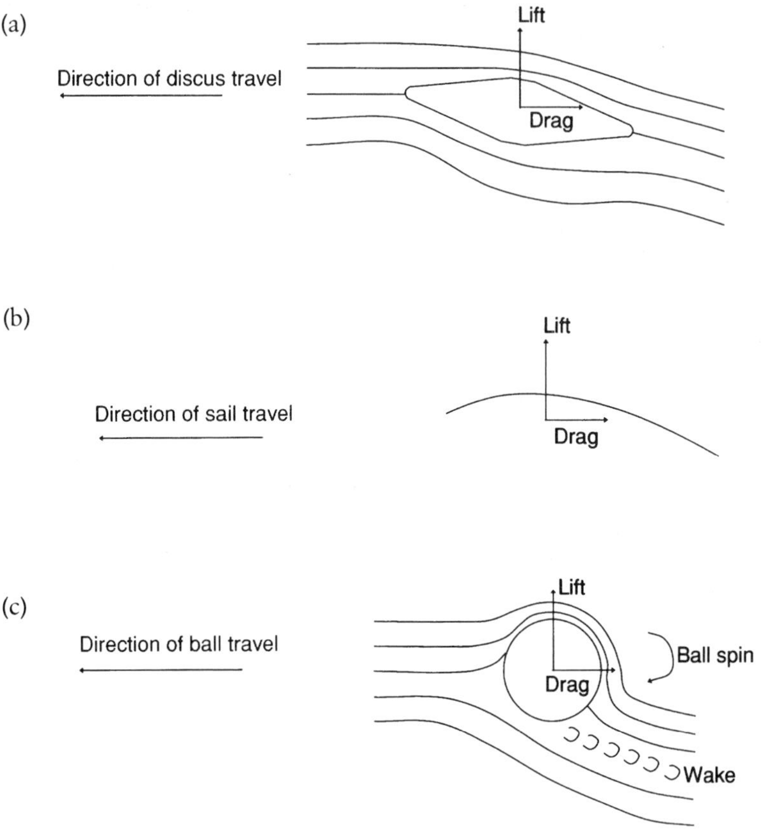

Figure 4.9 Generation of lift: (a) inclination of an axis of symmetry; (b) body asymmetry; (c) Magnus effect.

Lift forces on a flat surface

These occur when a surface is inclined at a small angle to the flow direction, in a similar way to the example of Figure 4.9(a) (this angle is known as the **angle of attack**). The flow beneath the surface is retarded, and the pressure becomes greater than the free stream pressure so that the body is pushed upwards. This 'dynamic lift' occurs in water sports and adds to the effect of buoyancy. In yachting, for example, as the yacht speed increases the dynamic lift enables the yacht to ride higher in the water. This reduces the wetted area of the hull of the yacht, so that the drag decreases and the speed can be further increased. This phenomenon is known as **planing;** the flatter the bottom of the boat the better the dynamic lift. This mechanism also generates the lift force on the skis of a water-skier. If both top and bottom surfaces of a body are submerged in the fluid then the flow above the body is accelerated. The fluid pressure above the body is lower than the free stream pressure so that the body is sucked upwards. The body is now behaving as a simple aerofoil or hydrofoil.

Basic foil theory

The ratio of lift to drag forces on a flat surface can be improved by slightly cambering the surface to produce the simplest foil shape, that of a sail (Figure 4.9(b)). Larger lift to drag ratios can be obtained by more conventional foil shapes, similar in cross-section to a glider wing. The lift and drag increase with angle of attack at relatively small values of this angle. At a sufficiently high angle of attack, known as the **stall angle**, the flow separates from the upper surface and the negative pressure region starts to collapse. The lift then decreases and the drag increases substantially. This effect is important in many sports. In sailing, for example, the required lift to drag ratios for the sails can be achieved by adjusting the angles of attack of the sails to the air flow past them.

It is generally agreed that swimmers use a mixture of lift and drag forces for propulsion (e.g. Schleihauf, 1974). A side view of a typical path of a front crawl swimmer's hand relative to the water is shown in Figure 4.10(a).

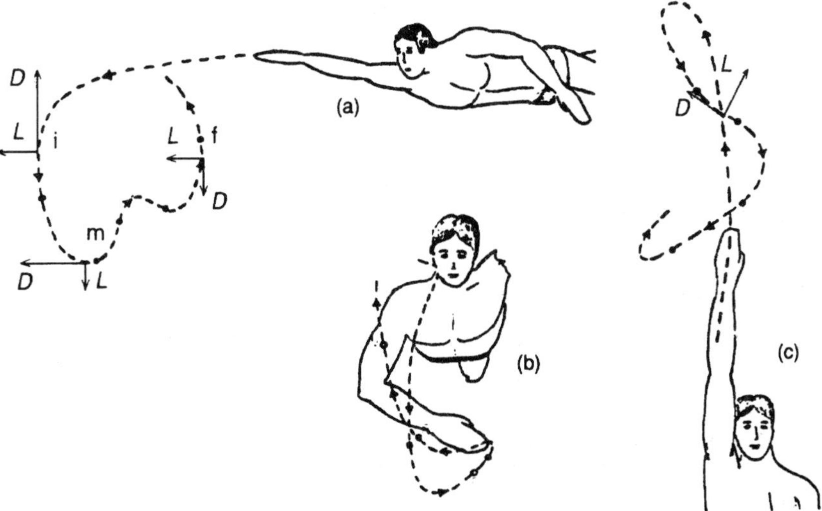

Figure 4.10 Typical path of a swimmer's hand relative to the water: (a) side view; (b) view from in front of the swimmer; (c) view from below the swimmer (adapted from Schleihauf, 1974).

In the initial and final portions of the pull phase of this stroke (marked i and f in Figure 4.10(a)) only the lift force (*L*) can make a significant contribution to propulsion. In the middle region of the pull phase (marked m), the side view would suggest that drag (*D*) is the dominant contributor. However, a view from in front of (Figure 4.10(b)) or below (Figure 4.10(c)) the swimmer shows an S-shaped pull pattern in the sideways plane. These sideways movements of the hands generate significant propulsion through lift forces perpendicular to the path of travel of the hand (Figure 4.10(c)). Swimmers need to develop a 'feel' of the water flow over their hands, and to vary the hand 'pitch' angle with respect to the flow of water to optimize the propulsive forces throughout the stroke.

The shape of oars and paddles suggests that they can behave as hydrofoils. The propulsive forces generated would then be a combination of both lift and drag components and not just drag. The velocity of the oar or paddle relative to the water is the crucial factor. If this is forwards at any time when the oar or paddle is submerged, the drag is in the wrong direction to provide propulsion. Only a lift force can then fulfil this function (see Figure 4.10). This explanation is now accepted in rowing, and the wing paddle in kayaking was developed to exploit this lift effect.

In both javelin and discus flight, the interplay between lift and drag forces is evident (see also Bartlett and Best, 1988). The discus is released with a small negative angle of incidence, which soon becomes positive owing to the ascending flight path. Lift forces are then high. Stall is delayed until late in flight, when it is beneficial, causing large drag forces that act against the weight of the discus and increase the range of the throw.

The Magnus effect

Many ball sports involve a spinning ball. Consider, for example, a ball moving through a fluid, and having backspin (Figure 4.9(c)). The top of the ball is moving in the same direction as the air relative to the ball, while the bottom of the ball is moving against the air stream. The rotational motion of the ball is transferred to the thin boundary layer adjacent to the surface of the ball. On the upper surface of the ball this 'circulation' imparted to the boundary layer reduces the difference in velocity across the boundary layer and delays separation. On the lower surface of the ball, the boundary layer is moving against the rest of the fluid flow (the free stream). This increases the velocity difference across the boundary layer and separation still occurs. The resulting wake has therefore been deflected downwards. Newton's laws imply a force (provided by the ball) acting downwards on the air and a reaction moving the ball away from the wake. This can also be discerned from the 'bunching' of the streamlines over the top of the ball, indicating a high-speed, low-pressure region (Bernoulli's principle). The higher pressure below the ball, where the speed is lower, causes an upward force on the ball. For a ball with backspin, the force acts perpendicular to the motion of the ball, i.e. it is a lift force.

Golf clubs are lofted so that the ball is undercut, producing backspin. This varies from approximately 50 Hz for a wood to 160 Hz for a 9 iron. The lift force generated may be sufficient to cause the initial ball trajectory to be curved slightly upwards. The lift force increases with the spin and substantially increases the length of drive compared to that with no spin (Cochran and Stobbs, 1968). The main function of golf ball dimples is to assist the transfer of the rotational motion of the ball to the boundary layer of air to increase the Magnus force and give optimum lift. A smooth ball would provide little Magnus force and would therefore have a poor carry, probably less than half that of a dimpled ball.

Backspin and topspin are used in games such as tennis and table tennis both to vary the flight of the ball and to alter its bounce.

In cricket, a spin bowler spins the ball so that it is rotating about the axis along which it is moving, and the ball only deviates when it contacts the ground. The spin bowler can also bowl so that the spin is partly or wholly imparted about the vertical axis and the ball has sidespin. The ball will then swerve in the air so that, for example, the skilled off-break bowler can float the ball away from a right-handed batsman, before breaking the ball into him. The swerve forces will be greater when the ball is rough, as the boundary layer attaches more firmly to the ball, as for the dimpled golf ball.

A baseball pitcher uses the Magnus effect to 'curve' the ball. At a pitching speed of 30 $m \cdot s^{-1}$ the spin imparted can be as high as 30 Hz, which gives a lateral deflection of 0.45 m in 18 m (Briggs, 1959). A soccer ball can be made to swerve in flight by moving the foot across the ball as it is kicked. This causes rotation of the ball about the vertical axis. If the foot is moved from right to left as the ball is kicked, the ball will swerve to the right. Slicing and hooking of a golf ball are caused (sometimes inadvertently) by sidespin imparted at impact. In cross-winds, the relative direction of motion between the air and the ball is changed. The correct sidespin can then increase the length of the drive.

A negative Magnus effect can also occur for a ball travelling below the critical Reynolds number. It happens when the boundary layer flow remains laminar on the side of the ball moving in the direction of the relative air flow, as the Reynolds number here remains below the critical value. On the other side of the ball, the rotation increases the relative speed between the air and the ball so that the boundary layer becomes turbulent. If this happens on a backspinning ball, laminar boundary layer flow will occur on the top surface and turbulent on the bottom surface. The wake will be deflected upwards, the opposite from the case discussed above, and the ball will plummet to the ground under the action of the negative lift force. Reynolds numbers in many ball sports are close to the critical value, and the negative Magnus effect may, therefore, be important. It has occasionally been proposed as an explanation for certain unusual ball behaviour in, for example, volleyball.

4.3 Fundamentals of energetics

4.3.1 FORMS OF ENERGY

There are various ways in which energy can be 'stored' in a thermodynamic system (see section 4.3.3) such as the sports performer. These are essentially properties of the system. Note that the SI unit of energy is the joule (J).

Mechanical energy includes kinetic and potential energy. **Kinetic energy** is energy of motion and is equal to the sum of all translational and rotational kinetic energies (see below). **Potential energy** is energy due to position. It can take the form of gravitational potential energy, often simply called potential energy, and that associated with deformation. Elastic potential energy (often called elastic strain energy as in this chapter) is caused by deformation of interatomic bonds within the strained material, as in a trampoline bed.

Thermal (internal) energy is due to the random motions of the molecules of a thermodynamic system. It is the product of a system's mass, mean specific heat and absolute temperature (the SI unit of absolute temperature is the kelvin, K).

Chemical energy is stored in chemicals in intermolecular and interatomic bonds. Chemical reactions can release energy (exothermic) or absorb energy (endothermic).

4.3.2 CONSERVATION OF MECHANICAL ENERGY

Consider a ball thrown vertically upwards (Figure 4.11(a)) with release speed v_0 and release height h_0, with no air resistance.

Uniform gravitational acceleration ($-g$) acts throughout the ball's flight. The ball in this example is known as a conservative system and its mechanical energy would vary as shown in Figure 4.11(b), and is given by:

$$T + V_g = \tfrac{1}{2}mv^2 + mgh = \text{constant} \tag{4.8}$$

Equation 4.8 is a statement of the principle of conservation of mechanical energy. In this example, the sum of the linear kinetic ($T = \tfrac{1}{2}mv^2$) and gravitational potential ($V_g = mgh$) energies is a constant.

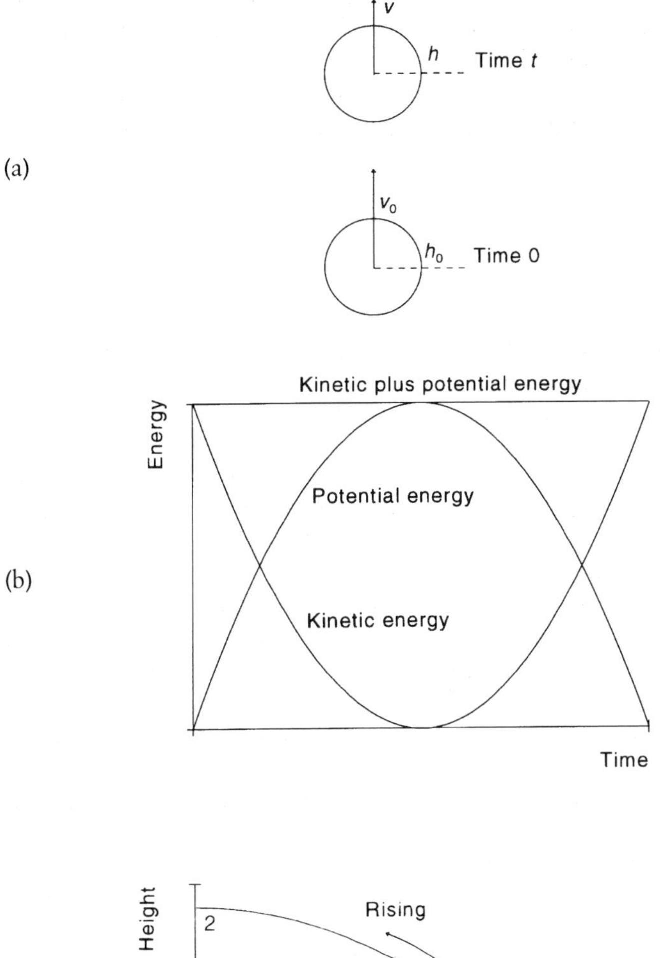

Figure 4.11 A conservative system: (a) ball thrown upwards; (b) mechanical energy variation during flight; (c) property diagram.

A trampolinist performing simple rebounds with no horizontal travel and no rotational manoeuvres would have energies as shown in Figure

4.11(b) for the airborne phase. During contact with the trampoline, the sum of the trampolinist's kinetic and potential energies would no longer be constant. To investigate this (and other more complex) aspects of energetics it is useful to define precisely some important concepts.

4.3.3 IMPORTANT CONCEPTS RELATED TO THE ENERGETICS OF THE SPORTS PERFORMER

The following concepts and definitions are taken from thermodynamics, the branch of science and engineering that deals with energy exchanges. Normal thermodynamic conventions will be used throughout this chapter.

A **thermodynamic system** is defined as the system, the behaviour of which is under investigation and which is separated from the surroundings by real (or imaginary) boundaries across which energy (in various forms) is transferred. Two types of system are associated with different processes.

- **Closed systems** are ones in which no mass is transferred across the system boundary and a non-flow process takes place. The trampoline bed in the above example is a closed system.
- **Open systems** experience mass transfer across the boundary. This is known as a flow process, which may be steady, with a constant mass flow rate, or non-steady. In general, the sports performer is an open system and the mass flows, principally caused by breathing, are non-steady.

A **thermodynamic process** is shown in Figure 4.11(c) and can, for present purposes, be defined as the path taken by a thermodynamic system during a change in state. The path taken refers to a diagrammatic representation of the process on a property diagram such as Figure 4.11(c), where the ball's speed and height are the two properties represented.

A **thermodynamic state** is a more or less temporary specification of the system's energy levels. An example would be the gravitational and linear kinetic energies of the ball (Figure 4.11(b)) at any instant. As these energy levels depend on the height and speed of the ball at that instant, the state can be represented by a point on the property diagram. States 1 and 2 in Figure 4.11(c) represent ball release and the maximum height respectively.

A **thermodynamic property** is a quantity used to define the state of a system, as above. A property depends only on the state, not on the process involved in arriving at that state from a previous state. Pressure, volume, mass, density, temperature, velocity (or speed), position (or height), (thermal) internal energy and enthalpy are properties. Work and heat transfer are not properties, as they depend on the process.

4.3.4 FORMS OF ENERGY TRANSFER

There are several ways in which energy can be transferred between a thermodynamic system, such as the sports performer, and its surroundings.

Work transfer

Work is the form of energy that crosses the boundary of a thermodynamic system because of the displacement of all or part of the system boundary in the presence of a force or torque. The convention of thermodynamics is that work transfer from the system to the surroundings is defined as positive, while work transfer to the system from the surroundings is negative. This is often expressed as: work done by a system on its surroundings is equal to the force exerted by that system on the surroundings multiplied by the distance moved in the direction of the force.

Heat transfer

Heat is the form of energy that crosses the system boundary because of a temperature difference between the system and the surroundings (conduction, convection and radiation) or because of a partial vapour pressure difference (evaporation). The convention of thermodynamics is that heat transfer from the system to the surroundings is defined as negative, while heat transfer to the system from the surroundings is positive. It is not usually possible to measure heat transfer during sport and exercise. This is one reason why sports biomechanics generally uses the impulse–momentum relationship (Chapter 3) rather than the work–energy one.

Energy transfer by mass flow

Mass can only cross the system boundary for open systems. This mass flow can then transport energy in various forms, including chemical energy, kinetic and pressure energy and thermal energy. When thermal energy is transferred across a system boundary by mass flow, the form of energy transferred is known as **enthalpy**. The enthalpy flow is the product of the mass (m) of substance entering (positive) or leaving (negative) the system, the specific heat at constant pressure (Cp) of the substance and its absolute temperature: $H = mCpT$. For the human system, the main mass transfers of chemical energy will be into the system though food and drink and out of the system through excretion of waste products. The main, although not only, thermal (enthalpy) flows will be of respiratory gases. Mass transfer of kinetic and pressure energy is probably of no significance for the whole body system. No mass transfer across the system boundary is associated with either heat or work transfer.

4.3.5 A TRAMPOLINIST IN CONTACT WITH A TRAMPOLINE

With the above definitions and conventions established, it is instructive to return to the trampolinist and trampoline problem. The assumption will be made that during contact with the trampoline bed, the trampolinist behaves as a rigid body.

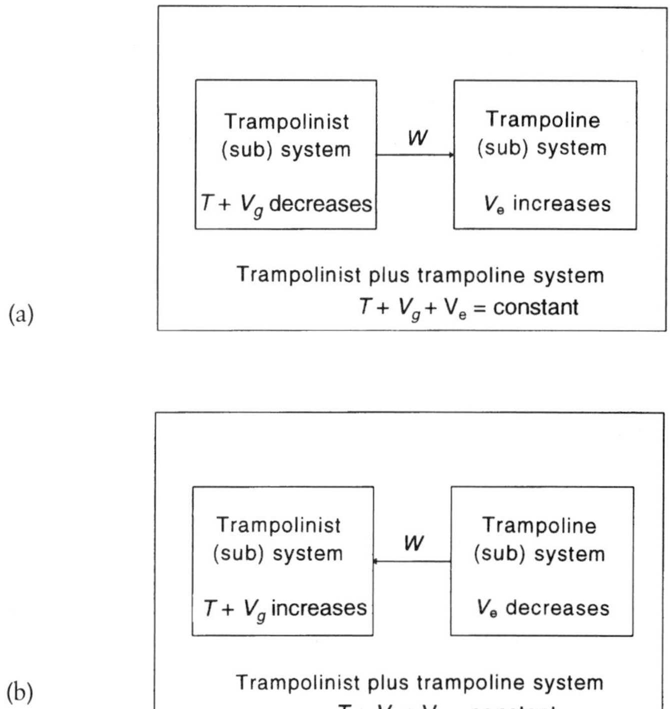

Figure 4.12 Energy exchanges for trampoline and trampolinist: (a) as bed is stretched; (b) as bed rebounds.

The trampoline plus trampolinist system

Assuming that the trampoline bed and springs are perfectly elastic, this is a conservative system (Figure 4.12). The energy lost by the trampolinist in deforming the trampoline is stored in the trampoline as elastic strain energy. Conversely, when the trampoline recoils it loses elastic strain energy to the trampolinist. Thus, the sum of the kinetic, gravitational potential and elastic strain energy of this (compound) system is constant and the principle of conservation of mechanical energy applies.

The trampolinist and trampoline (sub)systems

Neither of these is a conservative system. During trampoline contact equal but opposite forces act on these two subsystems. While the trampolinist is stretching the trampoline bed, work is being done by the trampolinist on the bed, this work transfer being positive for the trampolinist subsystem (Figure 4.12(a)). Hence, the trampolinist loses kinetic (T) and gravitational potential energy (V_g). The trampoline has work done on it, so that work transfer is negative for the trampoline subsystem, and gains elastic strain energy (V_e) during this period. The converse is true for both subsystems when the bed recoils (Figure 4.12(b)). Expressed mathematically, using Δ to represent a finite change over a specified time, and noting that the values for W will be positive or negative depending on whether they are done by or on the system under consideration:

For the trampolinist and trampoline:

$$\Delta T + \Delta V_g + \Delta V_e = 0 \tag{4.9}$$

For the trampolinist:

$$\Delta T + \Delta V_g + W = 0 \tag{4.10}$$

For the trampoline:

$$\Delta V_e + W = 0 \tag{4.11}$$

4.3.6 THE EQUATIONS OF ENERGY CONSERVATION

The trampolining examples of the previous section only considered work and mechanical energy. However, in sport, thermal and chemical energy storage are important as well as mechanical energy. Heat transfer and energy transfer by mass flow also require attention. It should be noted that, as with momentum, energy is conserved, providing all forms of energy transfer and storage are considered. Mechanical energy is only conserved in conservative systems. The various energy equations used in thermodynamics are essentially derived from the first law of thermodynamics, which will not be considered here. The energy equation for non-flow processes is known as the **non-flow energy equation** and is often used, with modifications, in the study of energetics in sport science. It is somewhat inappropriate as it ignores mass flow and has to be manipulated to account for motion of the system. A more versatile equation, which is not subject to either of these restrictions, is discussed in the following section.

4.3.7 THE NON-STEADY-FLOW ENERGY EQUATION

This is a thermodynamic equation for open systems with non-steady mass flows. Furthermore, it explicitly accounts for changes in the mechanical energy of the system during a thermodynamic process. The sports performer experiences both non-steady mass flows and changes in mechanical energy. This equation would appear, therefore, to be very appropriate for studying the energetics of sports performance. The general form of this equation applied to the sports performer as a whole (known as the **whole body system**) is:

$$Q - W + \Delta M = (T'' - T') + (V'' - V') + (U'' - U') + (C'' - C') \quad (4.12)$$

The left side of this equation refers to energy transfers across the system boundary during the process studied. W is the work done by the whole body system on the surroundings and Q is the heat transfer to the system from the surroundings during the process studied. ΔM is the net energy transferred by mass transfers to the system from the surroundings by all of the mechanisms discussed above. The right side of the equation refers to changes of energy levels within the human body (the system). V is the body's potential energy (gravitational and elastic), T its kinetic energy, U its stored thermal energy and C its stored chemical energy. The primes denote $''$ the properties after and $'$ the properties before the process studied.

The following assumptions will initially be made to simplify Equation 4.12 for application to the sports performer. The reader should seek to evaluate these assumptions (see Exercise 7).

- The kinetic energies of inspired and expired air are similar and negligibly small.
- No food intake nor waste product excretion occurs during the sporting activity.
- The thermal enthalpy of perspiration (the energy that depends on its temperature) is small. It should be noted that the important evaporation enthalpy (the energy required to cause the perspiration to evaporate) is contained in the heat transfer term, as evaporation occurs after the mass, and enthalpy, transfer across the system boundary.
- The enthalpy transfers in inspired and expired air are very similar, so that the net enthalpy transfer by breathing is negligible.
- No change in the mass of the body occurs.
- No change in the stored thermal energy (U) of the body occurs.
- There is no change in the stored elastic strain energy (V_e) of the body.

Then, Equation 4.12 simplifies to:

$$Q - W = (T'' - T') + (V_g'' - V_g') + (C'' - C') \quad (4.13)$$

Some important points should be noted at this juncture. Firstly, it must be reiterated that Equation 4.13 relates to the whole body system, where the system boundary is that of the sports performer. It could be adapted for other systems. Secondly, the study of energetics requires careful definition of the system and its boundary. Thereafter, the energy storage within that system and energy flows across the boundary are analysed. A failure to observe this simple rule has led to major misconceptions about human energetics in some biomechanics literature, one of which will be considered below. However, having carefully defined the system and its boundary, many processes can be studied. In the sprinting example in the following section, the end states are defined as the gun firing and the runner crossing the finish line. However, it would also be possible to study the energetics of a single stride by defining the end states as, for example, foot contact for one stride and foot contact for the following one. The system remains the same, the process differs, and so therefore do the energetics. Finally, in the complex multisegment movements of sport and exercise, the overall gravitational potential energy of the performer is exactly represented by the gravitational potential energy of the mass centre of the whole body. However, this is not the case for kinetic energy. The kinetic energy generally has both rotational and translational components. The total kinetic energy of the performer must be calculated by summing the rotational and translational kinetic energies of each independently moving body segment. That is:

$$T = \Sigma T_i = \Sigma \tfrac{1}{2} m_i v_i^2 + \Sigma \tfrac{1}{2} I_i \omega_i^2 \tag{4.14}$$

where I_i, m_i, v_i and ω_i are respectively the moment of inertia and mass and the magnitudes of the velocity and angular velocity of segment number i. If the translational kinetic energy of the whole body centre of mass is used to represent the total translational kinetic energy, a considerable underestimate will occur unless all the body segments are moving at the same speed, which is unlikely in sports movements.

4.3.8 SOME APPLICATIONS OF THE NON-STEADY-FLOW ENERGY EQUATION

Rest

$$T'' = T' = 0;\ V_g'' = V_g';\ W = 0. \text{ Hence:}$$

$$Q = C'' - C' \tag{4.15}$$

At rest, the heat lost by the body to the surroundings is equal to the loss of internal stored chemical energy.

Very-short-duration explosive activity

$$\text{Assume } Q = 0;\ C'' = C'. \text{ Hence:}$$

$$-W = (T'' + V_g'') - (T' + V_g') \tag{4.16}$$

It is possible to measure the work done in some activities providing the net effect of all external forces (F) is measured and the displacements ($\Delta \boldsymbol{r}$) that the forces generate are known. Then, the total work done:

$$W = \int \boldsymbol{F} \cdot \mathrm{d}\boldsymbol{r} \tag{4.17}$$

This can be calculated if a force platform is available, for example when performing a vertical jump. For this very-short-duration explosive activity, the work done by ground contact force during ground contact equals take-off kinetic energy plus take-off potential energy (as $T' = 0$ and V_g' can be defined as zero at the initial stance position). Note that in this example it is possible to calculate the work done by an external force on the system from the change in kinetic and potential (mechanical) energy. In such examples, it is also possible to define and measure the whole body power generated as power is the rate of doing work (Power = $\mathrm{d}W/\mathrm{d}t$). Consideration of the origins of the work–energy relationship (used in this example) from the force–momentum relationship allows the calculation of instantaneous whole body power as $P = \boldsymbol{F} \cdot \boldsymbol{v}$.

Bicycle ergometry

$V_g'' = V_g'$; $T'' = T'$. Hence:

$$Q - W = C'' - C' \tag{4.18}$$

In this equation, W is the work done by the subject, usually measured against a resistance. An interesting exercise for the reader (see Exercise 8) would be to apply the non-steady-flow energy equation to the combined system of rider and bicycle ergometer rather than just the rider, as here.

Sprinting

From firing of the gun to the end of the race: $T' = 0$; $V_g'' = V_g'$, approximately. Hence:

$$Q - W = T'' + C'' - C' \tag{4.19}$$

W is the net work transfer from the system to the surroundings. In this example, this is the work done against air resistance minus the work done by the ground contact force (minus the body weight) during the ground contact phase.

Trampolinist performing a tucked forward somersault

This is an extension of the example considered above. Assume now that the trampolinist takes off with an extended body position at the same speed as before, but with some somersaulting angular momentum (state ′). The trampolinist then tucks and is holding the tuck at the apex of the airborne phase (state ″). If the system was a conservative one, the increase in rotational kinetic energy from the extended to the tucked position would cause a loss of the other forms of mechanical energy and hence height achieved. This does not occur, because this, and all similar systems, are not conservative. The movement of the body segments needed to increase the rotational kinetic energy requires muscle activity within the body. This causes a reduction in the stored chemical energy and a consequent heat loss to the surroundings. Returning to the extended position also requires muscle activity and, therefore, a reduction of chemical energy stores and heat loss, even though the rotational energy now decreases. As $W = 0$:

$$Q = (T'' - T') + (V_g'' - V_g') + (C'' - C') \qquad (4.20)$$

4.3.9 EFFICIENCY

This is one of the most widely misused terms in the literature of sport and exercise science (for example, Gaesser and Brooks, 1975). Efficiency is a term used primarily in mechanics to denote the proportion of one form of energy converted to another. Usually efficiency is defined as some useful energy produced divided by the energy consumed to produce it. The useful energy produced is often the work done by the system on the surroundings. Values for the efficiency of the human whole body system are often obtained through the use of equations such as that derived above for bicycle ergometry. Here $C'' - C'$ is obviously the energy consumed, Q is wasted energy (heat loss) and W is a measure of energy produced (work done). Hence the relationship for efficiency (η)is as follows, although the 'usefulness' of the work produced in this example is open to question.

$$\eta = W/(C'' - C') \qquad (4.21)$$

Biomechanists and physiologists often use the terms **gross efficiency** for that calculated from Equation 4.21 and **net efficiency** for that calculated after 'correcting' for resting metabolism in the denominator of this equation. There are problems of a conceptual (as well as a practical) nature in the use of efficiency in this way. For the example of sprinting (see above) the reader is asked to consider whether it would be meaningful to use W (if indeed it could be measured) as the useful energy produced when the goal of the movement is generation of speed (expressed in T''). A further problem arises in treadmill running which, like bicycle ergometry, is widely used for fitness testing. In this activity, there is little air resistance and the work done by the runner is very small. The 'efficiency' of such activities therefore approaches zero.

4.3.10 MUSCLES AS THERMODYNAMIC SYSTEMS

Muscles are the only structures in the body able to generate force and do work. To study the thermodynamics of muscle requires a different thermodynamic system from the whole body one. This can be called the muscle power pack system. This consists of the muscle and its immediate source of chemical energy (ATP). It is important to note that muscles only do work on their surroundings, usually the body segments on which they act, when they shorten (concentric contractions). This is, by the thermodynamic convention above, positive work. When muscles lengthen while developing tension, work is done by the surroundings on the muscle. By thermodynamic convention, this is negative work, done on the muscle not by it. It is unfortunate that much of the scientific literature (e.g. Winter, 1990) refers to muscles 'doing' negative work, when it is the surroundings that do the work. When muscles develop tension but do not shorten, then no work is done. However, energy is still required to develop tension in order to support the activity of the actin–myosin cross-bridges, which continually detach and re-attach to maintain the tension (Alexander, 1992). Greater energy is needed if the muscle is doing work and the energy is reduced if the muscle is lengthening as the surroundings do work on it. Muscles are not very efficient at converting chemical energy into work, a value of about 25% being accepted as the maximum. The efficiency of a muscle is not constant, but differs at varying rates of shortening. The optimum occurs at around one-third of the maximum shortening speed (Alexander, 1992).

The relationship between the muscle power pack and whole body systems is easily seen for the single muscle power pack example of Figure 4.13(a). Here, the elbow flexors do work to raise the forearm–hand segment and the weight held in the hand.

For the muscle power pack system:

$$Q - W_{pp} = C'' - C' \tag{4.22}$$

For the surroundings of the muscle power pack system, considering only work and gravitational potential energy changes (the heat lost from the muscle Q will, of course, increase the thermal energy of the surroundings):

$$W_{pp} = V_g''_{\text{segment}} + V_g''_{\text{weight}} \tag{4.23}$$

For the whole body system:

$$Q - W_{wb} = C'' - C' + V_g''_{\text{segment}} \tag{4.24}$$

For the surroundings of the whole body system, simplified as above:

$$W_{wb} = V_g''_{\text{weight}} \tag{4.25}$$

The relationship between the two different systems is portrayed schematically in Figure 4.13(b).

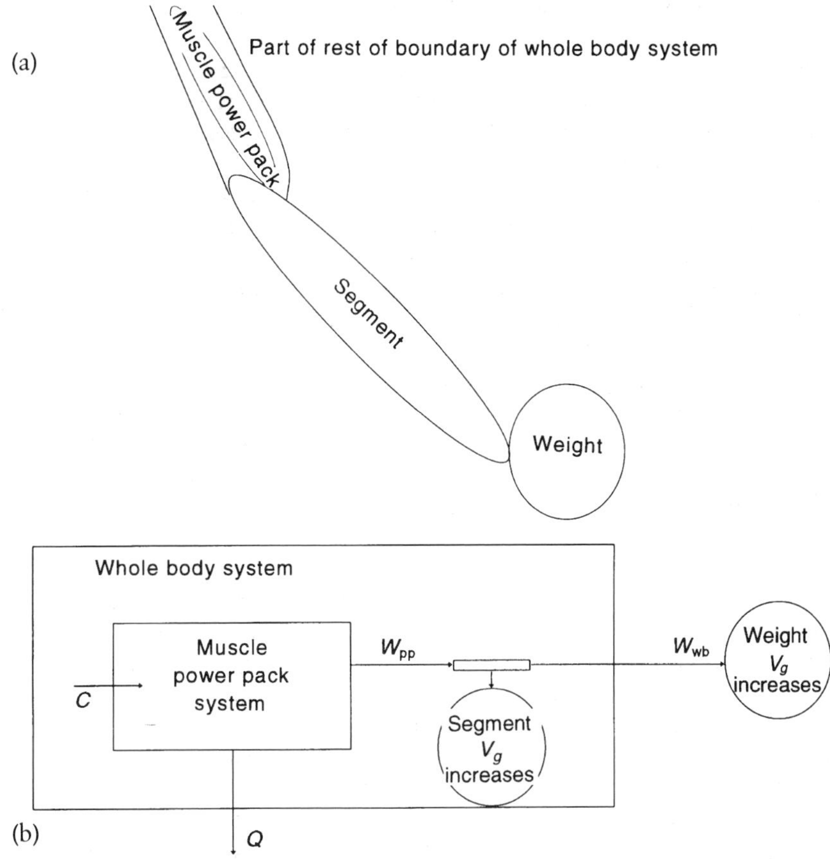

Figure 4.13 Whole body and muscle power pack systems: (a) example raising weight; (b) schematic representation of their relationship.

With care, it is possible to extend this approach to a compound system made up of all the body's muscle power packs. The work done by this compound system cannot be measured directly. It can be estimated, as in the above example, from changes in the mechanical energies of body segments. It is often referred to, somewhat misleadingly, as **internal work.** The work done by the whole body system is then referred to as **external work.** The total work done by the compound system of muscle power packs is estimated, as above, as the sum of the 'internal' and external work.

Some problems can occur with this approach. In cyclic movements, such as walking, running and cycling, positive work done by a particular muscle at some parts of the cycle will be cancelled out by negative work at other parts of the cycle. The net chemical energy is expended almost entirely in maintaining muscle tension, not in doing work (Alexander, 1992).

4.4 Energy transfers in sports motions

4.4.1 ENERGY TRANSFER RATES

At the start of this chapter, the cases of a ball thrown into the air and an airborne trampolinist performing simple rebounds were considered. Both of these are conservative systems for which the total mechanical energy, E, at any instant is the sum of its kinetic and gravitational potential energies (Equation 4.8 and Figure 4.10(b)). It is also possible to calculate the 'rate' at which energy is transferred from one form to the other, the **energy transfer rate**, expressed over a specified time interval and defined as:

$$R = \tfrac{1}{2}(|\Delta T| + |\Delta V_g|) \tag{4.26}$$

For a conservation system, the change in total mechanical energy is zero. Note that the ½ in Equation 4.26 is used to avoid adding increases in energy to corresponding decreases and moduli (||) – i.e. magnitudes but not signs – are taken to avoid $\Delta T + \Delta V_g$ summing to zero as $\Delta T = -\Delta V_g$.

Transfer rates can similarly be used for studying energy transfers in multisegmental movements. Because energy can be transferred from one form to another, there is no reason to expect mechanical energy conservation for the whole body model nor for individual body segments. To study such energy transfers in the human performer, it is necessary to define the forms in which energy can be stored which are of interest, and their interrelationships. It is also necessary to choose appropriate constraints on the various within and between segment energy transfers. Of considerable interest in human performance in sport and exercise is the possible existence of passive, non-metabolic energy transfers between the various forms of mechanical energy both within and between segments. It is noteworthy that the physiological costs of running at a given speed (the **running economy**) vary considerably between individuals. This is not only because of physiological differences, but also because of different running patterns causing different utilization of passive energy exchanges (Shorten, 1984). These passive energy exchanges are important as they can reduce metabolic energy requirements.

Passive energy transfers are likely within segments owing to the pendulum-like exchange of kinetic and gravitational potential energy, even for inanimate objects. In addition, energy transfers between segments of the same limb are justified on a multipendulum model and as muscle action between segments brings about changes in total energy with passive energy exchange as a secondary consequence. However, there is a lack of evidence in the scientific literature of passive transfers of energy between limbs (for example Pierrynowski, Winter and Norman, 1980; Shorten, 1984).

It is possible therefore to define rates of transfer of mechanical energy within a segment and between segments of the same limb in a similar way to Equation 4.26 but with the total energy change ($\Delta E = E_{in} - E_{out}$) now being non-zero. For energy transfers within segments:

$$|Rw_i| = ½(|\Delta T_i| + |\Delta V_i| - |\Delta E_i|) \quad (4.27)$$

Here Rw_i is the 'within segment', or intrasegmental, energy transfer rate for segment number i. For example, let 4 J of energy cross a segment boundary in unit time. Within the segment, let the gravitational potential energy decrease by 1 J, the elastic strain energy increase by 2 J, and the kinetic energy increase by 3 J. In this example, as two forms of potential energy storage have been considered, $|\Delta V_i|$ becomes $|\Delta V_{gi}| + |\Delta V_{ei}|$. (Figure 4.14).

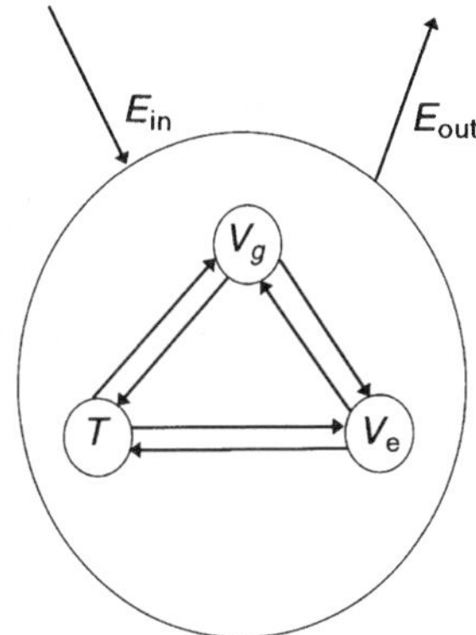

Figure 4.14 Single-segment energy transfer model.

Now $|\Delta T_i| + |\Delta V_i| = 1 + 2 + 3 = 6$ J, $|\Delta E_i| = 4$ J and $Rw = ½(6 - 4) = 1$ J. It is also possible that there is a transfer of energy between segments (Figure 4.15). Then a 'between segments', or intersegmental, energy transfer rate can be defined.

Energy transfers between body segments i and j can be computed by considering transfers of total segmental energies:

$$|Rb_{ij}| = ½(|\Delta E_i| + |\Delta E_j| - |\Delta E_{i+j}|) \quad (4.28)$$

Here Rb_{ij} is the between segment energy transfer rate for segments i and j. For example, in unit time for a two segment system, the total increase in stored energy for segment 1 (ΔE_1) is 10 J and the total decrease in stored energy for segment 2 (ΔE_2) is – 6 J. In this example, $|\Delta E_i| + |\Delta E_j| = 10 + 6 = 16$ J and $|\Delta E_{i+j}| = 10 - 6 = 4$ J. Then Rb_{12}, the energy transfer rate between segments 1 and 2 is $½(16 - 4) = 6$ J. The total transfer rate in such examples can be computed as the sum of $|Rw_i|$ and $|Rb_{ij}|$ for the various segments involved.

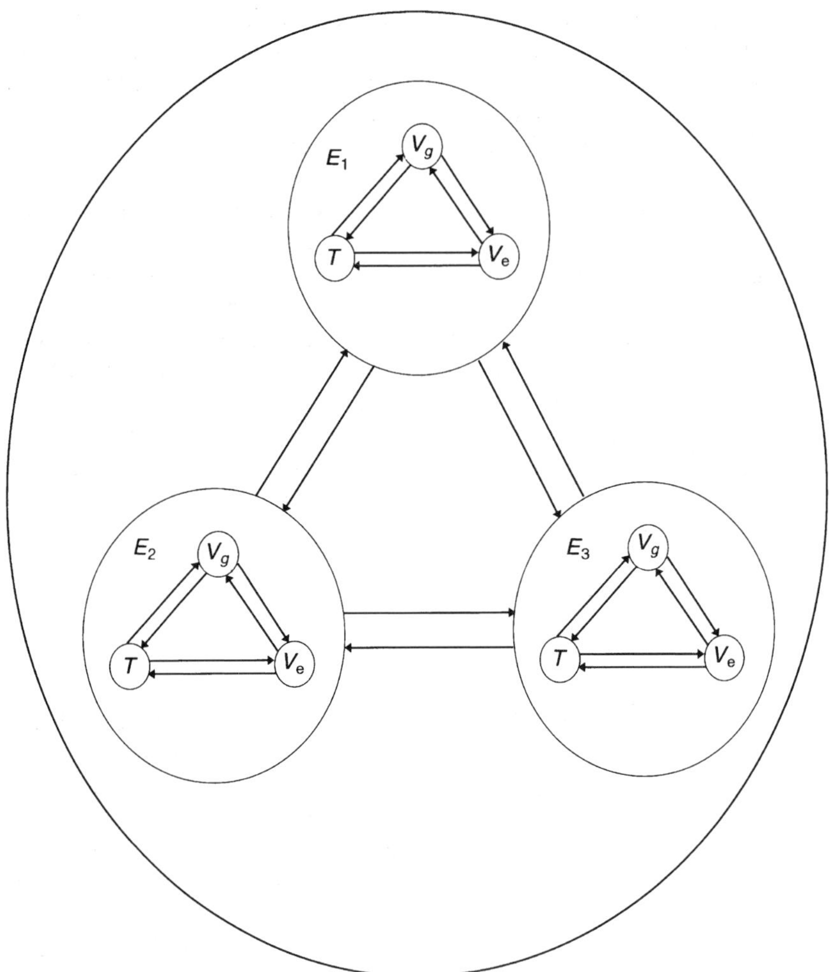

Figure 4.15 Multiple-segment energy transfer model.

4.4.2 PASSIVE ENERGY TRANSFERS AND HUMAN PERFORMANCE

The sports scientific literature provides many examples of studies of passive energy transfer and its relationship to performance in sport and exercise. For example, in treadmill running, approximately 70% of the total segmental energy has been reported to be transferred passively, of which about 80% is within and 20% between segments (mean values). About 70% of the increase in kinetic energy in the early part of the recovery phase in running (thigh flexion from the position of apparent hyperextension) is due to a loss in potential energy with 40% transfer being involved up to full thigh flexion. Furthermore, at the appropriate stage of recovery, up to 80% of the energy change is caused by transfer from the thigh to the calf (Shorten, 1984).

Alexander (1992) considered that about one-half of the work that would have had to have been done by the leg muscles in running was saved by the storage and release of elastic energy. The main factor was the passive stretch and recoil of the Achilles tendon, which returns about 93% of the energy it stores. The foot also stores elastic energy in the ligaments of the arch, and about 78% of this is returned on recoil. This example shows that the mechanism of the storage and recovery of elastic strain energy in the stretched elastic components of muscle and connective tissue must be included in the energy exchange model. As an further example, Shorten (1985) investigated this energy storage mechanism for the knee extensors and reported increases in within and between segment energy transfers with a reduction of 70 J in the whole body energy change. Unfortunately, it is very difficult to calculate the elastic potential (strain) energy of a segment, unlike the gravitational potential energy and the kinetic energy, which can be easily calculated.

It should be apparent that calculations of passive energy transfers are instructive in view of the reduced metabolic energy consumption which such transfers imply, and considering the principle of minimization of energy expenditure in human movement (see also Winter, 1990).

4.5 Summary

In this chapter the forces involved in moving through a fluid and the energetics of both linear and rotational motion were considered. Both topics are important for an understanding of human movement in sport and exercise. This included motion through a fluid, which always characterizes sport, and the forces which affect that motion. The principles and simple equations of fluid flow were outlined. The various forms of fluid drag were covered as well as the way in which differential boundary layer separation can cause a sideways force, as exemplified by cricket ball swing. The mechanisms of lift generation on sports objects were also explained. The chapter also covered the principles of thermodynamics and their application to the energetics of the sports performer. The non-steady-flow energy equation was introduced and how this could be simplified for sport and exercise was considered. The limitations of the concept of efficiency were outlined. Finally, an understanding was provided of the models of inter- and intrasegmental energy transfers and their application to movement in sport and exercise.

4.6 Exercises

1. Explain the differences between laminar and turbulent flow and give examples of when they occur in sport.
2. List the various types of fluid drag and give examples for each, other than those in this chapter, of a sport where it might be important.

3. An explanation was given in this chapter of the swing of a cricket ball. Choose a fluid dynamics phenomenon in any sport (such as propulsion in swimming, the floating serve in volleyball, the swerve of a soccer ball from a corner) and provide a brief, scientific explanation of why it occurs.
4. Measure the diameter (in metres) of each of the following sports balls. Calculate the Reynolds numbers for each ball travelling though air of density 1.23 kg·m^{-3} and dynamic viscosity 1.92×10^{-5} Pa·s. Then, from Figure 3.15, identify the type of fluid flow that will exist around the ball, establish its drag coefficient and calculate the drag force acting on it. Make very brief notes, based on the nature of the ball, its Reynolds number and Figure 3.15, of how the ball might behave in flight:
   a) soccer ball at a speed of 25 m·s^{-1};
   b) cricket ball at a speed of 60 m·s^{-1};
   c) golf ball at a speed of 65 m·s^{-1};
   d) tennis ball at a speed of 25 m·s^{-1};
   e) table tennis ball at a speed of 15 m·s^{-1}.
5. Give sports examples, other than those of section 4.2.3, of each of the three lift (or side force) generating mechanisms.
6. Golf balls, tennis balls and other sports balls have roughened surfaces to reduce air resistance. Alpine skiers wear very smooth clothing to reduce the air resistance. Are these contradictory, or are there different explanations in the two cases?
7. Relating to the application of the non-steady-flow energy equation to the human whole body system:
   a) Make a complete list of the energy flows by mass transfer into and out of the human body.
   b) Evaluate the simplifications made in section 4.3.7 to reduce the general form of the non-steady-flow energy equation, Equation 4.12, to the form of Equation 4.13.
8. Apply the non-steady-flow energy equation (Equation 4.13) to a combined system of rider and bicycle ergometer and contrast the result with that of the rider only, as analysed in section 4.3.8.
9. The data of Table 4.1 relate to energy levels in walking and have been adapted from Winter (1979).
   Complete the table including the total energies (E), total energy changes (ΔE) and within segment energy transfers (Rw) of both segments and the between segment energy transfers (Rb). Comment briefly on the results.
10. Explain how inter- and intrasegmental energy transfers occur and outline their importance in sport and exercise.

Table 4.1 Energy transfers in walking (in joules) – (adapted from Winter, 1979)

	Frame	*1*	*2*	*3*	*4*	*5*	*6*	*7*	*8*	*9*	*10*	*Notes:*
	T_{t1}	2.64	3.06	3.23	3.11	2.74	2.21	1.65	1.14	0.74	0.47	T_{t1} = Translational K.E.
S	T_{r1}	0.13	0.17	0.21	0.23	0.24	0.23	0.21	0.18	0.14	0.11	T_{r1} = Rotational K.E.
H	V_{g1}	0.35	0.25	0.17	0.12	0.10	0.09	0.09	0.10	0.09	0.09	V_{g1} = Gravitational P.E.
A	$\|\Delta T_{t1}\|$											From frame 1 to 2, etc.
N	$\|\Delta T_{r1}\|$											
K	$\|\Delta V_{g1}\|$											
	E_1											$E_1 = T_{t1} + T_{r1} + V_{g1}$
	$\|\Delta E_1\|$											
	Rw_1											$\frac{1}{2}(\|\Delta T_{t1}\| + \|\Delta T_{r1}\| + \|\Delta V_{g1}\| - \|\Delta E_1\|)$
T	T_{t2}	4.42	6.47	7.74	8.05	7.43	6.14	4.52	2.90	1.49	0.38	T_{r2} negligiable
H	V_{g2}	0.92	0.86	0.87	0.94	1.07	1.24	1.42	1.60	1.76	1.92	
I	$\|\Delta T_{t2}\|$											
G	$\|\Delta V_{g2}\|$											
H	E_2											$T_{t2} + V_{g2}$
	$\|\Delta E_2\|$											
	Rw_2											$\frac{1}{2}(\|\Delta T_{t2}\| + \|\Delta V_{g2}\| - \|\Delta E_2\|)$
	E_{12}											$E_1 + E_2$
	$\|\Delta E_{12}\|$											
	Rb_{12}											$\frac{1}{2}(\|\Delta E_1\| + \|\Delta E_2\| - \|\Delta E_{12}\|)$

4.7 References

Alexander, R. McN. (1992) *Exploring Biomechanics: Animals in Motion*, Scientific American Library, New York.

Bartlett, R. M. and Best, R. J. (1988) The biomechanics of javelin throwing: a review. *Journal of Sports Sciences*, **6**, 1–38.

Bown, W. and Mehta, R. D. (1993) The seamy side of swing bowling. *New Scientist*, **21 August**, 21–24.

Briggs, L. L. (1959) Effect of spin and speed on the lateral deflection (curve) of a baseball and the Magnus effect for smooth spheres. *American Journal of Physics*, **27**, 589–596.

Cochran, A. C. and Stobbs, J. (1968) *The Search for the Perfect Swing*, Heinemann, London.

Daish, C. B. (1972) *The Physics of Ball Games*, EUP, Cambridge.

Gaesser, G. A. and Brooks, G. A. (1975) Muscular efficiency during steady state exercise: effects of speed and work rate. *Journal of Applied Physiology*, **38**, 1132–1139.

Mehta, R. D. (1985) Aerodynamics of sports balls. *Annual Review of Fluid Mechanics*, **17**, 151–189.

Pierrynowski, M. R., Winter, D. A. and Norman, R. W. (1980) Transfers of mechanical energy within the total body and mechanical efficiency during treadmill walking. *Ergonomics*, **23**, 147–156.

Schleihauf, R. E. (1974) A biomechanical analysis of freestyle. *Swimming Technique*, **Fall**, 89–96.

Shorten, M. (1984) Mechanical energy models and the efficiency of human movement. *International Journal of Modelling and Simulation*, **3**, 15–19.

Shorten, M. (1985) Mechanical energy changes and elastic energy storage during treadmill running, in *Biomechanics IX-B* (eds D. A. Winter, R. W. Norman, R. P. Wells *et al.*), Human Kinetics, Champaign, IL, pp.313–318.

Wellicome, J. F. (1967) Some hydrodynamic aspects of rowing, in *Rowing: a Scientific Approach*, (eds G. P. Williams and A. C. Scott), Kaye and Ward, London, pp. 22–63.

Winter, D. A. (1979) *Biomechanics of Human Movement*, Wiley-Interscience, New York.

Winter, D. A. (1990) *Biomechanics and Motor Control of Human Movement*, Wiley-Interscience, New York.

4.8 Further reading

Alexander, R. McN. (1992) *Exploring Biomechanics: Animals in Motion*, Scientific American Library, New York: chapter 2 provides fascinating insights into some energetic aspects of walking and running.

Daish, C. B. (1972) *The Physics of Ball Games*, EUP, Cambridge: chapters 6 and 7 deal with the aerodynamics of balls non-mathematically. The mathematically inclined reader will also benefit from reading chapters 12 and 13.

Mehta, R. D. (1985) Aerodynamics of sports balls. *Annual Review of Fluid Mechanics*, **17**, 151–189.

Winter, D. A. (1990) *Biomechanics and Motor Control of Human Movement*, Wiley-Interscience, New York, chapter 5: a different approach to human energetics from that of section 4.3 and well worth reading.

Part Two

Techniques for Recording and Analysing Sports Movements

Introduction

The techniques for data collection, processing and analysis are essential components of biomechanical studies of sports movements. These techniques include: two- and three-dimensional video and cinematography and automated opto-electronic motion analysis systems; other ways of measuring displacement; dynamometry; electromyography; accelerometry; pressure measurement and measurement of body segment parameters (anthropometry). These will be covered in the following chapters. An emphasis will be placed on the uses of these techniques in sports biomechanics and their limitations, on the accuracy and reliability of the data obtained from them and on practical considerations concerning their use.

Chapter 5 covers the use of cinematography and video analysis in the study of sports movements, including the equipment and methods used, and the importance of cinematography and video in the qualitative and quantitative analysis of sports techniques. The necessary features of cine and video equipment for recording movements in sport are considered as well as the advantages and limitations of two- and three-dimensional recording of sports movements. The possible sources of error in recorded movement data are outlined and experimental procedures are described that will minimize recorded error in two- and three-dimensional movements. The need for, and the ways of performing, smoothing and filtering of kinematic data are covered, along with the requirement for accurate body segment inertia parameter data and how these can be obtained. Finally, a convention for the specification of three-dimensional segment orientations and some aspects of error analysis are introduced.

In Chapter 6, the use of the force platform in sports biomechanics is covered, including the equipment and methods used, the processing of force platform data and some examples of the use of force measurements in sport. The importance of the measurement of the contact forces on the sports performer is considered. The advantages and limitations of the two main types of force platform and the measurement characteristics required for a force platform in sports biomechanics are outlined. The procedures for calibrating a force platform are described, along with those used to record forces in practice. The different ways in which force platform data can be processed to obtain other movement variables is also covered, along with some applications of these data. The chapter concludes with a consideration of some of the reported research in sports biomechanics that has used force measurements.

Chapter 7 deals with the use of electromyography in the study of muscle activity in sports biomechanics. It includes the equipment and methods used, the processing of EMG data and the important relationship between EMG and muscle tension. Consideration is given to the importance of the EMG signal in sports biomechanics and why the recorded EMG differs from the physiological EMG signal. The chapter also covers the advantages and limitations of the three types of EMG electrodes suitable for use in sports biomechanics, the main characteristics of an EMG amplifier and other EMG equipment. The processing of the raw EMG signal is considered in terms of its time-domain descriptors and the EMG power spectrum and the measures used to define it. Finally, the chapter overviews the research that has been conducted into the relationship between EMG and muscle tension, and provides an understanding of the importance of this relationship.

In Chapter 8, some of the other techniques used in the recording and analysis of sports movements are dealt with, including their advantages and limitations. The uses and limitations of single-plate multiple image photography for recording movement in sport are considered. The use of automated opto-electronic motion analysis systems is described and the advantages and limitations, for sports movements, of three types of opto-electronic system are also covered. The use of electrogoniometry to record joint motion is outlined, along with the use of accelerometry, and the limitations of both of these techniques for sports movements are addressed. The value of contact pressure measurements in the study of sports movements is covered, including the relative advantages and limitations of pressure insoles and pressure platforms and of the three types of pressure transducer used in sports biomechanics. Some examples are provided of the ways in which pressure transducer data can be presented to aid analysis. Finally, brief consideration is given to the restrictions on the use of direct tendon force measurement in sport and how isokinetic dynamometry can be used to record the net muscle torque at a joint.

Cinematography and video analysis 5

This chapter is intended to provide an understanding of the use of cinematography and video analysis in the study of sports movements, including the equipment and methods used. After reading this chapter, you should be able to:

- understand the importance of cinematography and video analysis in the study of sports movements;
- undertake a qualitative video analysis of a sports technique of your choice and plan how you would carry out a quantitative video analysis of that technique;
- understand the important features of cine and video equipment for recording movements in sport;
- outline the advantages and limitations of two- and three-dimensional recording of sports movements;
- list the possible sources of error in recorded movement data;
- describe and implement experimental procedures that would minimize measurement inaccuracy in a study of an essentially two-dimensional movement;
- appreciate how these procedures can be extended and modified to record a three-dimensional movement;
- understand the need for, and the ways of performing, smoothing and filtering of kinematic data;
- appreciate the need for accurate body segment inertia parameter data, and ways in which these can be obtained;
- describe a three-dimensional convention for specifying segment position and orientation.

5.1 The use of cine and video analysis in sports biomechanics

5.1.1 INTRODUCTION

Sports biomechanics primarily focuses on two main areas, the enhancement of performance and the reduction of injury. A mixture of experimental and theoretical approaches is used to seek answers to such questions as: what is the best running technique to minimize energy expenditure; how should the sequence of body movements be organized in a javelin throw to maximize the distance thrown; why are lumbar spine injuries so common among fast bowlers in cricket?

The Biomechanics Guidelines of the British Association of Sport and Exercise Sciences (Bartlett, 1992) identify two levels of experimental biomechanical analysis of sport. Qualitative analysis involves a careful, if somewhat subjective, description of sports movements; quantitative analysis requires detailed measurement and evaluation of the measured data. This chapter will concentrate on the latter, although the former should not be neglected. It should precede any quantitative analysis in order to build up a comprehensive understanding of the sport or event studied and to highlight the variables that should be measured.

The quantitative experimental approach often takes one of two forms. A cross-sectional study, for example, could evaluate a sports movement by comparing the techniques of different sports performers. This can lead to a better understanding of the biomechanics of the skill studied, and help in diagnosis of faults in technique. An alternative, less frequently used approach is to compare several trials of the same individual, for example at the same sports competition, such as a series of high jumps by one athlete. This would be done to identify the performance parameters which, for that person, correlate with success. In a longitudinal study, the same person, or group, is analysed over time in order to improve their technique. This will probably involve feedback and modification to their movement pattern. Both cross-sectional and longitudinal approaches are relevant to the sports biomechanist, although conclusions drawn from a cross-sectional study of several athletes cannot be generalized to a single athlete or *vice versa*. An experimental study may be used in conjunction with a theoretical approach such as computer simulation. Both of these approaches are part of the overall scientific method of sports biomechanics proposed by Yeadon and Challis (1993) and illustrated in Figure 5.1.

The main techniques for recording and studying sports movements are cinematography and video. A great strength of both cinematography and video recording is that they enable the investigator to film not only in a controlled laboratory setting, but also in competition. The two techniques also minimize any possible interference with the performer.

5.1.2 LEVELS OF BIOMECHANICAL ANALYSIS OF SPORTS MOVEMENTS

Cine or video recording of sports movements is normally undertaken for one or more of the following reasons (see also Bartlett, Challis and Yeadon, 1992).

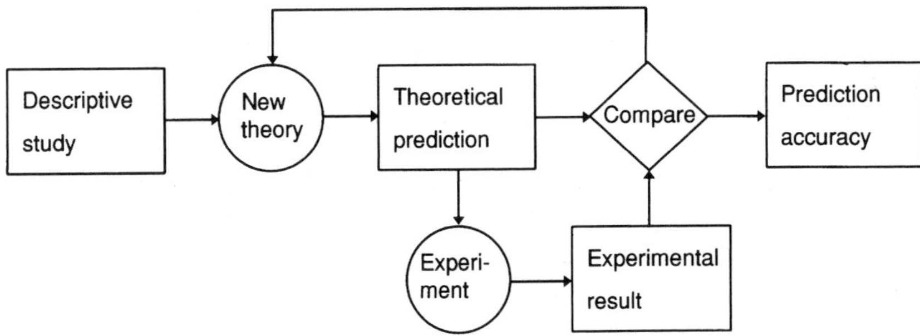

Figure 5.1 Schematic diagram of scientific method (adapted from Yeadon and Challis, 1993).

Qualitative analysis

This involves detailed study of the movement (technique), usually at slow speed or frame by frame, with minimal disturbance to the performer. It may be extended to a semi-quantitative level by taking simple timings and displacement measures. This can be done easily from a suitably displayed image without sophisticated equipment. For example, the durations of phases of the movement can be estimated by counting frames. Also, basic displacement measures, such as stride length, can be estimated using some simple length-measuring device, providing a real-life scale has been filmed.

Quantitative kinematic analysis

This will often involve the biomechanist having to digitize a large amount of data. This process of 'coordinate digitization' involves the identification of a number of body landmarks, such as estimated joint axes of rotation. The coordinates of each point are usually recorded and stored in computer memory by use of a digitizing pen or pointer or a cursor controlled by a computer mouse or similar device. After digitizing a movement sequence, linear and angular displacements can be calculated and presented as a function of time (for example Figure 5.2(a)).

Some additional data processing will normally be performed to obtain centre of mass displacements. Velocities and accelerations will also probably be obtained from the displacement data (e.g. Figure 5.2(b),(c)). Such analysis may also involve the identification of values of some of these variables at important instants in the movement to allow inter- or intraperformer comparisons. These values, often called **performance parameters**, are usually defined at the key events that separate the phases of sports movements, such as foot strike in running, release of a discus and bar release in gymnastics.

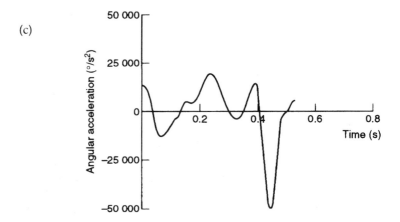

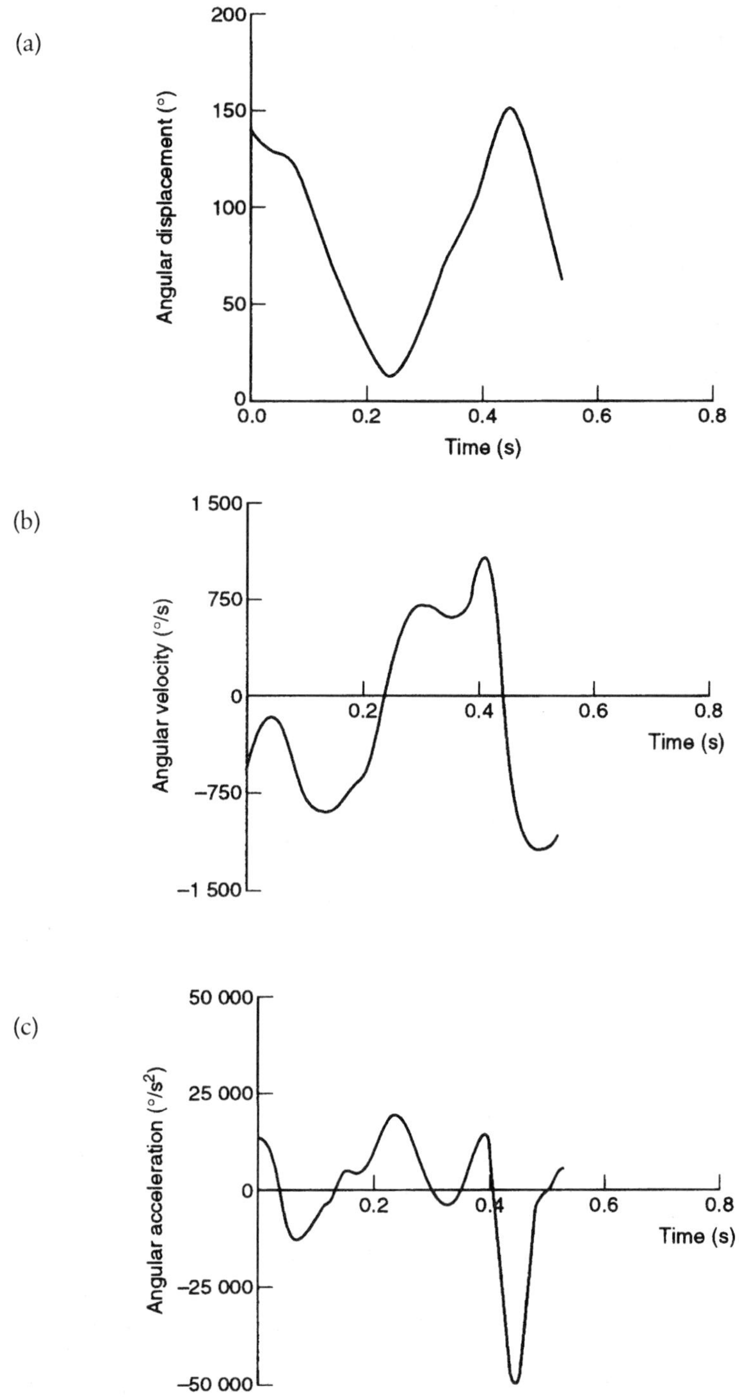

(a)

(b)

(c)

Figure 5.2 Time functions of upper arm to trunk angle for a fast bowler from start of delivery stride until after ball release: (a) displacement; (b) velocity; (c) acceleration.

Computer visualization of the movement will also be possible. This can be in the form of stick figure sequences (Figure 5.3).

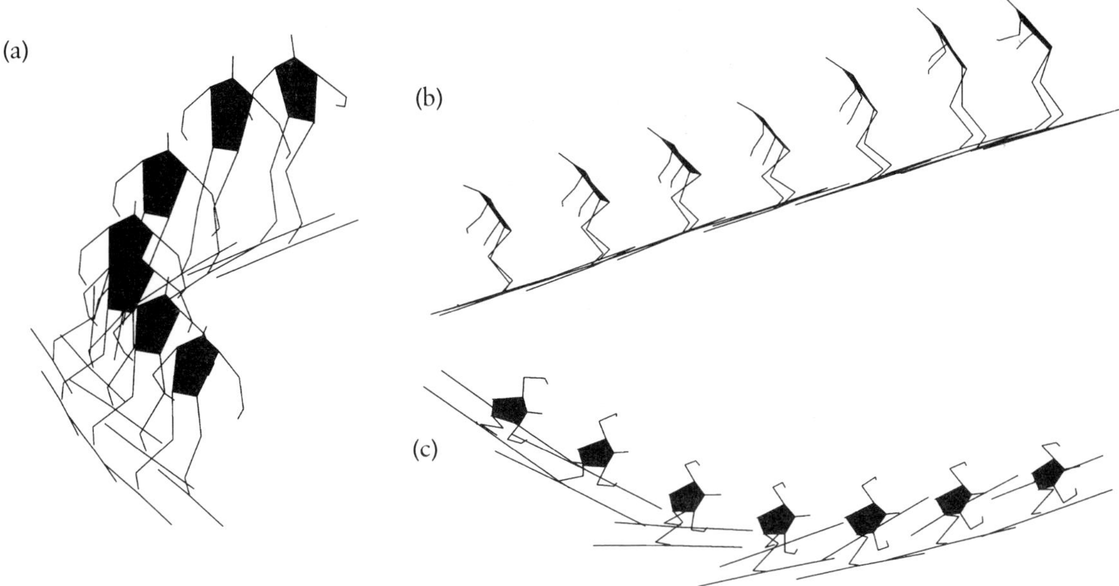

Figure 5.3 Stick figure sequences of a skier: (a) front view; (b) side view; (c) top view.

These are quick and easy to produce but have ambiguities with respect to whether limbs are in front of or behind the body (Figure 5.4(a)).

In three-dimensional analysis, this can be partially overcome by filling in the body and using hidden line removal (Figure 5.4(b)). Full solid-body modelling (Figure 5.4(c)) is even more effective in this respect, but computationally very time-consuming. Solid-body models can be made more realistic through the use of shading and surface rendering, but at an even greater time cost (e.g. Calvert and Bruderlin, 1995).

Kinetic analysis (inverse dynamics)

This involves the calculation of kinetic variables for joints and body segments to try to understand the underlying processes that give rise to the observed motion patterns. The method of inverse dynamics (for example Gagnon, Robertson and Norman, 1987) is used to calculate joint reaction forces and moments. This is usually done in combination with external force measurements from, for example, a force platform. The method of inverse dynamics is valuable in sports biomechanics research to provide an insight into the musculoskeletal dynamics that generate the observed characteristics of sports techniques. However, there are many calculations needed to produce these data, and an assessment of the measurement and data processing errors involved is important.

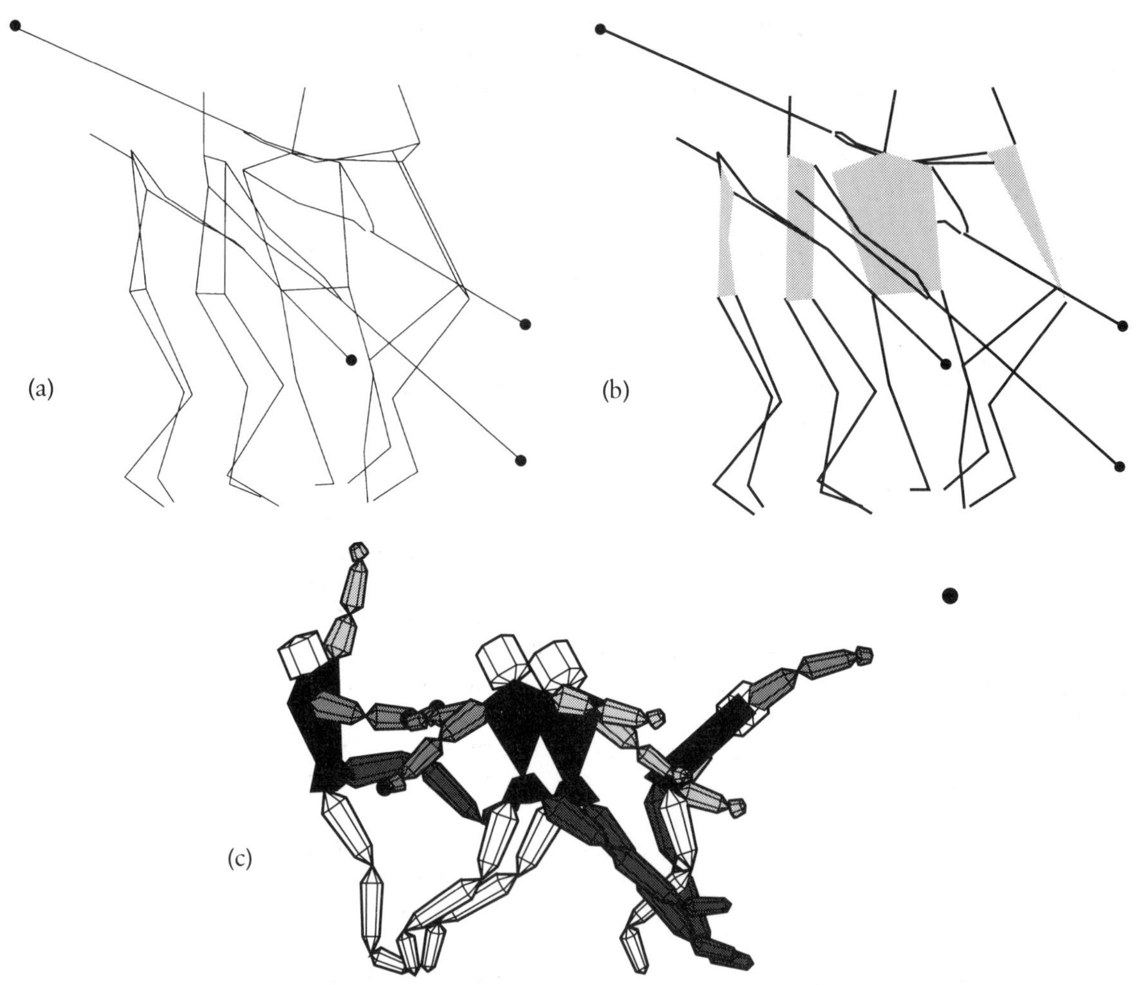

Figure 5.4 Computer visualization: (a) stick figure of hammer thrower; (b) as (a) but with body shading and hidden line removal; (c) solid body model of cricket fast bowler.

5.2 Recording the movement

5.2.1 CINE OR VIDEO?

Cinematography has long been used because of excellent picture quality and the availability of coordinate digitizer tablets having very good resolution (the number of individual positions on the tablet which return different image coordinates). The cameras used normally take up to 500 pictures per second (they have frame rates up to 500 Hz). The limitations of cinematography include the lack of immediacy of the results, because of film processing, the large quantities of expensive film that are often used and the need for a working knowledge of photography.

Video recording overcomes these limitations. Recent improvements in video technology have included an increased availability of electronically shuttered video cameras. In these cameras, electronic signals are applied to the sensor to control the time over which the incoming light is detected. This reduces the 'smearing' (blurring) of the image that occurs when recording moving objects with a non-shuttered video camera that has an exposure time of 1/50 s. Other important developments include high-quality slow-motion and 'freeze-frame' playback devices that allow the two fields that make up an interlaced video frame to be displayed sequentially. A further improvement has been the increased resolution of video image capture boards, which allow fields of video to be captured and stored in a computer for further processing, such as coordinate digitization. Standard (50 fields/25 frames per second) video equipment has therefore become an attractive alternative to cinematography because of its price, immediacy and accessibility. Also, video offers far better opportunities than does cine for automated image processing (Chapter 8). The major drawbacks of video are the resolution of the image, which restricts digitizing accuracy when compared with high-resolution cine digitizing tablets (e.g. Kerwin and Templeton, 1991), and the high cost of video equipment to match the higher frame rates of cinematography.

Cine and video are both sampling processes – i.e. the movement is captured for a short time interval and then no further changes in the movement are recorded until the next field or frame. The number of such pictures taken per second is therefore called the **sampling rate** or **sampling frequency**. For the recording stage of movement analysis, this will correspond to the field rate or frame rate. The overall sampling rate for the analysis may be less than this if not every field or frame is digitized. An important limitation of cine-based and most standard video-based motion analysis systems is the vast amount of manual coordinate digitization required for quantitative analysis. This is especially the case when a three-dimensional analysis is being undertaken, as at least two images have to be digitized for each field or frame to be analysed. To overcome this drawback, opto-electronic systems have been developed, some of which use video technology, which automatically track markers attached to the body (Chapter 8).

5.2.2 RECORDING THE IMAGE – CAMERAS AND LENSES

As cinematography involves taking a sequence of still photographs, some familiarity with photographic equipment and techniques is necessary to obtain the best results. The main equipment requirement for a cinematographic study is a good quality cine camera. This should be capable of taking pictures at the frame rate required for the investigation with a minimum of optical distortion. The size of the film used (usually 35 mm, 16 mm or 8 mm) is important, as larger sizes provide better picture resolution. The high cost of 35 mm cine cameras mitigates against their use, despite their large picture size, while 8 mm cameras provide a picture size which is too small and they are often of inferior optical quality. In practice, 16 mm cameras are generally used, as they provide a good compromise between economy and image size.

Motor-driven cameras should be used because of their consistent interframe interval; spring-driven cameras should only be used for qualitative analysis purposes. A camera with an adjustable frame rate up to 500 Hz will normally be suitable for biomechanical studies of sports techniques. A higher frame rate might be necessary for the study of impacts, such as a golf club hitting a golf ball. In such cases, a rotating prism camera will be needed, in which light is directed on to continuously moving film by a continuously rotating prism (Figure 5.5(a)).

For cameras operating up to 500 Hz, film is moved past a rotating shutter, which controls the entry of light. A pin-registered camera (Figure 5.5(b)) should be used to minimize film movement in the camera gate. In such a camera, the film is transported into the gate while the shutter is closed by a pair of pins (the retraction pins), which locate in the perforations on the edges of the film. A separate pair of pins (the registration pins) holds the film in place while the picture is taken.

A camera with an internal timing light, which records timing marks on one edge of the film, is convenient. Otherwise some other method of calibrating the frame rate must be used. If two or more cameras are to be used in a three-dimensional study, then provision for phase-locking of the cameras is useful, as it allows the cameras to take synchronized pictures. Preferably the camera should have a variable shutter with a good choice of openings (see Figure 5.6, where the shutter factor is 360% divided by the shutter opening). This enables the shutter to be closed, subject to lighting conditions, to control the exposure time and therefore reduce blurring when fast movements are recorded. This is more cost-effective than the alternative use of excessively high frame rates, as the exposure time = 1/(frame rate × shutter factor). An exposure time of 1/500 s, suitable for many fast movements, could be achieved, for example, with the following combinations of frame rate and shutter factor: 50 Hz and factor 10, 100 Hz and factor 5, 250 Hz and factor 2.

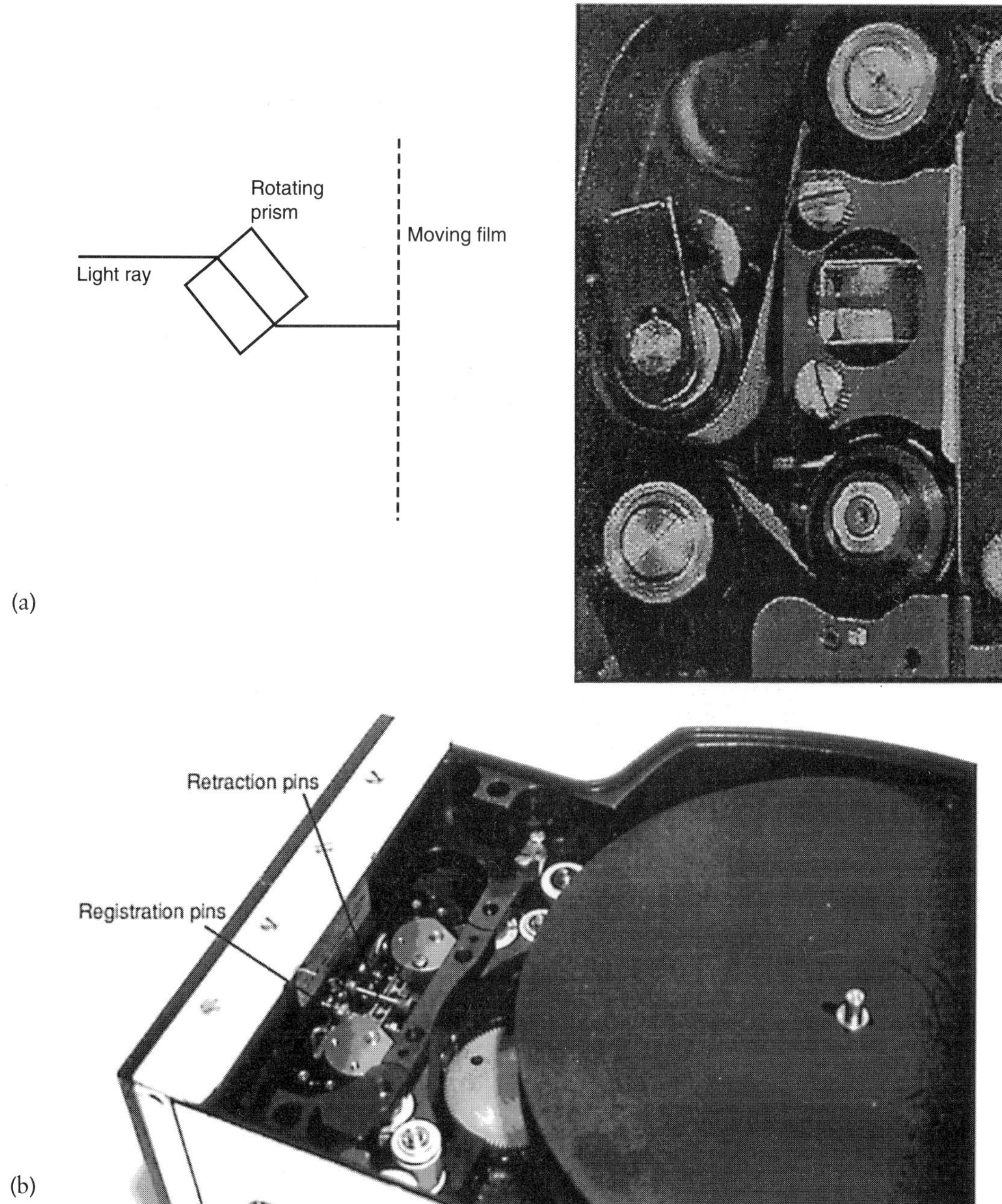

Figure 5.5 Cine camera principles: (a) rotating prism camera; (b) pin-registration camera.

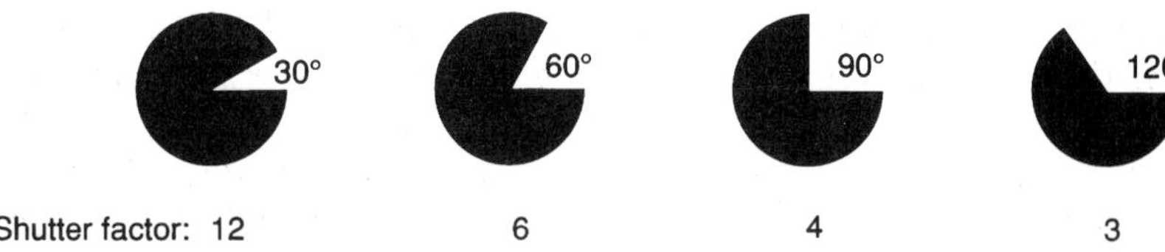

Figure 5.6 Camera shutter openings.

Only high-quality lenses should be used. Fixed focal length lenses usually cause less image distortion, called lens errors, than variable focal length (zoom) lenses. However, a high-quality telephoto zoom lens is ideal for allowing filming from variable, large distances. Filming from well away from the activity using such a lens minimizes perspective error (see below) while maintaining a large image of the performer. It also reduces any disturbance to the performer. A lens is needed with a suitable range of aperture settings for the lighting conditions and depth of field requirements of a particular study. The aperture is the opening at the front of the camera which admits light, in a way analogous to the iris of the human eye. It can usually be adjusted, or 'stopped down', by means of a diaphragm. The aperture is normally expressed as an *f* number (or *f* stop), which is the ratio of the focal length of the lens to the aperture diameter. An *f* stop of 2 admits about twice as much light as one of 2.8 but has a more restricted depth of field (that range of distances from the lens over which the image will be sharply focused).

Traditionally, black and white film was preferred to colour, because of its sharper image definition and lower cost. However, improvements in colour film and the increased cost of, and fewer processing laboratories for, black and white film have overturned this. The choice of film sensitivity depends on the lighting levels involved and the frame rates to be used, as well as likely problems with graininess (lack of resolution) on the projected image. Film sensitivity is expressed by the ISO number such that the ISO number doubles as the sensitivity of the film to light doubles. The need to avoid excessive graininess makes film sensitivities greater than ISO 400 unsuitable (Bartlett, Challis and Yeadon, 1992). Reversal film (similar to slide film) has generally been preferred to negative film because of the sharper images produced. The disadvantage of reversal film are that copies cannot easily be made and that few commercial laboratories now process such film.

For video recording of sports movements for biomechanical analysis, video cameras capable of similar frame rates to those of pin-registered cine cameras would ideally be used. However, the 50 pictures per second that can be obtained with playback systems that display in sequence the two fields that make up a video frame is a sufficiently high sampling rate for some sports movements. Such video cameras are of varying quality and different formats. Hi-Band U-matic and Betacam are far superior to VHS and are generally considered to be better than super VHS (S-VHS). Modern, solid-state video cameras detect the image using an array of light sensors precisely etched into silicon and are claimed to have zero geometric distortion, unlike cine film (Greaves, 1995). It is essential to use electronically shuttered video cameras to obtain good quality, unblurred picture 'freezing' for analysis, as the 40 ms scan time of a standard video frame is excessive (Bartlett, Challis and Yeadon, 1992). Many video cameras now have electronic shutters with several settings to accommodate to various light levels. If two or more cameras are to be used in a three-dimensional study, provision for gen-locking the cameras allows the recording of images to be synchronized.

Although camcorders are convenient and not expensive, they do not normally allow the interchange of lenses, which can be useful, and they cannot be gen-locked. The latter is a limitation to their use for three-dimensional analysis. Although low-distortion lenses can be obtained for video cameras for biomechanical measurement (Allard, Blanchi and Aïssaoui, 1995), the quality of most standard video lenses is not as high as the more expensive lenses normally used in cinematography. This can result in image distortion, particularly at wide-angle settings and this can lead to increased errors in digitized coordinates.

No frame rate calibration is needed with video. However, unlike the wide range of sampling rate settings on most pin-registered cine cameras, 50 Hz video cameras have only a fixed field rate. Even high-speed (100 Hz and above) video cameras normally only provide a few alternative sampling rates. High-speed video cameras are also more expensive than pin-registered cine cameras. They can achieve higher field rates by splitting each field into several pictures, but this decreases the vertical resolution. Alternatively, for the highest field rates (up to around 5000 Hz) the recording heads rotate faster and the tape moves past them at greater speeds.

5.2.3 DISPLAYING THE IMAGE – CINE PROJECTORS AND VIDEO PLAYERS

The optical quality of the cine projector used is very important. Ideally a pin-registered projector should be used to minimize movement of the film in the gate. Such projectors can be obtained at reasonable cost. The projector should give an image at least 25 times the size of the film (Dainty *et al.*, 1987) and must permit stop frame operation for both qualitative and quantitative analysis. The projected image should be bright enough to be clearly viewed. For a given projector there is an inverse relationship between image size and brightness (Bartlett, Challis and Yeadon, 1992). The projector should also be firmly fixed in its operating position to reduce day-to-day variability and the need for frequent realignment.

The video playback system should provide absolutely still images on 'freeze-frame', both for qualitative and quantitative analysis, and flicker-free slow-motion replay. A playback system that displays each of the two video fields that make up a video frame (thus providing 50 pictures per second) rather than only one field per frame (and hence only 25 pictures per second) is a requirement for quantitative biomechanical analysis of most sports. Playback systems that display the two fields combined into a single frame are unsatisfactory for biomechanical analysis (Bartlett, Challis and Yeadon, 1992).

5.2.4 OBTAINING BODY COORDINATES

A coordinate digitizer is required for any detailed quantitative analysis. The coordinates of important points on the image are obtained from a digitizing tablet for film. For video, the image is stored by an image capture board ('frame grabber') and displayed on a video or computer monitor for digitizing. A cine digitizing tablet should have a minimum screen resolution of 0.5 mm, and preferably better than 0.1 mm (Dainty *et al.*, 1987). The resolution of video digitized coordinates is limited by the number of individual spots of light, or picture elements (**pixels**), that are captured and generated by the image capture board (1024 horizontally by 1024 vertically would be a very good specification). This is the main reason for the greater digitizing errors usually reported in comparisons of cine and video digitized coordinates (e.g. Kerwin and Templeton, 1991). Projection of the video image on to a digitizing tablet using a high-quality video projector or interpolating the position of the digitizing cursor between pixels have been used as solutions to this problem. The calibration and reliability of the digitizer used should be checked regularly, and the digitizer should be connected to a computer equipped with appropriate software for data processing (Bartlett, Challis and Yeadon, 1992). Many reputable software systems are now commercially available for cine and video digitizing and data processing.

5.2.5 TWO-DIMENSIONAL OR THREE-DIMENSIONAL ANALYSIS?

A decision must be made as to whether a two-dimensional or three-dimensional analysis is required. Both have advantages and disadvantages as summarized by the following.

Two-dimensional recording and analysis:

- Is simpler and cheaper as fewer cameras and other equipment are needed.

- Requires movements to be in a preselected movement plane (the plane of motion or plane of performance). It can yield acceptable results for essentially planar movements but it ignores movements out of the chosen plane. This can be important even for an event which might appear essentially two-dimensional, such as the long jump (Yeadon and Challis, 1993).
- Is conceptually easier to relate to than three-dimensional coordinates.
- Requires less digitizing time and has fewer methodological problems, such as the transformation of image to movement plane coordinates.

Three-dimensional recording and analysis:

- Has more complex procedural considerations.
- Can show the body's true spatial motions and is closer to the reality of the movements studied.
- Requires more equipment and is thus more expensive. Although it is possible by intelligent placement of mirrors to record several images on one camera, this is rarely practical in sports movements.
- Has increased computational complexity associated with three dimensional reconstructions, and software time synchronization of the results from cameras that are not physically time-synchronized (phase-locked or gen-locked).
- Allows intersegment angles to be calculated accurately, without viewing distortions. It also allows the calculation of other angles which cannot be easily obtained from a single camera view in many cases – an example is the hip axis to shoulder axis angle (Figure 5.7), which can be visualized from above even if the two cameras were horizontal. It poses a problem with respect to the convention used for segment orientation angles (see later), which two-dimensional analysis sidesteps.
- Enables the reconstruction of simulated views of the performance (e.g. Figure 5.3(c)) other than those seen by the cameras, an extremely useful aid to movement analysis and evaluation.

5.2.6 PROBLEMS AND SOURCES OF ERROR IN MOTION RECORDING

The requirement of recording human motion in sport can be formally stated as: to obtain a record that will enable the accurate measurement of the position of the centre of rotation of each of the moving body segments and of the time lapse between successive pictures. The following problems and sources of error can be identified in two-dimensional recording of sports movements.

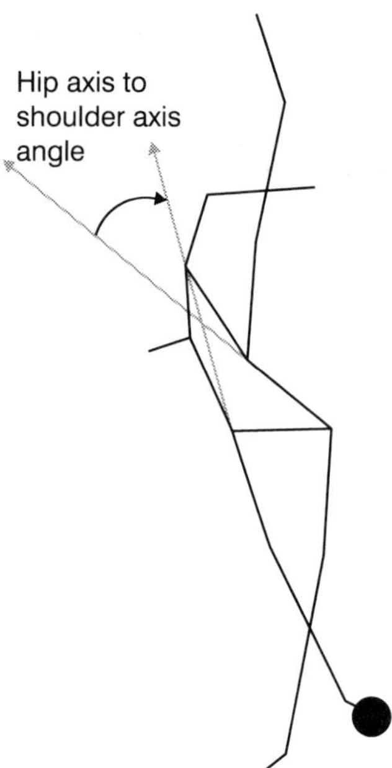

Figure 5.7 Hip axis to shoulder axis angle (reconstructed view from above).

- The three-dimensionality of the position of centres of rotation requires the analysis of movements recorded from one camera to be done with care.
- Any non-coincidence of the plane of motion (the plane of performance) and the plane perpendicular to the optical axis of the camera (the photographic plane) is a source of error if calibration is performed with a simple scaling object in the plane of motion.
- Perspective and parallax errors need attention. Perspective error is the apparent discrepancy in length between two objects of equal length, such as left and right limbs, when one of the limbs is closer to the camera than the other. It occurs for movements away from the photographic plane. The term is also sometimes used to refer to the error in recorded length for a limb or body segment which is at an angle to the photographic plane and therefore appears to be shorter than it really is. Associated with this error is that caused by viewing away from the optical axis, such that, across the plane of motion, the view is not always side-on, as at the positions marked (*) in Figure 5.8(b). This is sometimes referred to as **parallax error**. The combined result of these optical errors is that limbs nearer to the camera appear bigger and appear to travel further than those further away. This causes errors in the digitized coordinates.

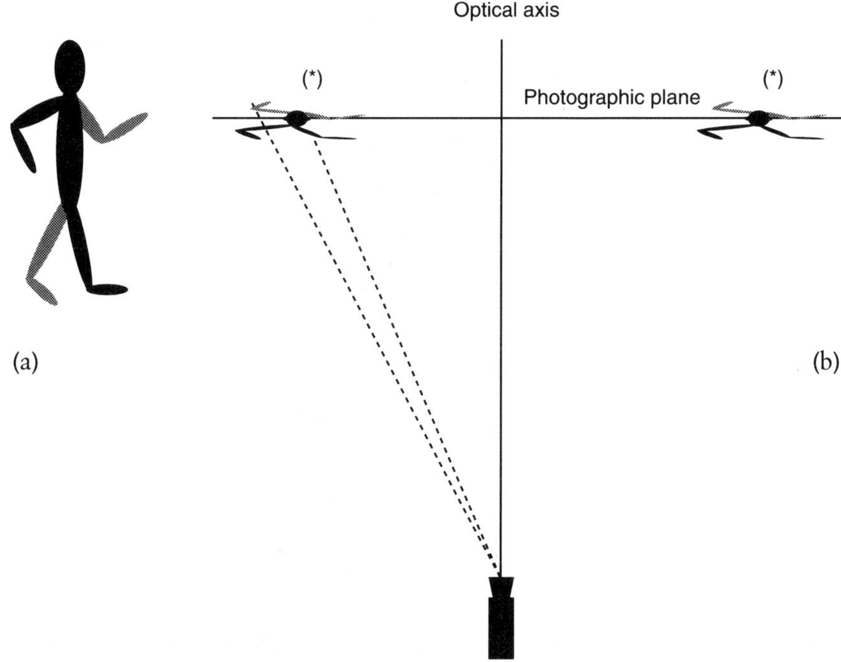

Figure 5.8 Errors from viewing movements away from the photographic plane and optical axis of the camera: (a) side view; (b) as seen from above.

- Lens distortions, and film distortions such as stretching and imperfect registration, may be a source of error.
- Locations of joint axes of rotation are only estimates, based on the positions of superficial skin markers or identification of anatomical landmarks. Use of skin markers can both help and hinder the biomechanist, as they move with respect to the underlying bone and to one another. The digitizing of such markers, or estimating the positions of axes of rotation without their use, is probably the major source of random error (or noise) in recorded joint coordinates. Locating joint axes of rotation is especially difficult when the joint is obscured by other body parts or by clothing.
- The accuracy of the sampling rate is no problem for video, but needs calibrating for cine cameras. Much biomechanical software will require a constant value of the frame rate.
- Other possible sources of error include: the sharpness of the projected image; locating cine film in the camera and projector gates; camera vibration; digitizing errors, related to coordinate digitizer resolution and human digitizer errors; computer round-off errors.

For three-dimensional analysis, there are other potential error sources, although several of those above are partly or wholly overcome.

- Relating the two-dimensional video or film image coordinates to the three-dimensional movement space ('real world' or object) coordinates may be a source of error. Several methods of doing this will be considered below, but they all present problems. Use of an array of calibration points, such as a 'calibration frame' (e.g. Figure 5.9), is probably the most common method. Errors within the frame volume can be accurately assessed, while those outside the frame will be greater and more difficult to assess (see Wood and Marshall, 1986). Errors will increase with the ratio of the size of the movement space to that of the image.

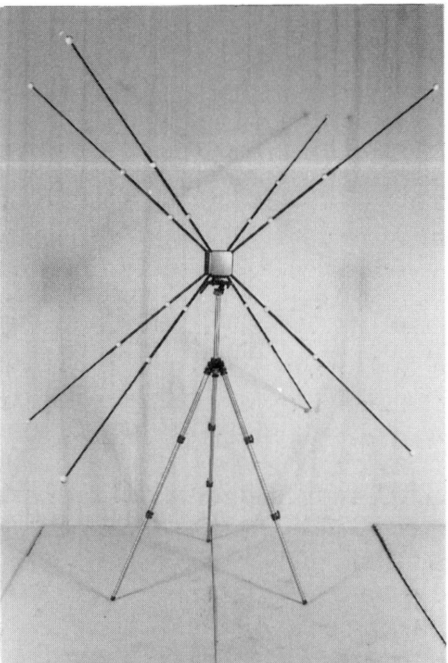

Figure 5.9 A typical calibration frame.

- All the calibration points must be clearly visible on the images from both cameras; they must also have three-dimensional coordinates that are known to a good accuracy.
- Placements of cameras must relate to the algorithm chosen for reconstruction of the movement space coordinates. Errors will be caused by deviations from these requirements.

In summary, digitized coordinate data will be contaminated with measurement inaccuracies or errors. These will be random (noise), systematic, or both. All obvious systematic errors (such as those caused by lens distortion and errors in calibration objects) should be identified and removed, for example by calibration or software corrections. Any remaining sources of systematic error will then be very small or of low frequency and will, therefore, have little effect on velocities and accelerations. The remaining random noise in the displacement data, expressed as relative errors, has been estimated as within 1% for a point in the photographic plane for two-dimensional cinematography and within 2% for a point in the calibration volume for three-dimensional cinematography (Dainty *et al.*, 1987). Random errors must be minimized at source by good experimental procedures. Any remaining noise should be removed, as far as possible, from the digitized data before further data processing. These two aspects will be covered in the next two sections. Also, more consideration needs to be given to the estimation of errors in digitized coordinate data and their effects on derived values (e.g. Payton and Bartlett, 1995).

5.3 Experimental procedures

5.3.1 TWO-DIMENSIONAL RECORDING PROCEDURES

The following steps (adapted from Bartlett, Challis and Yeadon, 1992) have often been considered necessary to minimize errors recorded during filming, thereby improving the accuracy of all derived data (e.g. Miller and Nelson, 1973; Smith, 1975). The following steps can still often prove useful especially for students who are unfamiliar with cine and video recording and analysis. They also allow for a simple linear transformation from image to movement plane coordinates using simple scale information recorded in the field of view.

- The camera should be mounted on a stationary, rigid tripod pointing towards the centre of the plane of motion and no panning of the camera should be used.
- The camera should be sited as far from the action as possible to reduce perspective error. A telephoto zoom lens should be used to bring the performer's image to the required size. The lens should be carefully adjusted to focus the image. This is best done, for zoom lenses, by zooming in on the performer, focusing and then zooming out to the required field of view. The field of view should be adjusted to coincide with the performance area that is to be recorded. This maximizes the size of the performer on the projected image and increases the accuracy of digitizing. For particularly long movements (e.g. long and triple jumps), consideration should be given to using two or more synchronized cameras to cover the filming area (Figure 5.10).
- The plane of motion should be perpendicular to the optical axis of the camera. This can be done in various ways, that include the use of spirit levels, plumb lines and 3–4–5 triangles.
- A length scale (for example a metre rule) and vertical reference (for example a plumb line) must be included in the field of view. For

video recording, horizontal and vertical scalings are required as video cameras scale horizontal and vertical coordinates differently (that is the 'aspect ratio' is not unity). This is also recommended for cine studies. The length scale must be positioned in the plane of motion. Its length should be at least that required to give a scaling error, when digitized, of no more than 0.5% (Dainty *et al.*, 1987).

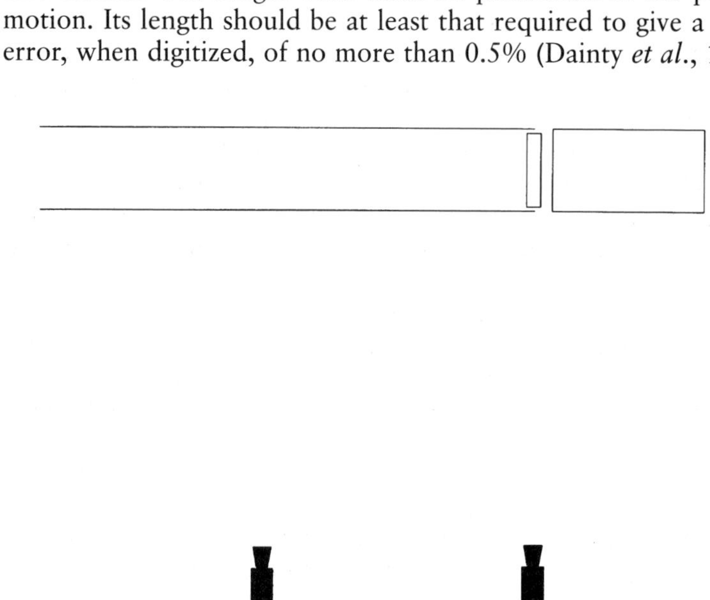

Figure 5.10 Possible camera placements for movement such as long jump.

- Ideally the background would consist of a square grid to provide calibration and accuracy checks across the whole field of view. This is rarely practical if filming in sports competitions, and is unnecessary providing strict experimental procedures are adopted throughout. The background should be as uncluttered as possible, plain and non-reflective.
- If using cine, provision must be made to assess the true frame rate from information obtained during the filming sequence, using a clock or internal timing marks. Procedures that calibrate the frame rate at some other time, such as by filming dropped objects, should be avoided and cine camera dial settings should not be relied on (Dainty *et al.*, 1987). The acceleration and deceleration times required for a cine camera to reach and slow down from its set frame rate should be established beforehand. Allowance should be made for these when filming, to ensure a constant frame rate during the motion sequences that will be analysed.
- The use of colour contrast markers on the appropriate body landmarks is sometimes recommended. The use of several non-collinear markers on each body segment can help to establish the instanta-

neous centre of rotation of each joint for each frame (e.g. Ladin, 1995). In much sports biomechanics practice, positions are marked on the skin. These correspond to an axis through the appropriate joint centre, as seen from the camera, when the subject is standing in a specific position and posture (e.g. Plagenhoef, 1971): these markers are sometimes referred to as joint centre markers. The use of any skin markers will not normally be possible in competition, and has some potential disadvantages when digitizing (see below). If using skin markers, then marking the skin directly is less problematical than using adhesive markers, which often fall off. If axes of rotation are being estimated from anatomical landmarks, skin markers are not essential, although their use probably saves time when digitizing.

- A sufficiently high frame rate should be used. Plagenhoef (1971) recommended minimum rates of 24 Hz for swimming, 64 Hz for the tennis serve and 80 Hz for a golf swing. The frame rate (sampling frequency) needs to be at least twice the maximum frequency in the signal (not twice the maximum frequency of interest) to avoid aliasing (Figure 5.11). This is a phenomenon often seen when wheels on cars and stage coaches, for example, appear to revolve backwards on film. The temporal resolution (the inverse of the sampling rate) improves the precision of both the displacement data and its time derivatives (Lanshammer, 1982). For accurate time (temporal) measurements, or to reduce errors in velocities and accelerations, higher frame rates may be needed (see also section 5.4.1).
- Lighting should be adequate to film at the required frame rate. If a choice is available, then natural daylight is usually preferred. If artificial lighting is used, floodlights mounted with one near the optical axis of the camera and one to each side at 30° to the plane of motion give good illumination. In a cine study, if tungsten lighting is used, then either tungsten balanced film or a tungsten filter with daylight film should be used. Careful attention must also be paid to light-meter readings when choosing the film sensitivity, the camera *f* stop, frame rate and shutter opening.
- Whenever possible, information should be incorporated within the camera's field of view, identifying important features such as the name of the performer and date. The 'take number' is especially important when filming is used in conjunction with other data acquisition methods.
- The recording of the movement should be as unobtrusive as possible. The performer may need to become accustomed to performing in front of a camera in an experimental context. The number of experimenters should be kept to the bare minimum in such studies. In controlled studies, away from competition, as little clothing as possible should be worn by the performers, to minimize errors in

locating body landmarks, providing that this does not affect their performance. In recording movements that are not in the public domain, written informed consent should be obtained from all participants in the study.

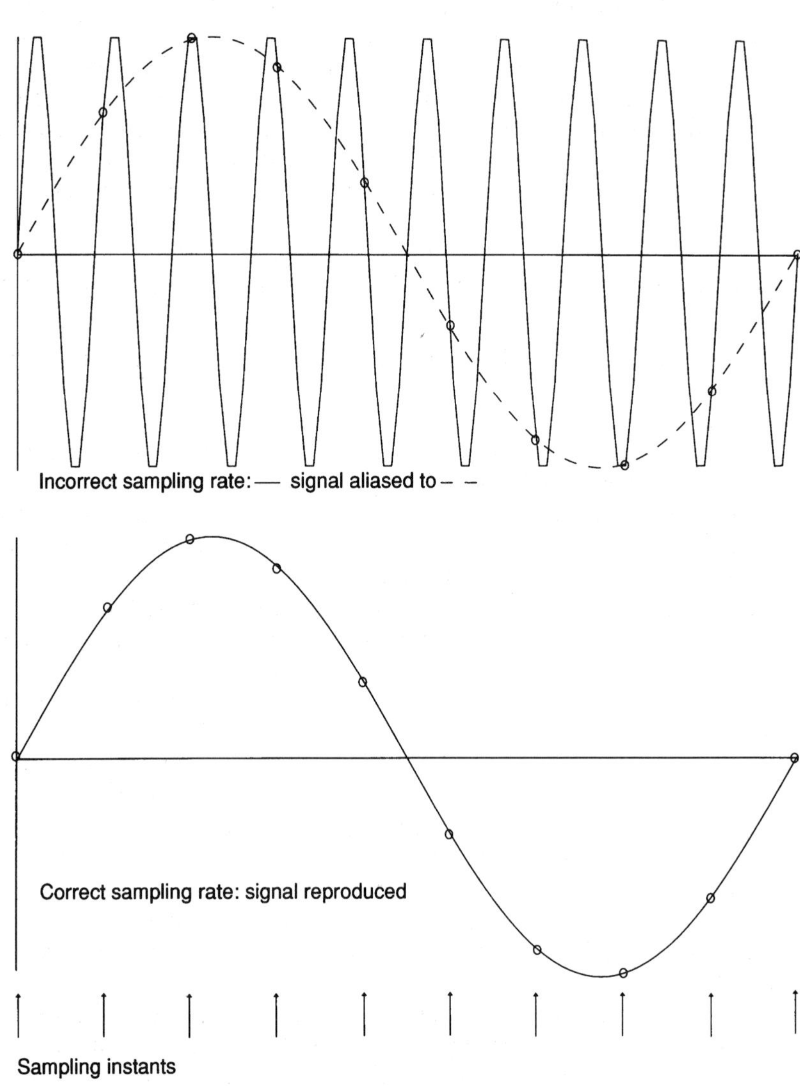

Figure 5.11 Aliasing.

The above protocol imposes some unnecessary restrictions on two-dimensional cine or video analysis. It can lead to severe practical limitations on camera placements, particularly in competitions, because of the requirement that the optical axis of the camera is perpendicular to the plane of motion. A more flexible camera placement can be obtained, for example by the use of a two-dimensional version of the direct linear transformation (see below). This overcomes some of the camera location problems that arise in competition because of spectators, officials and advertising hoardings. It requires the use of a more complex transformation from image to movement plane coordinates. Many of the above procedural steps then become redundant, but others, similar to those used in three-dimensional analysis, are introduced (Bartlett, Challis and Yeadon, 1992). This more flexible approach can also be extended to allow for camera panning (see Yeadon, 1989a). The reader should assess (Exercise 4) which of the above steps could not be ensured if filming at a top-level sports competition.

When projecting and digitizing the film, attention should be paid to the following points (adapted from Bartlett, Challis and Yeadon, 1992):

- If a cine projector is used, it must be carefully set up and fixed if possible. Checks should be carried out to ensure that the projector's optical axis is perpendicular to the plane of the digitizing tablet. This should ensure that the latter coincides with the image plane.
- If joint centre markers are used, careful attention must be paid to their movement relative to underlying bones. If segments move in, or are seen in, a plane other than that for which the markers were placed, the marker will no longer lie along the axis through the joint as seen by the camera. A thorough anatomical knowledge of the joints and the location of their axes of rotation with respect to superficial landmarks throughout the range of segmental orientations is needed to minimize errors. It should also be noted that many joint axes of flexion–extension are not exactly perpendicular to the sagittal plane often filmed for two-dimensional analysis.
- The alignment and scaling of the projected image must be checked. Independent horizontal and vertical scaling must be performed for video analysis and is also recommended for cine.
- For cine, the true frame rate must be obtained from the information included during the filming sequence. Its uniformity throughout the sequence to be analysed should also be checked. A consistent frame rate makes analysis easier, although software compensations for variable interframe intervals can be made.
- At least one recorded sequence should be digitized several times to check on operator reliability (consistency). It is also recommended that more than one person should digitize a complete sequence to check on operator objectivity. The validity of the digitizing can only be established by frequent checks on digitizer accuracy, by careful adherence to good experimental protocols, by analysis of a relevant and standard criterion sequence and by carrying out checks on all calculations. Film of a falling object is sometimes used as a criterion sequence as the object's acceleration is known.

- Projected image movement may occur, especially when cine cameras or projectors without registration pins have been used. This will introduce errors into the data unless corrected for by digitizing a series of control (or realignment) points. These must have been fixed in the filming (movement) space and, ideally, should be visible in each projected frame of film. A minimum of two such points, and preferably four or more, has been recommended (Kerwin, 1988). Image movement can also arise with video equipment, especially between adjacent fields, unless a time-base corrector is used. If this problem is found to occur, a similar use of control points will be necessary.

5.3.2 THREE-DIMENSIONAL RECORDING PROCEDURES

Some, but not all, of the considerations of the previous section also apply to three-dimensional recording and analysis. The major requirements of three-dimensional analysis are discussed in this section (for further details see for example Van Gheluwe, 1978; Yeadon and Challis, 1994). It is recommended that readers should gain good experience in all aspects of two-dimensional quantitative analysis before attempting three-dimensional analysis.

For reconstruction of three-dimensional movement space coordinates of a point, at least two cameras are needed. The cameras should ideally be phase-locked for cine or directly gen-locked for video to provide shutter synchronization and identical frame rates. For cine cameras, identical frame rates are only obtained for phase-locked cameras when they have run up to their set frame rate. The shutters should then be synchronized to within a reasonably small error. Cine cameras that have been phase-locked should incorporate a timing mechanism to check when the frame rates are synchronized. Furthermore, synchronization may also be needed to link the events being recorded by the two cameras (**event synchronization**). This can be done, for example, by a visual event, such as bursting a balloon. Alternatively, simple pulse generators, or switches, can be used to record simultaneous marks on the edge of the film in cine cameras (usually the opposite edge from the one with timing light marks). For video cameras, event synchronization can be achieved, for example, by the use of synchronized character generators.

Synchronization of cameras requires a physical connection in the form of cabling between the cameras. This will not always be possible, especially when recording movements during sports competitions. In such cases, a timing device, such as a digital clock, can be included in the field of view, visible from all cameras. Synchronization then needs to be performed mathematically at a later stage. Obviously some error is involved in this process. If this method is used then it is recommended (Dainty *et al.*, 1987) that the slowest camera should provide the timing baseline. It may also not be possible, again especially in competition, to include a timing device in the fields of view of the cameras (e.g. Yeadon, 1989a). Some synchronization method will then need to be used, based on information available from the recordings.

From two or more sets of image coordinates, some method is needed to reconstruct the three-dimensional movement space coordinates. Several algorithms can be used for this purpose and the choice of the algorithm may have some procedural implications. Most of these algorithms involve the explicit or implicit reconstruction of the line (or ray) from each camera that is directed towards the point of interest, such as a skin marker. The location of that point is then estimated as that which is closest to the two rays (Yeadon and Challis, 1994).

The simplest algorithm (e.g. Martin and Pomgrantz, 1974) requires two cameras to be aligned with their optical axes perpendicular to each other (Figure 5.12(a)). The cameras are then largely independent and the depth information from each camera is used to correct for perspective error for the other. The alignment of the cameras in this technique is difficult, although the reconstruction equations are very simple. This technique is generally too restrictive for use in sports competitions, where flexibility in camera placements is beneficial and sometimes essential (Bartlett, Challis and Yeadon, 1992).

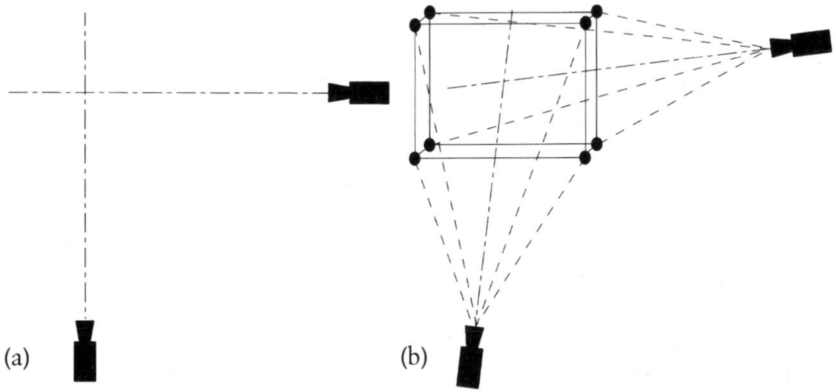

Figure 5.12 Three-dimensional camera alignments: (a) with perpendicularly intersecting optical axes; (b) DLT camera set-up – note that the rays from the calibration spheres are unambiguous for both cameras (for clarity only the rays from all the upper or lower spheres are traced to one or other camera).

Flexible camera positions can be achieved with the most commonly used reconstruction algorithm, the 'direct linear transformation (DLT)'. This transforms the image to movement space coordinates by camera calibration involving 11 (or more) independently treated transformation parameters (C_1–C_{11}) for each camera (Abdel-Aziz and Karara, 1971). The simplest form of the pair of transformation equations, for sports biomechanics purposes, for each camera is:

$$C_1 + C_2 X + C_3 Y + C_4 Z + C_5 xX + C_6 xY + C_7 xZ + x = r_x$$
$$C_8 + C_9 X + C_{10} Y + C_{11} Z + C_5 yX + C_6 yY + C_7 yZ + y = r_y \quad (5.1)$$

The algorithm requires a minimum of six calibration points with known three-dimensional coordinates X,Y,Z and measured image coordinates x,y to establish the DLT (transformation) parameters, or coefficients, C_1-C_{11}, for each camera independently. The DLT parameters incorporate the optical parameters of the camera and linear lens distortion factors. Because of the errors in sports biomechanical data, the DLT equations (Equation 5.1) incorporate residual error terms (r_x,r_y). The equations can then be solved directly by minimization of the sum of the squares of the residuals. Once the DLT parameters have been established for each camera, the unknown movement space coordinates (X,Y,Z) of other points, such as skin markers, can then be reconstructed using the DLT parameters and the image coordinates (x,y) for both cameras. Up to 11 additional DLT parameters can also be included, if necessary, to allow for symmetrical lens distortion (five parameters), asymmetrical lens distortions caused by decentring of the lens elements (two parameters) and non-linear components of film distortion (four parameters). Karara and Abdel-Aziz (1974) found no improvements in accuracy by incorporating non-linear lens distortions, although this might not apply to low-priced, wide-angle video lenses. The DLT algorithms impose several experimental restrictions (adapted from Bartlett, Challis and Yeadon, 1992).

- There is a need for an array of calibration (or control) points, the coordinates of which are accurately known with respect to three mutually perpendicular axes. This is usually provided by some form of calibration frame (e.g. Figures 5.9 and 5.12(b)) or similar structure. The accuracy of the calibration coordinates is paramount, as it determines the maximum accuracy of other measurements. Where a calibration frame is not suitable (for example in filming kayaking, ski jumping, skiing, or javelin flight just after release) separate calibration poles can be used. The coordinates of these will usually need to be established using surveying equipment. Use of calibration poles is generally more flexible and allows for a larger calibration volume than does a calibration frame. The greater the number of calibration points, the stronger and more reliable is the reconstruction (Karara, 1980). It is often convenient to define the reference axes to coincide with directions of interest for the sports movement being investigated. The usual convention is for the x axis to correspond with the main direction of horizontal motion.

- All the calibration points must be visible to each camera and their image coordinates must be clearly and unambiguously distinguishable (as in Figure 5.12(b)). Calibration posts or limbs of calibration frames should not, therefore, overlap or nearly overlap, when viewed from any camera. Although an angle of 90° between optical axes might be considered ideal, deviations from this can be tolerated if kept within a range of about 60–120°. The cameras should also be placed so as to give the best views of the performer.
- Accurate coordinate reconstruction can only be guaranteed within the space (the calibration or control volume) defined by the calibration (or control) points. These should therefore be equally distributed within or around the volume in which the sports movement takes place. Errors depend on the distribution of the calibration points (e.g. Challis and Kerwin, 1992) and increase if the performer moves beyond the confines of the control volume (Wood and Marshall, 1986). This is the most serious restriction on the use of the DLT algorithm. It has led to the development of other methods which require smaller calibration objects (e.g. Woltring, 1980; Dapena, Harman and Miller, 1982; Ball and Pierrynowski, 1988). They all have greater computational complexity than the relatively straightforward DLT algorithms.

To circumvent the problem of a small image size, which would prevent identification of body landmarks if the control volume was very large, panning cameras can be used. Dapena (1978) developed a three-dimensional reconstruction technique allowing two cameras to rotate freely about their vertical axis (panning). Yeadon (1989a) reported a method that allowed for the cameras to pan and tilt (rotation about vertical and horizontal axes). In both of these techniques, the cameras must be in known positions (Yeadon and Challis, 1994). These approaches have also been extended to allow for variation of the focal lengths of the camera lenses during filming.

It is obviously necessary to check the validity of these methods. This can be done by calculating the rms error between the reconstructed and known three-dimensional coordinates of points, preferably ones that have not been used to determine the DLT parameters (Allard, Blanchi and Aïssaoui, 1995). Furthermore, the success with which these methods reproduce three-dimensional movements can be checked, for example by filming the three-dimensional motion of a body segment of known dimensions or a rod thrown into the air (Dainty *et al.*, 1987). In addition, reliability and objectivity checks should be carried out on the digitized data.

5.4 Data processing

The data obtained from digitizing, either before or after transformation to three-dimensional coordinates, are often referred to as 'raw' data. Many difficulties arise when processing raw kinematic data and this can lead to large errors. As noted in the previous section, some errors can be minimized by careful equipment selection and rigorous attention to experimental procedures. However, the digitized coordinates will still contain random errors (noise).

The importance of this noise removal can be seen from the general equation of harmonic motion:

$$r = a_0 + \sum_{i=1}^{n} a_i \sin i\omega t \quad + \quad \sum_{i=1}^{n} b_i \cos i\omega t \tag{5.2}$$

where a_i and b_i are the amplitudes of the sine and cosine components of the signal respectively, ω is the fundamental frequency, in radians per second, and i is the harmonic number. To specify the frequency (f) in Hz, we note that $\omega = 2\pi f$. Equation 5.2 can be differentiated to give velocity (v), which in turn can be differentiated to give acceleration (a), using the differentation rules: $a_i \mathrm{d}(\sin i\omega t)/\mathrm{d}t = i\omega a_i \cos i\omega t$ and $b_i \mathrm{d}(\cos i\omega t)/\mathrm{d}t = -i\omega b_i \sin i\omega t$, and substituting $\omega = 2\pi f$, as in the simplified Equations 5.3 and 5.4. Let us choose an extremely simplified representation of a recorded sports movement to illustrate this, expressed by the equation:

$$r \quad = \quad 2\sin 2\pi\, 2t \quad + \quad 0.02 \sin 2\pi\, 20t \tag{5.3}$$

The first term on the right-hand side of Equation 5.3 ($2 \sin 2\pi\, 2t$) represents the signal, i.e. the motion being observed. The amplitude of this signal is 2 metres and its fundamental frequency is a 2 Hz (that is $i = 1$, $f = 2$ Hz). The second term is the noise; this has an amplitude of only 1% of the signal (this would be a low value for many sports biomechanics studies) and a frequency of 20 Hz, 10 times ($i = 10$) that of the signal. The difference in frequencies is because human movement generally has a low frequency content and noise is at a higher frequency. Figure 5.13(a) shows the signal with the noise superimposed; it should be noted that there is little difference between the noise-free and noisy displacement. Using the rules of the previous paragraph, Equation 5.3 can be differentiated to give:

$$v \quad = \quad 8\pi \cos 2\pi\, 2t \quad + \quad 0.8\pi \cos 2\pi\, 20t \tag{5.4a}$$

$$a \quad = \quad -32\pi^2 \sin 2\pi\, 2t \quad - \quad 32\pi^2 \sin 2\pi\, 20t \tag{5.4b}$$

The noise amplitude in the velocity is now 10% ($0.8\pi/8\pi \times 100$) of the signal amplitude. The noise in the acceleration data has the same amplitude ($32\pi^2$), which is an intolerable error. Unless the errors in the displacement data are reduced (by smoothing or filtering), they will lead to considerable inaccuracies in velocities and accelerations and any derived data. This will be compounded by any errors in body segment data.

(a)

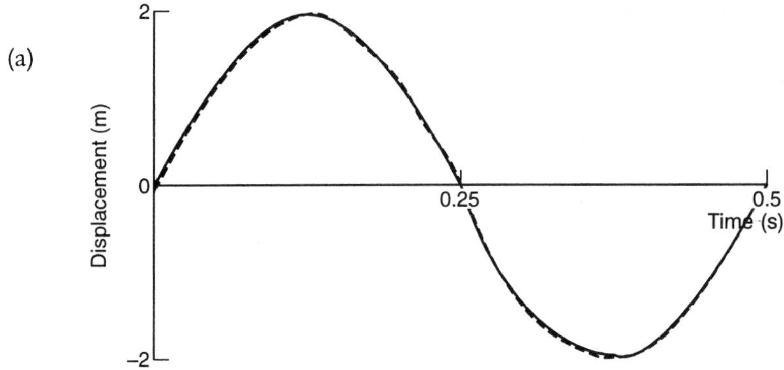

(b)

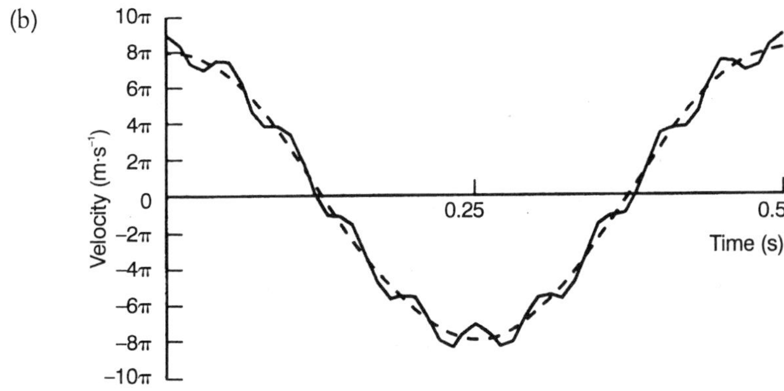

(c)

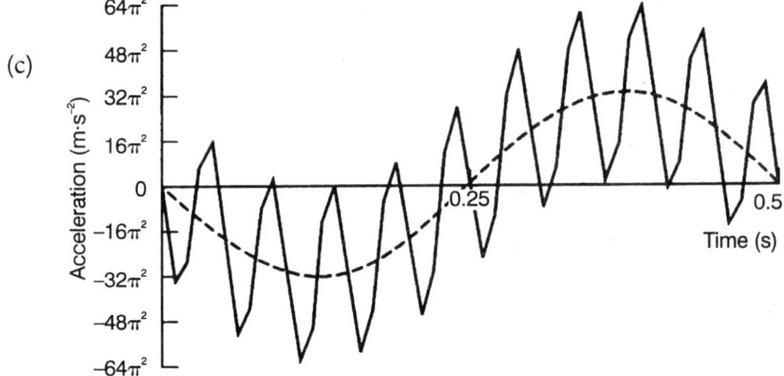

Figure 5.13 Simple example of noise-free (dashed lines) and noisy (continuous lines) data: (a) displacement; (b) velocity; (c) acceleration.

5.4.1 DATA SMOOTHING, FILTERING AND DIFFERENTIATION

Much attention has been paid to the problem of removal of noise from discretely sampled data in biomechanics (e.g. Wood, 1982). Solutions are not always, or entirely, satisfactory, particularly when transient signals, such as those caused by foot strike or other impacts, are present (Woltring, 1995). Noise removal is normally performed after reconstruction of the movement coordinates from the image coordinates as, for three-dimensional studies, each set of image coordinates does not contain full coordinate information. However, the noise removal should be performed before calculating other data, such as segment orientations and joint forces and moments. The reason for this is that the calculations are highly non-linear, leading to non-linear combinations of random noise. This can adversely affect the separation of signal and noise by low-pass filtering (Woltring, 1995).

An introduction to the process of data smoothing can be gained through an extension of the finite difference technique introduced in Chapter 2. Central finite difference smoothing can be performed by taking a weighted average of the data point being smoothed (x_0) and the two most adjacent data points to each side (x_{-2}, x_{-1}, x_1, x_2), for example $x_{\text{smooth }0} = (x_{-2} + 2x_{-1} + 2x_0 + 2x_1 + x_2)/8$. The interested reader can investigate the use of this technique on noisy data, such as that of Table 2.2. Such techniques tend not to smooth the data sufficiently if accelerations are required. This section will therefore consider three techniques that are suitable for noise removal if the resulting data are to be differentiated to obtain satisfactory velocities and accelerations.

Digital low-pass filters

These are widely used to remove, or filter, high-frequency noise from digital data. Butterworth filters (of order $2n$ where n is a positive integer) are often used in sports biomechanics, because they have a flat passband (Figure 5.14), i.e. the band of frequencies that is not affected by the filter. However, they have relatively shallow cut-offs. This can be improved by using higher-order filters, but round-off errors in computer calculations can then become a problem. They also introduce a phase shift, which must be removed by a second, reverse filtering, which increases the order of the filter and further reduces the cut-off frequency.

Butterworth filters are recursive, i.e. they use filtered values of previous data points as well as noisy data values to obtain filtered data values. This makes for faster computation but introduces problems at the ends of data sequences, where filtered values must be estimated. This can mean that extra frames must be digitized and included in the data processing, but some frames at each end of the sequence then have to be discarded after filtering. This involves unwelcome extra work for the sports biomechanist. Other solutions have been proposed for padding the ends of the data sets (e.g. Smith, 1989).

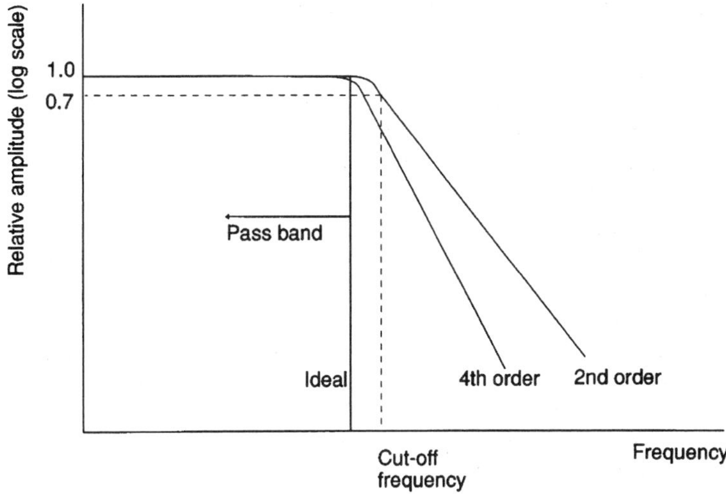

Figure 5.14 Low-pass filter frequency characteristics.

The main decision for the user, as with Fourier series truncation, is the choice of cut-off frequency (see below). The filtered data are not obtained in analytic form, so a separate numerical differentiation process must be used (e.g. Lees, 1980), similar to the finite difference calculation of velocity in Chapter 2. Although Butterworth filtering appears to be very different from spline fitting (see below), the two are, in fact, closely linked (Woltring, 1995).

Fourier series truncation

This involves, firstly, the transformation of the noisy data into the frequency domain by means of a Fourier transformation. This, in essence, replaces the familiar representation of displacement as a function of time (time domain, Figure 5.15(a)) by a series of sinusoidal waves of different frequencies (as in the very simple example of Equation 5.3 and in Figure 5.16(a)).

The frequency domain representation of the data is then presented as amplitudes of the sinusoidal components at each frequency (the harmonic frequencies, Figure 5.15(b)) or as a continuous curve. Figure 5.16(a) shows the frequency domain representation of the simplified data of Figure 5.13(a) and Equation 5.3. The data are then filtered to remove high-frequency noise. This is done by reconstituting the data up to the chosen cut-off frequency and truncating the number of terms in the series from which it is made up. For the simplified data of Equation 5.3, if the cut-off frequency was, say, 4 Hz, then the second term on the right-hand side would be rejected as noise. The reconstituted (filtered) data would be obtained as shown in Equation 5.5.

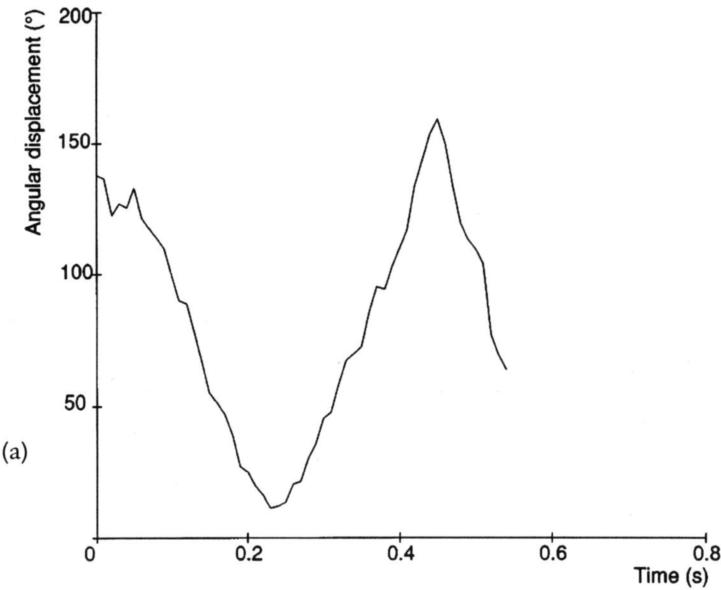

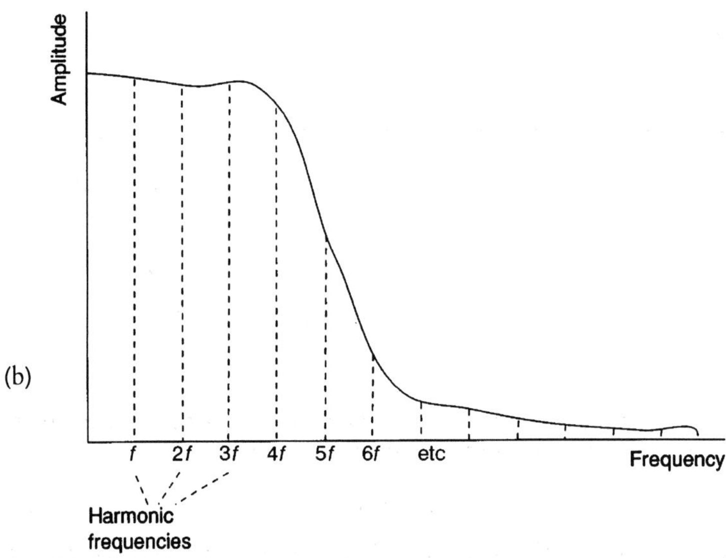

Figure 5.15 Displacement data represented in: (a) the time domain; (b) the frequency domain.

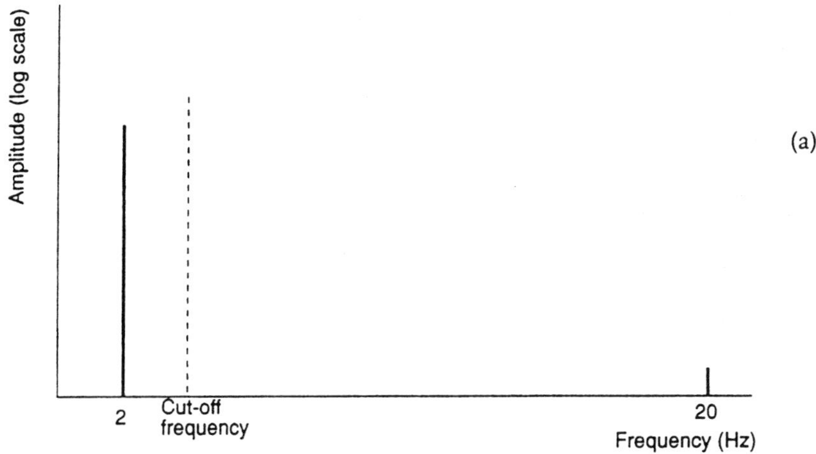

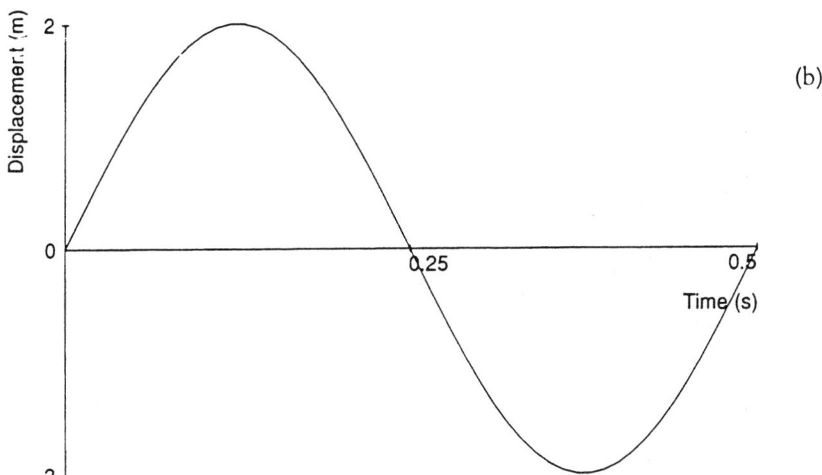

Figure 5.16 Simple example of noisy data: (a) in the frequency domain; (b) in the time domain after filtering by Fourier series truncation.

$$r = 2 \sin 4\pi t \qquad (5.5)$$

In this simplified case, the noise would have been removed perfectly, as the time domain signal of Figure 5.16(b) demonstrates. There would be no resulting errors in the velocities and accelerations. The major decision here concerns the choice of cut-off frequency, and similar principles to those described below can be applied. Unlike digital filters, the cut-off can be infinitely steep, as in Figure 5.14 (these are sometimes called ideal filters), but this is not necessarily the case (Hatze, 1981). The filtered data are analytic in form (i.e. they can be represented as an equation similar to Equation 5.5) and can be differentiated analytically (as for Equations 5.3 and 5.4). This technique requires the raw data points to be sampled at equal time intervals, as also do digital low-pass filters (see above).

Quintic spline curve fitting

Many techniques used for the smoothing and differentiation of data in sports biomechanics involve the use of spline functions. These are a series of polynomial curves through one or more points, joined (or pieced) together at points called **knots**. This smoothing technique, which is performed in the time domain, can be considered to be the numerical equivalent of drawing a smooth curve through the data points. Indeed, the name 'spline' derived from the flexible strip of rubber or wood used by draftsmen for drawing curves. Splines are claimed to represent the smooth nature of human movement while rejecting the normally distributed random noise in the digitized coordinates (e.g. McLaughlin, Dillman and Lardner, 1977).

Many spline techniques are derived from the work of Reinsch (1967), and have a knot at each data point, obviating the need for the user to choose optimal knot positions. The user has simply to specify a weighting factor for each data point and select the value of the smoothing parameter, which controls the extent of the smoothing. Reinsch (1967) proposed that the weighting factor should be the inverse of the estimate of the variance of the data point. This is easily established in sports biomechanics by repeated digitization of a film or video sequence. The use of different weighting factors for different points can be useful, especially where points are obscured from the camera, and hence have a higher variance than ones that can be seen clearly. Inappropriate choices of the smoothing parameter can cause problems of over-smoothing (Figure 5.17(a),(b)) and under-smoothing (Figure 5.17(c),(d)).

Generalized cross-validated quintic splines do not require the user to specify the error in the data to be smoothed, but instead automatically select an optimum smoothing level (Woltring, 1995). Computer programmes for spline smoothing are available in various software packages and allow a choice of automatic or user defined smoothing levels (e.g. Woltring, 1986). Generalized cross-validation can accommodate data points sampled at unequal time intervals. Splines can be differentiated analytically (as was done with Equations 5.3 and 5.4). Quintic splines are continuous up to the fourth derivative, which is a series of interconnected straight lines. This provides accurate generation of the second derivative, acceleration.

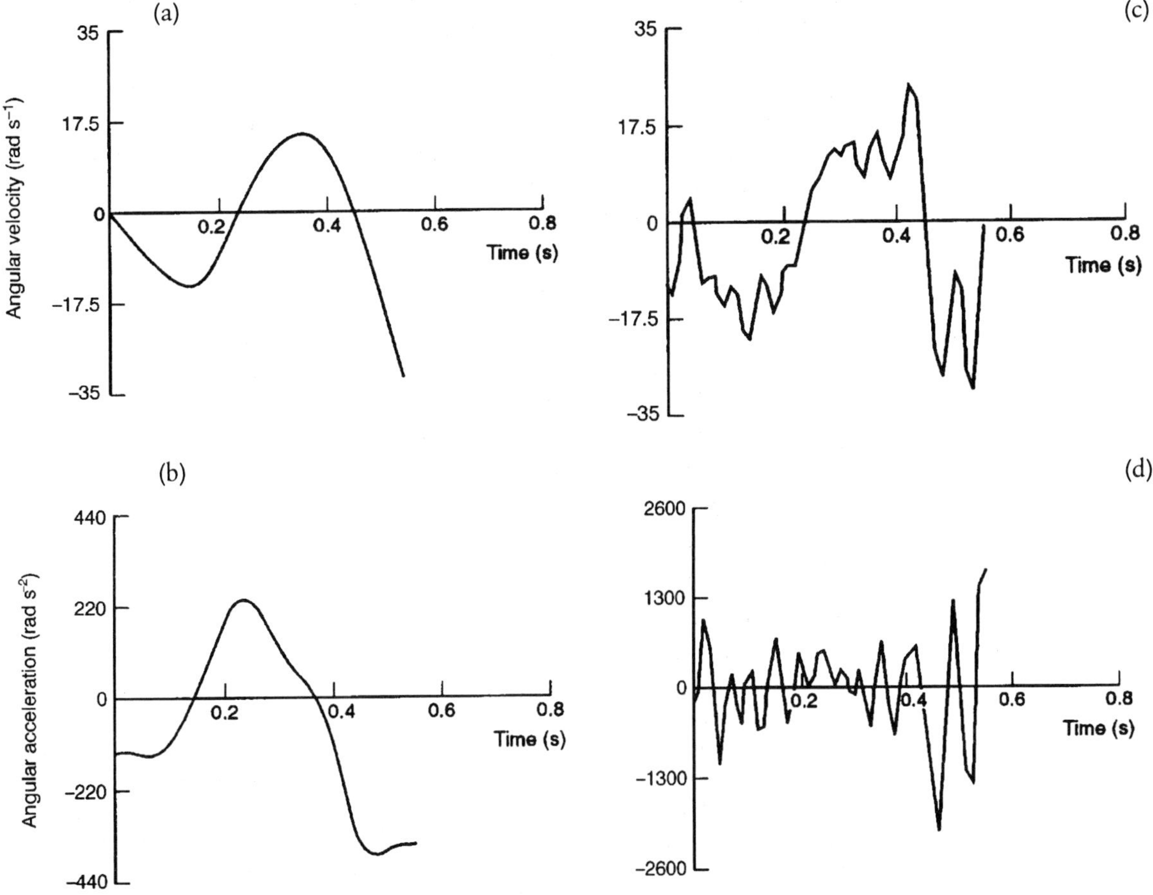

Figure 5.17 Over-smoothing: (a) velocity; (b) acceleration, and under-smoothing: (c) velocity; (d) acceleration. The optimum smoothing is shown in Figure 5.2(b),(c).

Choice of noise removal technique

Quintic splines appear to produce more accurate first and second derivatives than most other techniques which are commonly used in sports biomechanics (e.g. Challis and Kerwin, 1988). The two filtering techniques (Fourier truncation and digital filters) were devised for periodic data, where the pattern of movement is cyclical (as in Figure 5.13(a)). Sporting activities of a cyclic nature (such as running) are obviously periodic, and some others can be considered quasi-periodic. Problems may be found in attempting to filter non-periodic data, although these may be overcome by removing any linear trend in the data prior to filtering (this makes the first and last data values zero). The Butterworth filter often creates less problems here than Fourier truncation but neither technique deals completely satisfactorily with constant acceleration motion, as for the centre of mass when the performer is airborne.

The major consideration for the sports biomechanist using smoothing or filtering routines, as suggested above, is a rational choice of filter cut-off frequency or spline smoothing parameter. A poor choice can result in some noise being retained (e.g. if the filter cut-off frequency is too high) or some of the signal being rejected (e.g. if the cut-off frequency is too low). As most human movement is at a low frequency, a cut-off frequency of between 4 and 8 Hz is often used. Lower cut-off frequencies may be preferable for slow events such as swimming, and higher ones for impacts (Antonsson and Mann, 1985) or other rapid energy transfers (Best and Bartlett, 1988). The cut-off frequency should be chosen to include the highest frequency of interest in the movement. As filters are often implemented as the ratio of the cut-off and the sampling frequencies, an appropriate choice of the latter needs to have been made at an earlier stage.

The need for data smoothness demands a minimum ratio of the sampling to cut-off frequencies of 4:1 and preferably one as high as 8:1 or 10:1 (Lanczos, 1957). The frame rate used when filming or video recording, and the digitizing rate (the sampling rate), must allow for these considerations. Attempts to base the choice of cut-off frequency on some objective criterion have not always been successful. The best approach is probably to compare the root mean square difference between the noisy data and that obtained after filtering at several different cut-off frequencies with the standard deviation obtained from repetitive digitization of the same anatomical point. The cut-off frequency should then be chosen so that the magnitudes of the two are similar (e.g. Kerwin, 1988).

The use of previously published filter cut-off frequencies (e.g. Winter, Sidwell and Hobson, 1974) or manual adjustment of the smoothing parameter (Vaughan, 1982) are not recommended. Instead, a technique should be used that involves a justifiable procedure (such as Hatze, 1981; Woltring, 1986 and Yeadon, 1989b) to take into account the peculiarities of each new set of data (Yeadon and Challis, 1994). Furthermore, it is often necessary to use a different smoothing parameter or cut-off frequency for the coordinates of the different points recorded. This is especially necessary when the frequency spectra for the various points are different. Finally, it should be remembered that no automatic noise removal algorithm will always be successful, and that the smoothness of the processed data should always be checked.

5.4.2 BODY SEGMENT INERTIA PARAMETERS

Various body segment inertia parameters are used in sports biomechanical analysis. The mass of each body segment and the segment centre of mass position are used in calculating the position of the whole body centre of mass (Chapter 3). These values, and segment moments of inertia, are used in calculations of net joint forces and moments using the method of inverse dynamics. The most accurate and valid values available for these inertia parameters should obviously be used. Ideally, they should be obtained from, or scaled to, the sports performer being studied. The values of body segment parameters used in sports biomechanics have been obtained from cadavers (e.g. Dempster, 1955), and from living persons (e.g. Zatsiorsky, 1983), including measurements of the performers being filmed. A useful comparison of imaging techniques, gamma-ray scanning and mathematical modelling methods for estimating segment moments of inertia is provided by Mungiole and Martin (1990).

Cadaver studies have provided very accurate segmental data. However, limited sample sizes throw doubt on the extrapolation of these data to a general sports population. They are also highly questionable in terms of the unrepresentative nature of the samples in respect of sex, age and morphology. Problems also arise from the use of different dissection techniques by different researchers and losses in tissue and body fluid during dissection and degeneration associated with the state of health preceding death (Dainty *et al.*, 1987). Segment mass may be expressed as a simple fraction of total body mass or, more accurately, in the form of a linear regression equation with one or more anthropometric variables. Even the latter may cause under- or overestimation errors of total body mass as large as 4.6% (Miller and Morrison, 1975).

Studies on living people have been very limited. Body segment data have been obtained using gamma-ray scanning (e.g. Zatsiorsky, 1983) or from imaging techniques, such as computerized tomography and magnetic resonance imaging (e.g. Martin et al., 1989). These may eventually supersede the cadaver data which are still too often used (Bartlett, Challis and Yeadon, 1992).

Obtaining body segment parameter data from sports performers may require sophisticated equipment and a great deal of the performer's time. The immersion technique is simple, and can be easily demonstrated by any reader with a bucket, a vessel to catch the overflow from the bucket, and some calibrated measuring jugs or weighing device (see Figure 5.18 and Exercise 8). It provides accurate measurements of segment volume and centre of volume, but requires a knowledge of segmental density to calculate segment mass. Also, as segment density is not uniform throughout the segment, the centre of mass does not coincide with the centre of volume.

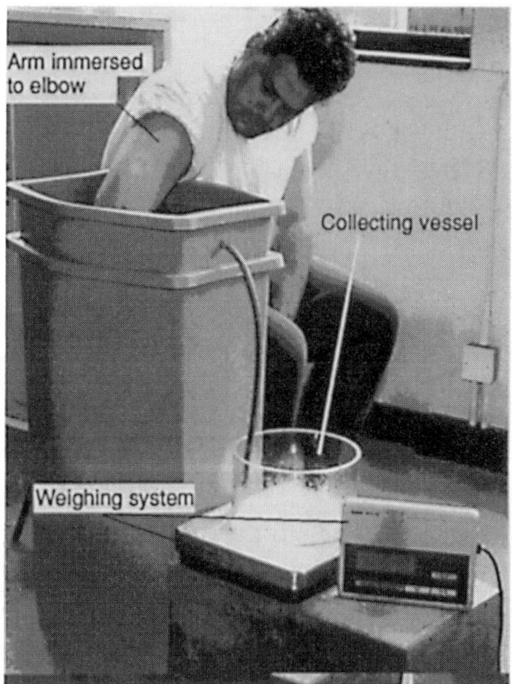

Figure 5.18 Simple measurement of segment volume.

Several 'mathematical' models exist which calculate body segment parameters from standard anthropometric measurements, such as segment lengths and circumferences. Some of these models result in large errors even in estimates of segment volumes (e.g. by 15% for the forearm in the model of Hanavan, 1964). The model of Hatze (1980) requires well over 200 anthropometric measurements, which would take at least an hour or two to complete. All these models require density values from other sources (usually cadavers) and most of them assume constant density throughout the segment, or throughout large parts of the segment.

The greatest problems in body segment data occur for moments of inertia. There are no simple yet accurate methods to measure segmental moments of inertia for a living person. Many model estimations are either very inaccurate or require further validation. A relative error of 5% in segmental moments of inertia may be quite common (Dainty *et al.*, 1987). Norms or linear regression equations are often used, but these should be treated with caution as the errors involved in their use are rarely fully assessed. It may be necessary to allow for the non-linear relationships between segmental dimensions and moment of inertia values (Yeadon and Morlock, 1989).

5.4.3 SEGMENT ORIENTATIONS

For two-dimensional analysis, joint angles are usually defined and calculated simply as the angle between two lines representing the proximal and distal body segments. This will not be the true joint angle if the segments move out of the photographic plane. A similar definition could be used to specify the angle between line representations of segments in three dimensions for simple hinge joint. Generally the process is more complex, and there are many different conventions for defining the orientation angles of two articulating rigid bodies and of specifying the orientation angles of the human performer as a whole. Most of these conventions have certain problems, one of which is to have a joint angle convention that is intelligible to the sports performer and coach to whom the results are communicated (Yeadon and Challis, 1994).

The International Society of Biomechanics convention involves:

- specification of a global coordinate system (*XYZ*) fixed in the ground (Figure 5.19);
- specification of segment linear positions by the location of the segment's centre of mass relative to the global coordinate system – a local coordinate system (*x*,*y*,*z*) is fixed at the segment's centre of mass (Figure 5.19);
- specification of segment attitudes (angles) with respect to the global coordinate system as an ordered series of rotations about the local coordinate system's *z* axis (flexion–extension), *y* axis (medial–lateral rotation) and *x* axis (abduction–adduction);
- specification of the relative orientations of adjacent body segments using a joint rotation convention which, although conceptually more complex than segment attitudes, is effectively the same and preserves the anatomical meaningfulness (see Nigg and Cole, 1994).

5.4.4 DATA ERRORS

Uncertainties, also referred to as errors, in the results of biomechanical data processing, especially for computation of kinetic variables, can be large. This is mainly because of errors in the body segment data and linear and angular velocities and accelerations, and the combination of these errors in the inverse dynamics equations. If such computations are to be attempted, scrupulous adherence to good experimental protocols is essential. A rigorous assessment of the applicability of the body-segment data used, and of the appropriateness and accuracy of the data processing techniques is also necessary (Bartlett, Challis and Yeadon, 1992). The topic of error analysis is a very important one and the value of sports biomechanical measurements cannot be assessed fully in the absence of a quantification of the measurement error (Yeadon and Challis, 1994). The accuracy of the measuring system and the precision of the measurements should be assessed separately.

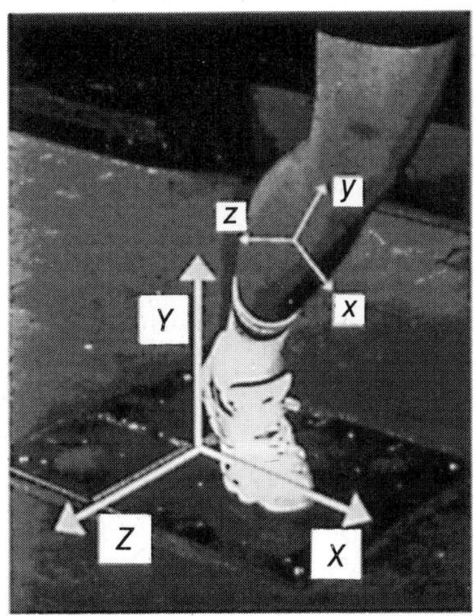

Figure 5.19 Global and local coordinate systems.

The formulae of Lanshammer (1982) can provide precision estimates for smoothed and differentiated data:

$$\sigma_k^2 \geq \frac{\sigma^2 T \omega_s^{2k+1}}{\pi(2k+1)} \tag{5.6}$$

In this equation, σ_k is the standard deviation of the normally distributed, random noise in the estimated kth derivative, σ is the standard deviation of the noise in the measured displacement data, T is the sampling interval (the inverse of the sampling rate in Hz). The term ω_s is the bandwidth of the signal (in rad·s^{-1}). This would be 10π rad·s^{-1} for a movement with a maximum signal frequency of 5 Hz. If such a signal was sampled at 100 Hz (T = 1/100 s) – this is 20 times the maximum frequency – then the standard deviation of the noise in the filtered and differentiated acceleration data (k = 2) would be 140 times that of the noise in the raw displacement data. It would be even worse for lower sampling rates. Error propagation in calculations can be estimated using standard formulae (e.g. Taylor, 1982) as reported, for example, by Payton and Bartlett (1995).

5.5 Summary

In this chapter, the use of cinematography and video analysis in the study of sports movements was covered, including the equipment and methods used, and the importance of cinematography and video in the qualitative and quantitative analysis of sports techniques. The necessary features of cine and video equipment for recording movements in sport were considered as were the advantages and limitations of two- and three-dimensional recording of sports movements. The possible sources of error in recorded movement data were outlined and experimental procedures described that would minimize recorded errors in two- and three-dimensional movements. The need for, and the ways of performing, smoothing and filtering of kinematic data were covered, along with the requirement for accurate body segment inertia parameter data and how these can be obtained. The chapter concluded with a brief consideration of a convention for the specification of three-dimensional segment orientations and some aspects of error analysis.

5.6 Exercises

1. Explain why cinematography and video analysis are important in the study of sports techniques.
2. Obtain a video recording of a sports movement of your choice. Study the recording carefully, frame by frame. Identify major features of the technique, such as key events that separate the various phases of the movement. Write a short description of the essential features of the movement. Also, identify the displacements, angles, velocities and accelerations that you would need to include in a quantitative analysis of this technique. If you have access to a video camera, you may wish to perform this exercise for the movement you will use in exercise 10.
3. List the possible sources of error in recorded movement data, and identify which would lead to random and which to systematic errors.
4. Briefly describe the procedures that would minimize the recorded error in a study of an essentially two-dimensional movement. Assess which of these steps could not be ensured if filming at a top-level sports competition. Briefly explain how these procedures would be modified for recording a three-dimensional movement.
5. List the important features of cine and video cameras, image display equipment and coordinate digitizers. List the advantages and disadvantages of cine and video analysis of sports movements.
6. Outline the advantages and limitations of two-dimensional and three-dimensional recording of sports movements.
7. Explain the need for smoothing or filtering of kinematic data, and outline the differences between spline smoothing, digital low-pass filters and Fourier series truncation.

8. Carry out an experiment to determine the volume of (a) a hand, (b) a forearm segment. You will need a bucket or similar vessel large enough for the hand and forearm to be fully submerged. You will also need a bowl, or similar vessel, in which the bucket can be placed, to catch the overflow of water, and calibrated containers to measure the volume of water. Repeat the experiment at least three times, and then calculate the mean volume and standard deviation for each segment. How reproducible are your data?
9. Outline the need for accurate body segment parameter data and ways in which such data can be obtained.
10. Plan an experimental session in which you would record an essentially two-dimensional sports movement, such as a long jump, running or simple gymnastics vault. You should carefully detail all the important procedural steps (see section 5.3.1). If you have access to a suitable video camera, record several trials of the movement from one or more performers. If you have access to a video digitizing system, then digitize at least one of the sequences you have recorded. If you have access to analysis software, then plot stick figure sequences and graphs of relevant kinematic variables (these should have been established by a qualitative analysis of the movement similar to that performed in Exercise 2). Make some brief notes on the important results.

5.7 References

Abdel-Aziz, Y. I. and Karara, H. M. (1971) Direct linear transformation from comparator coordinates into space coordinates in close range photogrammetry, in *Proceedings of the Symposium on Close Range Photogrammetry*, American Society of Photogrammetry, Falls Church, USA.

Allard, P., Blanchi, J.-P. and Aïssaoui, R. (1995) Bases of three-dimensional reconstruction, in *Three-Dimensional Analysis of Human Movement*, (eds P. Allard, I. A. F. Stokes and J.-P. Blanchi), Human Kinetics, Champaign, IL, pp.19–40.

Antonsson, E. K. and Mann, R. W. (1985) The frequency content of human gait. *Journal of Biomechanics*, **18**, 39–47.

Ball, K. A. and Pierrynowski, M. R. (1988) A modified direct linear transformation (DLT) calibration procedure to improve the accuracy of 3D reconstruction for large volumes, in *Biomechanics XI*, (eds G. de Groot, A. P. Hollander, P.A. Huijing and G. J. van Ingen Schenau), Free University Press, Amsterdam, pp.1045–1050.

Bartlett, R. M. (ed.) (1992) *Biomechanical Analysis of Performance in Sport*, British Association of Sports Sciences, Leeds.

Bartlett, R. M., Challis, J. H. and Yeadon M. R. (1992) Cinematography/video analysis, in *Biomechanical Analysis of Performance in Sport*, (ed. R. M. Bartlett), British Association of Sports Sciences, Leeds, pp.8–23.

Best, R. J. and Bartlett, R. M. (1988) A critical appraisal of javelin high-speed cinematography using a computer flight simulation program. *Ergonomics*, **31**, 1683–1692.

Calvert, T. W. and Bruderlin, A. (1995) Computer graphics for visualization and animation, in *Three-Dimensional Analysis of Human Movement*, (eds P. Allard, I. A. F. Stokes and J.-P. Blanchi), Human Kinetics, Champaign, IL, pp.101–123.

Challis, J. H. and Kerwin, D. G. (1988) An evaluation of splines in biomechanical data analysis, in *Biomechanics XI*, (eds G. de Groot, A. P. Hollander, P.A. Huijing and G. J. van Ingen Schenau), Free University Press, Amsterdam, pp.1057–1061.

Challis, J. H. and Kerwin, D. G. (1992) Accuracy assessment and control point configuration when using the DLT for photogrammetry. *Journal of Biomechanics*, **25**, 1053–1058.

Dainty, D. A., Gagnon, M., Lagasse, P.P. *et al.* (1987) Recommended procedures, in *Standardizing Biomechanical Testing in Sport*, (eds D. A. Dainty and R. W. Norman), Human Kinetics, Champaign, IL, pp.73–100.

Dapena, J. (1978) Three-dimensional cinematography with horizontal panning cameras. *Science et Motricité,* **1**, 3–15.

Dapena, J., Harman, E. A. and Miller, J. (1982) Three-dimensional cinematography with control objects of unknown shape. *Journal of Biomechanics*, **15**, 11–19.

Dempster, W. T. (1955) Space requirements of the seated operator, WADC-TR-55-159, Aerospace Medical Research Laboratory, Wright-Patterson Air Force Base, Dayton, OH.

Gagnon, M. Robertson, R. and Norman, R. W. (1987) Kinetics, in *Standardizing Biomechanical Testing in Sport*, (eds D. A. Dainty and R. W. Norman), Human Kinetics, Champaign, IL, pp.21–57.

Greaves, J. O. B. (1995) Instrumentation in video-based three-dimensional systems, in *Three-Dimensional Analysis of Human Movement*, (eds P. Allard, I. A. F. Stokes and J.-P. Blanchi), Human Kinetics, Champaign, IL, pp.41–55.

Hanavan, E. P. (1964) A mathematical model of the human body, AMRL-TDR-63-18, Aerospace Medical Research Laboratory, Wright-Patterson Air Force Base, Dayton, OH.

Hatze, H. (1980) A mathematical model for the computational determination of parameter values of anthropometric body segments. *Journal of Biomechanics*, **13**, 833–843.

Hatze, H. (1981) The use of optimally regularized Fourier series for estimating higher order derivatives of noisy biomechanical data. *Journal of Biomechanics*, **13**, 833–843.

Karara, H. M. (1980) Non-metric cameras, in *Developments in Close Range Photogrammetry*, (ed. K. B. Atkinson), Applied Science Publishers, London, pp.63–80.

Karara, H. M. and Abdel-Aziz, Y. I. (1974) Accuracy aspects of non-metric imageries. *Photogrammetric Engineering*, **40**, 1107–1117.

Kerwin, D. G. (1988) Digitizing: computer interfacing, digitization and data storage protocols, in *Proceedings of the Sports Biomechanics Section of the British Association of Sports Sciences*, (ed. R. M. Bartlett), British Association of Sports Sciences, Loughborough.

Kerwin, D. G. and Templeton, N. (1991) Cine-film and video: an assessment of digitization accuracy. *Journal of Sports Sciences*, **9**, 402.

Ladin, Z. (1995) Three-dimensional instrumentation, in *Three-Dimensional Analysis of Human Movement*, (eds P. Allard, I. A. F. Stokes and J.-P. Blanchi), Human Kinetics, Champaign, IL, pp.3–17.

Lanczos, C. (1957) *Applied Analysis*, Pitman, London.

Lanshammer, H. (1982) On precision limits for derivatives numerically calculated from noisy data. *Journal of Biomechanics*, **15**, 459–470.

Lees, A. (1980) An optimised film analysis method based on finite difference techniques. *Journal of Human Movement Studies*, **6**, 165–180.

McLaughlin, T. M., Dillman, C. J. and Lardner, T. J. (1977) Biomechanical analysis with cubic spline functions. *Research Quarterly*, **48**, 568–582.

Martin, P. E., Mungiole, M., Marzke, M. W. and Longhill, J. M. (1989) The use of magnetic resonance imaging for measuring segment inertial properties. *Journal of Biomechanics*, **22**, 367–376.

Martin, T. P. and Pomgrantz, M. B. (1974) Validation of a mathematical model for correction of photographic perspective error, in *Biomechanics IV*, (eds R. C. Nelson and C.A. Morehouse), University Park Press, Baltimore, MD, pp.469–751.

Miller, D. I. and Morrison, W. E. (1975) Prediction of segmental parameters using the Hanavan human body model. *Medicine and Science in Sports*, **7**, 207–212.

Miller, D. I. and Nelson, R. C. (1973) *Biomechanics of Sport*, Lea & Febiger, Philadelphia, PA.

Mungiole, M. and Martin, P. E. (1990) Estimating segment inertial properties: comparison of magnetic resonance imaging with existing methods. *Journal of Biomechanics*, **23**, 1039–1046.

Nigg, B. M. and Cole, G. K. (1994) Optical methods, in *Biomechanics of the Musculoskeletal System*, (eds B. M. Nigg and W. Herzog), John Wiley, Chichester, pp.254–286.

Payton, C. J. and Bartlett, R. M. (1995) Estimating propulsive forces in swimming from three-dimensional kinematic data. *Journal of Sports Sciences*, **13**, 447–454.

Plagenhoef, S. (1971) *Patterns of Human Movement*, Prentice-Hall, Englewood Cliffs, NJ.

Reinsch, C. H. (1967) Smoothing by spline functions. *Numerical Mathematics*, **10**, 177–183.

Smith, A. J. (1975) Photographic analysis of movement, in *Techniques for the Analysis of Human Movement*, (eds D. W. Grieve, D. I. Miller, D. Mitchelson *et al.*), Lepus, London, pp.3–32.

Smith, G. (1989) Padding point extrapolation techniques for the Butterworth digital filter. *Journal of Biomechanics*, **22**, 967–971.

Taylor, J. R. (1982) *An Introduction to Error Analysis: the Study of Uncertainties in Physical Measurements*, University Science Books, Mill Valley, CA.

Van Gheluwe, B. (1978) Computerized three-dimensional cinematography for any arbitrary camera set-up, in *Biomechanics VI-A*, (eds E. Asmussen and K. Jorgensen), Human Kinetics, Champaign, IL, pp.343–348.

Vaughan, C. L. (1982) Smoothing and differentiating of displacement data: an application of splines and digital filtering. *International Journal of Biomedical Computing*, **13**, 375–382.

Winter, D. A., Sidwell, H. G. and Hobson, D. A. (1974) Measurement and reduction of noise in kinematics of locomotion. *Journal of Biomechanics*, **7**, 157–159.

Woltring, H. J. (1980) Planar control in multi-camera calibration for 3-D gait studies. *Journal of Biomechanics*, **13**, 39–48.

Woltring, H. J. (1986) A Fortran package for generalized cross-validatory spline smoothing and differentiation. *Advances in Engineering Software*, **8**, 104–113.

Woltring, H. J. (1995) Smoothing and differentiation techniques applied to 3-D data, in *Three-Dimensional Analysis of Human Movement*, (eds P. Allard, I. A. F. Stokes and J.-P. Blanchi), Human Kinetics, Champaign, IL, pp.79–99.

Wood, G. A. (1982) Data smoothing and differentiation procedures in biomechanics, in *Exercise and Sport Sciences Reviews*, Vol. 10, (ed. R. L. Terjung), Franklin Institute Press, New York, pp.308–362.

Wood, G. A. and Marshall, R. N. (1986) The accuracy of DLT extrapolation in three-dimensional film analysis. *Journal of Biomechanics*, **19**, 781–785.

Yeadon, M. R. (1989a) A method for obtaining three-dimensional data on ski jumping using pan and tilt cameras. *International Journal of Sport Biomechanics*, **5**, 238–247.

Yeadon, M. R. (1989b) Numerical differentiation of noisy data. Communication to the XII Congress of the International Society of Biomechanics, Los Angeles, CA.

Yeadon, M. R. and Challis, J. H. (1993) *Future Directions for Performance Related Research in Sports Biomechanics*, Sports Council, London.

Yeadon, M. R. and Challis, J. H. (1994) The future of performance related sports biomechanics research. *Journal of Sports Sciences*, **12**, 3–32.

Yeadon, M. R. and Morlock, M. (1989) The appropriate use of regression equations for the estimation of segmental inertia parameters. *Journal of Biomechanics*, **22**, 683–689.

Zatsiorsky, V. M. (1983) Biomechanical characteristics of the human body, in *Biomechanics and Performance in Sport*, (ed. W. Baumann), Verlag Karl Hofmann, Schorndorf, pp.71–83.

5.8 Further reading

Allard, P. Stokes, I. A. F. and Blanchi, J-P. (eds) (1994) *Three-Dimensional Analysis of Human Movement*, Human Kinetics, Champaign, IL: chapters 1–3 and 5–6 contain material that extends the treatment of three-dimensional analysis contained in this book. The material is generally very interesting, although somewhat advanced in places (especially in Chapter 5). Chapter 15 deals with the application of three-dimensional analysis to sports, and is fairly readable if the mathematical parts are omitted.

Bartlett, R. M., Challis, J. H. and Yeadon M. R. (1992) Cinematography/video analysis, in *Biomechanical Analysis of Performance in Sport*, (ed. R. M. Bartlett), British Association of Sports Sciences, Leeds, pp.8–23: this contains some useful advice on the reporting of a cine or video study, which you could adopt for a formal experimental report of exercise 10.

There are many review articles that summarize biomechanical studies of specific sports and which usually contain information mostly derived from cine and video studies. Such reviews can be found, for example, in *Medicine and Sports Science Reviews*, volumes 1–24 (1971–1996), and the *Journal of Sports Sciences*. Copies of these can be found in most university libraries or obtained through interlibrary loan. You should seek to read one such review article of a sport (or event) in which you are interested.

6 Force platforms and external force measurement

This chapter is intended to provide an understanding of the use of the force platform to measure contact forces between the sports performer and the ground (or other surfaces), including the equipment and methods used. After reading this chapter, you should be able to:

- appreciate why the measurement of the external contact forces acting on the sports performer is important in sports biomechanics;
- understand the characteristics of a force platform that influence the accuracy of the measurements taken;
- evaluate the suitability of any force platform system for use in sports biomechanics applications;
- understand the advantages and limitations of the two main types of force platform;
- outline how a force platform can be calibrated and the procedures to be adopted when using a force platform in investigations of sports movements;
- appreciate the types of information that can be obtained from force platform recordings and how this information can be applied;
- understand and evaluate reported biomechanical research which has used the force platform (or other force transducers) in the study of sports movements.

6.1 Introduction and equipment considerations

A force platform (also called a force plate) is used to measure the force exerted on it by a subject or sports performer. By Newton's third law of motion, this force has the same magnitude as, but opposite direction from, the reaction force exerted on the subject by the platform (Figure 6.1).

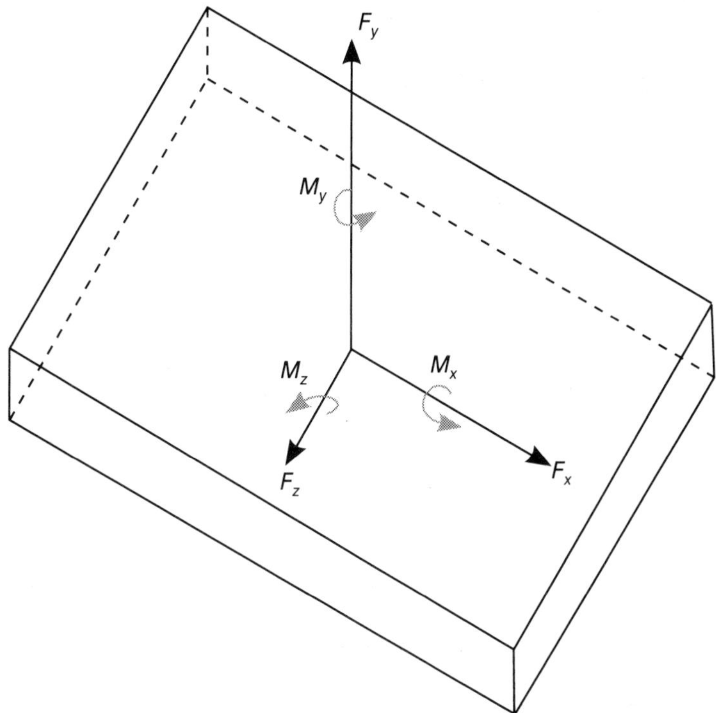

Figure 6.1 Force and moment components which act on the sports performer.

Force platforms are most commonly used in sports biomechanics to measure the contact forces between a sports performer and the ground (the **ground contact force**). Force platforms can be obtained in a variety of sizes. The most commonly used ones have a relatively small contact area, for example 600 mm × 400 mm (Kistler type 9281B11) or 508 mm × 643 mm (AMTI model OR6-5-1) and weigh between 310 and 410 N. They are normally bolted to a base plate which is set in concrete. The measured forces can be used, for example, to evaluate the footstrike patterns of runners or the balance of archers. They can also serve as inputs for joint moment and force calculations (e.g. Winter, 1983) and they can help identify injury potential (e.g. Mason, Weissensteiner and Spence, 1989). The forces can be further processed to provide other information (section 6.3). Force platforms provide basically whole-body measurements, with no explicit information on forces between body segments. In biomechanical evaluation of the sports performer, they would normally be used in conjunction with cinematography or an alternative motion analysis technique (Bartlett, 1992). This allows the biomechanist to link the kinematics and kinetics of the movement.

Forces of interaction between the sports performer and items of sports equipment can also be measured using other force transducers, usually purpose-built or adapted for a particular application. Such transducers have been used, for example, to measure the forces exerted by a rower on an oar (Smith and Spinks, 1989). Many of the principles discussed in this chapter for force platforms also apply to force transducers in general. Further consideration of these, usually specialist, devices will not be undertaken in this chapter.

One limitation of force platforms is that they do not show how the applied force is distributed over the contact surface, for example the shoe or the foot. This information can be obtained from pressure platforms, pads and insoles, which are becoming increasingly available commercially, and which will be considered in Chapter 8.

6.1.1 GENERAL EQUIPMENT CONSIDERATIONS

Force platforms used for the evaluation of sporting performance are sophisticated electronic devices and are generally very accurate. Essentially, they can be considered as weighing systems that are responsive to changes in the displacement of a sensor or detecting element (Ramey, 1975). They incorporate some form of force transducer, which converts the force into an electrical signal. The transducers are mounted on the supports of the rigid plate which forms the platform surface (usually one support at each of the four corners of a rectangle). One transducer is used, at each support, to measure each of the three force components (one vertical and two horizontal). The signals from the transducers are amplified and may undergo other electrical modification. The amplified and modified signals can be recorded, for example by using an oscillograph, or converted to digital format for computer processing. A complete force platform system (e.g. Figure 6.2(a)) can be represented schematically in the form of Figure 6.2(b). The components of such a system are considered in the following sections.

6.1.2 THE DETECTOR–TRANSDUCER

The detector–transducer (Figure 6.3) detects the force to be measured and converts (or transduces) it into an electrical signal proportional to the force applied.

Basically, any force-sensitive mechanical element linked to a compatible electrical transducer could be used. Most commercial platforms use either piezoelectric or strain gauge transducers.

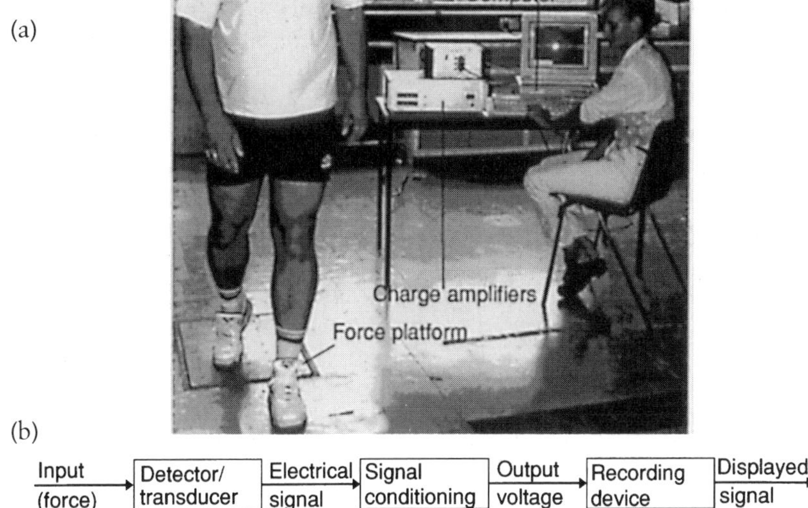

Figure 6.2 Force platform system: (a) photograph; (b) schematic representation.

Figure 6.3 Detector-transducer stage: (a) strain gauge type; (b) piezoelectric type.

Strain gauge platforms

These consist of electric resistance strain gauges, made of a material the resistance of which changes with its deformation (or strain). The strain gauges are mounted on a sensor, such as an axially loaded cylinder, which deforms slightly when a force is applied to it. The strain gauge configuration has to provide temperature compensation, as strain gauges respond to changes in temperature as well as strain. Strain gauge transducers may need more frequent calibration checks than piezoelectric platforms. They are considered easier to install, as they are less sensitive to pre-loading errors caused by the four mountings not being exactly in the same plane. They may be more suitable than piezoelectric platforms for static or quasi-static applications. Strain gauge force platforms are generally less expensive than piezoelectric ones. The AMTI platforms are probably the best known examples of this type of force platform.

Piezoelectric platforms

These rely on the development of an electrical charge on certain crystals (e.g. quartz) when subject to an applied force. Some limitations may be experienced for static or quasi-static applications because of **drift**, i.e. a change in output without a change in the force applied. Provision can be made to account for drift, for example by resetting zero output before recording. These platforms are, however, well suited for the rapidly changing forces that characterize most sports movements. They are less affected by changes in temperature than strain gauge platforms, but are considered to be more difficult to install. They have a widely adjustable range, such that the same platform can be used to record both impact (as in triple jumping) and the forces on the ground caused by the heart beating (Nigg, 1994). The Kistler platforms are the best known piezoelectric platforms and are probably still among the most widely used force platforms in sports biomechanics.

6.1.3 SIGNAL CONDITIONING AND RECORDING

For a strain gauge system, signal conditioning will usually incorporate temperature compensation and an amplifier. For a piezoelectric platform a charge amplifier will be used, which may incorporate circuitry to improve the static measuring capability by compensating for drift. The main function of this part of the system is to produce an amplified voltage output suitable for recording.

A force platform must be used with a suitable recorder (Figure 6.4) if the signal is to be adequately reproduced.

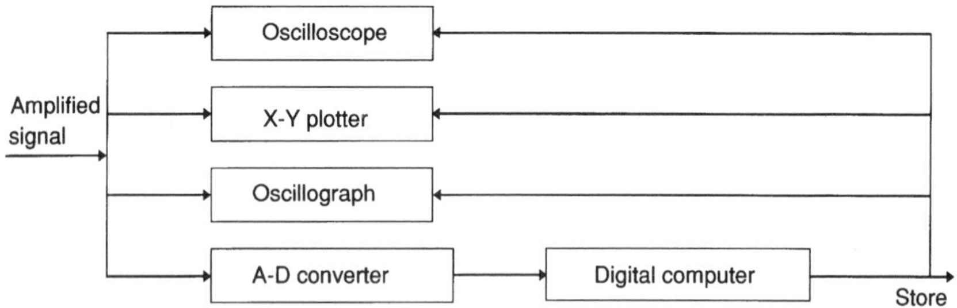

Figure 6.4 Possible ways of recording the force platform output signals.

For a recording system that produces a continuous trace (an analogue device), such as a pen recorder, the recorder should possess a natural frequency (see below) greater than five times that of the highest signal frequency of interest and a damping ratio (see below) of about 0.5–0.8. This will ensure adequate transient and steady state response.

To produce a signal in a form suitable for digital computer processing, an analog-to-digital converter is used. The signal is then sampled at discrete time intervals, expressed as the **sampling rate** or **sampling frequency**. The Nyquist sampling theorem (Chapter 5) requires a sampling frequency greater than twice that of the highest signal frequency. It should be remembered that, although the frequency content of much human movement is low, many force platform applications involve impacts and hence a higher frequency content. A sampling frequency as high as 500 Hz or 1 kHz may, therefore, be appropriate. When choosing or setting up an analog-to-digital converter, it is also important to note whether the signals from all channels are sampled sequentially (**multiplexing**), simultaneously or quasi-simultaneously (**burst sampling**) (Bartlett, 1992). In non-simultaneous sampling, errors can arise in calculations, such as the point of force application, made from force channels sampled at slightly different times, unless software corrections are performed.

6.1.4 OPERATIONAL CHARACTERISTICS OF A FORCE PLATFORM SYSTEM

Accurate (valid) and reliable force platform measurements depend on adequate system sensitivity, a low force detection threshold, high linearity, low hysteresis, low cross-talk and the elimination of cable interference, electrical inductance and temperature and humidity variations. The platform must be sufficiently large to accommodate the movement under investigation. Extraneous vibrations must be excluded. Mounting instructions for force platforms are specified by the manufacturers. As a general rule, the platform is usually sited on the bottom floor of a building in a large concrete block. If mounted outdoors, a large concrete block sited on pebbles or gravel is usually a suitable base and attention must be given to problems of drainage (Bartlett, 1992). The main measurement characteristics of a force platform are considered below. In addition, a good temperature range (–20 to 70°C) and a relatively light weight may be important (Dainty *et al.*, 1987). Also, any variation in the recorded force with the position on the platform surface at which it is applied should be less than 2–3% in the worst case (Biewener and Full, 1992).

Linearity

This is expressed as the maximum deviation from linearity as a percentage of full scale deflection (FSD). For example, in Figure 6.5(a), linearity would be expressed as $y/Y \times 100\%$.

Although good linearity is not essential for accurate measurements, as a non-linear system can be calibrated, it is useful and does make calibration easier. A suitable figure for a force platform for use in sports biomechanics would be 0.5% FSD or better.

Hysteresis

This exists when different input–output relationships occur, depending on whether the input force is increasing or decreasing (Figure 6.5(b)). Hysteresis can be caused, for example, by the presence of deforming mechanical elements in the transducer stage. It is expressed as the maximum difference between the output voltage for the same force, increasing and decreasing, divided by the full scale output voltage. It should be low, a suitable figure being 0.5% FSD or less.

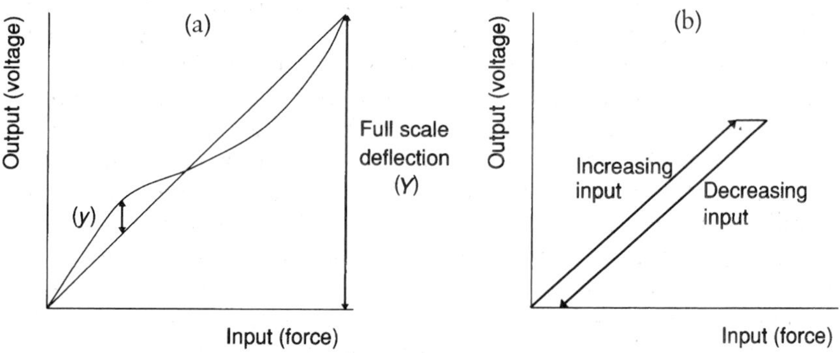

Figure 6.5 Force platform characteristics: (a) linearity; (b) hysteresis (simplified).

Range

The range of forces that can be measured must be adequate for the application and the range should be adjustable. If the range is too small for the forces being measured, the output voltage will saturate (reach a constant level) as shown in Figure 6.6. Suitable maximum ranges for most sports biomechanics applications would be –10 kN to +10 kN on the two horizontal axes and –10 kN to +20 kN on the vertical.

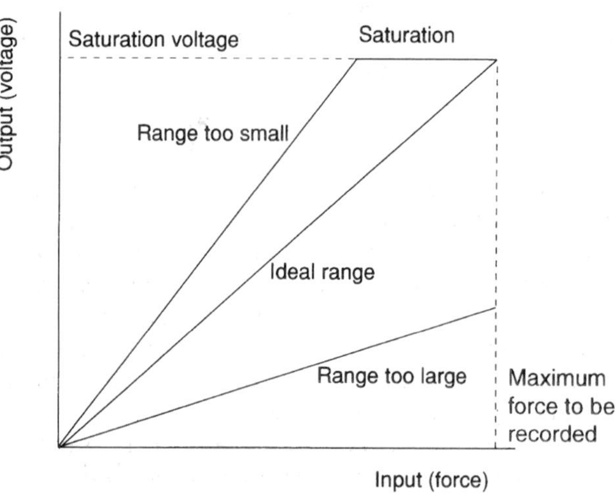

Figure 6.6 Output saturation and effect of range on sensitivity.

Sensitivity

This is the change in the recorded signal for a unit change in the force input, or the slope of the idealized linear voltage–force relationships of Figure 6.6. The sensitivity decreases with increasing range. Good sensitivity is essential as it is a limiting factor on the accuracy of the measurement. In most modern force platform systems, an analog-to-digital converter (A–D converter) is used and is usually the main limitation on the overall system **resolution.** An 8-bit A–D converter divides the input into an output that can take one of 256 (2^8) discrete values. The resolution is (100/255)%, which is approximately 0.4%. A 12-bit converter will improve this figure to about 0.025%. In a force platform system with adjustable range, it is essential to choose the range that just avoids overloading in order to achieve optimum sensitivity. For example, consider that the maximum vertical force to be recorded is 4300 N, an 8-bit A–D converter is used, and a choice of ranges of 10 000 N, 5000 N and 2500 N is available. The best available range is 5000 N, and the force would be recorded approximately to the nearest 20 N (5000/255). The percentage error in the maximum force is then only about 0.5% (20 × 100/4300). A range of 10 000 N would double this error to about 1%. A range of 2500 N would cause saturation, and the maximum force recorded would be 2500 N, an underestimate of over 40% (compare with Figure 6.6).

Cross-talk

Force platforms are normally used to measure force components in more than one direction. The possibility then exists of forces in one component direction affecting the forces recorded by the transducers used for the other components. The term cross-talk is used to express this interference between the recording channels for the various force components. Cross-talk must be small and a figure of better than 3% has been reported as necessary (Dainty *et al.*, 1987).

Dynamic response

Forces in sport almost always change rapidly as a function of time. The way in which the measuring system responds to such rapidly changing forces is crucial to the accuracy of the measurements and is called the dynamic response of the system. The considerations here relate mostly to the mechanical components of the detector–transducer and any slow display elements, such as a pen recorder. A representation of simple sinusoidally varying input and output signals of a single frequency (ω) as a function of time is presented in Figure 6.7.

The ratio of the amplitudes (maximum values) of the output to the input signal is called the **amplitude ratio**. The time by which the output signal lags the input signal is called the **phase lag**. This is often expressed as the phase angle (ϕ) which is the time lag multiplied by the signal frequency.

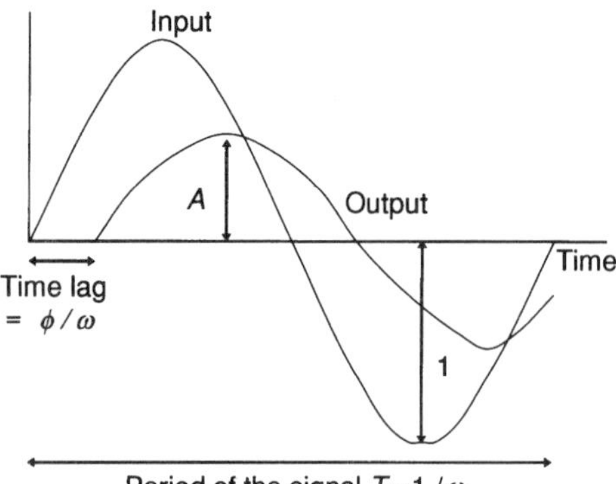

Figure 6.7 Representation of force input and recorded output signals as a function of time.

In practice, the force signal will contain a range of frequency components, each of which could have different phase lags and amplitude ratios. The more different these are across the range of frequencies present in the signal, the greater will be the distortion of the output signal. This will reduce the accuracy of the measurement. What is required, therefore, is:

- that all the frequencies present in the force signal should be equally amplified; this means that there should be a constant value of the amplitude ratio A (system calibration can allow A to be considered as 1, as in Figure 6.8);
- that values of the phase lag (or phase angle ϕ) should be small.

A force platform is a second-order measuring system consisting, essentially, of a mass (m), a spring, of stiffness k, and a damping element (c). The steady-state frequency response characteristics of such a system are usually represented by the unique series of non-dimensional curves of Figure 6.8. The response of such a system to an instantaneous change of the input force is known as its transient response and can be represented as in Figure 6.9.

In these two Figures:

- ζ is the damping ratio of the system ($c/(4\ km)^{1/2}$);
- ω_n is the natural frequency ($(k/m)^{1/2}$), i.e. the frequency at which the platform will vibrate if struck and then allowed to vibrate freely;
- ω is the signal frequency (as in Figure 6.7).

To obtain a suitable transient response, the damping ratio needs to be around 0.5–0.8 to avoid under- or overdamping. A value of 0.707 is considered ideal (as in Figure 6.9), as this ensures that the output follows the input signal in the shortest possible time.

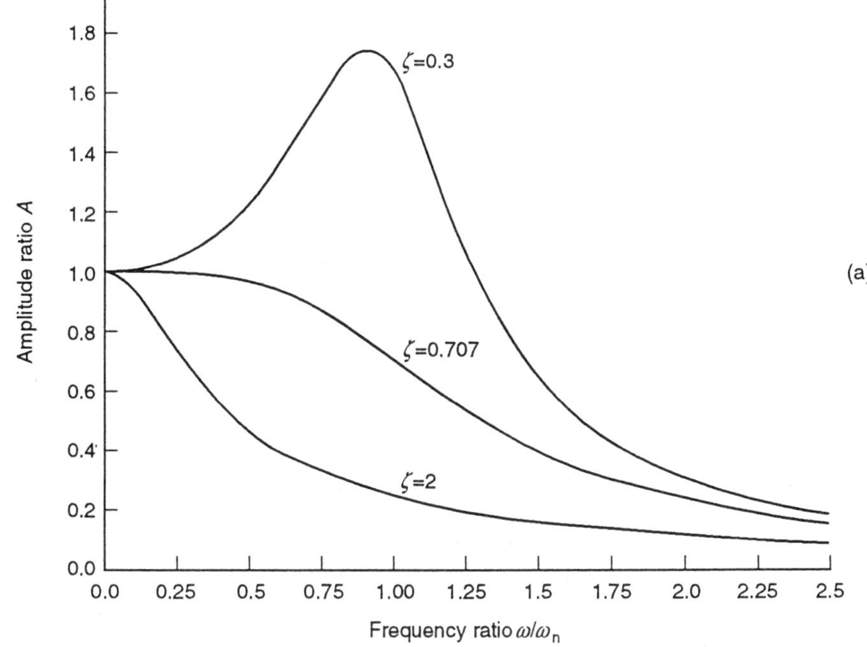

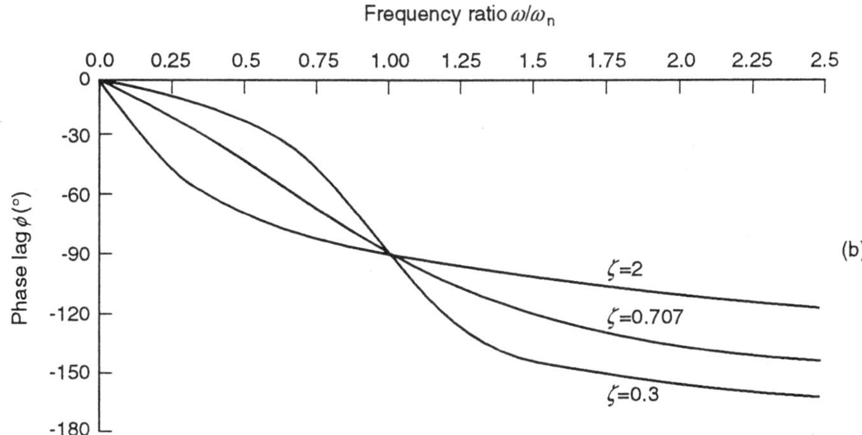

Figure 6.8 Steady state frequency response characteristics of a second order force platform system: (a) amplitude plot; (b) phase plot.

Careful inspection of Figure 6.8 allows the suitable frequency range to be established for steady-state response in order to achieve a near-constant amplitude ratio (Figure 6.8(a)) and a small phase lag (Figure

6.8(b)), with $\zeta = 0.707$. The frequency ratio ω'/ω_n, where ω' is the largest frequency of interest in the signal, must be small, ideally less than 0.2. Above this value, the amplitude ratio (Figure 6.8(a)) starts to deviate from a constant value and the phase lag becomes large and very frequency dependent, causing errors in the output signal. For impact forces, Ramey (1975) considered that the natural frequency should be 10 times the equivalent frequency of the impact, which can exceed 100 Hz in sports activities. The natural frequency must, therefore, be as large as possible to record the frequencies of interest. The platform's structure must be relatively light (m) with high stiffness (k) to give a high natural frequency ($(k/m)^{1/2}$). (This consideration relates not only to the platform but also to its mounting). A high natural frequency, of around 1000 Hz, would be desirable for most applications in sports biomechanics. Among the highest natural frequencies specified for commercial platforms are: for piezoelectric platforms 850 Hz, all three channels; for strain gauge platforms 1000 Hz for the vertical channel, 550 Hz for the two horizontal channels, and 1500 Hz (vertical) and 320 Hz (horizontal). However, the value specified for a particular platform may not always be found in practice. For example, Kerwin and Chapman (1988) reported natural frequencies between 288 and 743 Hz in some vibration modes for a commercial force platform with a quoted natural frequency of at least 850 Hz.

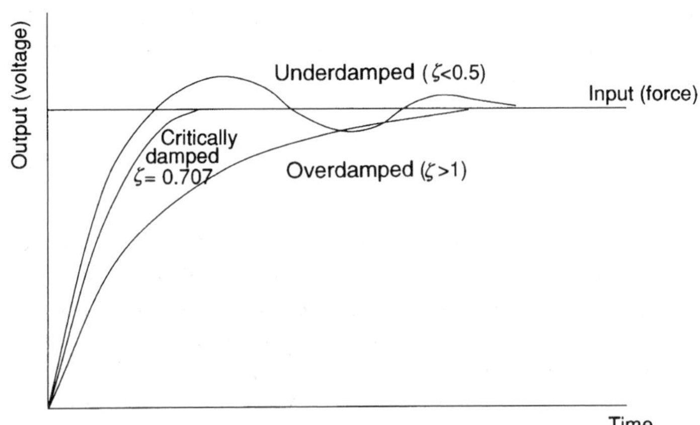

Figure 6.9 Transient response characteristics of a second order force platform system.

6.2 Experimental procedures

6.2.1 GENERAL

A force platform, if correctly installed (mounted) and used with appropriate auxiliary equipment, is generally simple to use. When used with an analog-to-digital converter and computer, the timing of data collection is important. This can be achieved, for example, by computer control of the collection time, by the use of photoelectric triggers or by sampling only when the force exceeds a certain threshold. When using a force platform with cinematography or video recording, synchronization of the two will be required. This can be done in several ways, such as causing the triggering of the force platform data collection to illuminate a light in the field of view of the cameras.

The sensitivity of the overall system will need to be adjusted to prevent saturation while at the same time ensuring the largest possible use of the equipment's range. This can often be done by trial-and-error adjustments of either amplifier or recorder gains: this would obviously be done before the main data collection (Bartlett, 1992). If the manufacturers have recommended warm-up times for the system amplifiers, then these must be carefully observed.

Care must be taken in the experimental protocol, if the performer moves on to the platform, to ensure that foot contact occurs with little, preferably no, targeting of the platform by the performer. This may, for example, require the platform to be concealed by covering the surface with a material similar to that of the surroundings of the platform. Obviously the external validity of an investigation will be compromised if changes in movement patterns occur in order to contact the platform. Also important for external (ecological) validity, especially when recording impacts, is matching the platform surface to that which normally exists in the sport being studied. The aluminium surfaces usual for force platforms are unrepresentative of most sports surfaces.

6.2.2 CALIBRATION

It is essential that the system should be capable of calibration in order to minimize systematic errors. The overall system will require regular calibration checks, even if this is not necessary for the platform itself. Calibration of the amplifier output as a function of force input will usually be set by the manufacturers and may require periodic checking. The vertical channel is easily calibrated under static loading conditions by use of known weights. If these are applied at different points across the platform surface, the variability of the recorded force with its point of application can also be checked. The horizontal channels can also be statically calibrated, although not so easily. One method (Biewener and Full, 1992) involves attaching a cable to the platform surface, passing the cable over a frictionless air pulley at the level of the platform surface, and adding weights to the free end of the cable. Obviously this cannot be done while the platform is installed in the ground flush with the surrounding surface. There appears to be little guidance provided to users on the need for (or regularity of) dynamic calibration checks on force platform systems. The tendency of piezoelectric transducers to drift may mean that zero corrections are required, and strain gauge platforms may need more frequent calibration checks than do piezoelectric ones.

Biewener and Full (1992) presented a full protocol for platform calibration using generally available equipment. This provided full details of static calibration and checks on the variability of the recorded force over the platform surface. Cross-talk was checked by recording the outputs from the two horizontal channels when only a vertical force was applied to the platform. A similar procedure can be used for assessing cross-talk on the vertical channel if horizontal forces can be applied. Positions of the point of force application were checked by placing weights on the platform at various positions and comparing these with centre of pressure positions calculated from the outputs from the individual vertical force transducers. As errors in these calculations are especially problematic when small forces are being recorded, small as well as large weights should be included in such checks. Finally, Biewener and Full (1992) recommended that the natural frequency should be checked by lightly striking the platform with a metal object and using an oscilloscope to show the ringing of the platform at its natural frequency. This should be carried out, of course, in the location in which the platform is to be used.

6.3 Data processing

Processing of force platform signals is relatively simple and accurate, compared with most data in sports biomechanics. The example data of Figure 6.10 were obtained from a standing broad jump.

The three orthogonal components of the ground contact force (Figures 6.1, 6.10(a)) are easily obtained by summing the outputs of individual transducers. As the platform provides whole body measurements, these forces can be easily converted to the three components of centre of mass acceleration, as $F = m\,a$ (Figure 6.10(b)), after subtracting the performer's weight from the vertical force component. The coordinates of the point of application of the force, the centre of pressure (Figures 6.10(c), 6.11(b)), on the platform working surface can also be calculated (see Nigg, 1994). The accuracy of the centre of pressure calculations in particular depends on careful calibration of the force platform (Bobbert and Schamhardt, 1990). This accuracy becomes highly suspect at the beginning and end of any contact phase, when the calculation of centre of pressure involves the division of small forces by other small forces.

The moment of the ground contact force about the (vertical) axis perpendicular to the plane of the platform, the frictional torque (Figure 6.10(d)) can be calculated. With appropriate knowledge of the position vector of the performer's centre of mass, in principle at one instant only, the component moments of the ground contact force (Figure 6.10(d)) can also be calculated about the two horizontal and mutually perpendicular axes passing through the performer's mass centre and parallel to the platform plane.

From the force–time and moment–time data, integration can be performed to find overall or instant-by-instant changes in centre of mass momentum (and hence velocity) and whole body angular momentum. Absolute magnitudes of these variables at all instants (Figure 6.10(e),(f)) can be calculated only if their values are known at least at one instant. These values could be obtained from cinematography or another motion analysis system. They are easily obtained if the performer is at rest on the platform at some instant, when linear and angular momentum are zero. If absolute velocities are known, then the changes in position (displacement) can be found by integration. In this case, absolute values of the position vector of the centre of mass with respect to the platform coordinate system (Figure 6.10g) can be obtained if that position vector is known for at least one instant. Again, that value could be obtained from cinematography or another motion analysis system. Alternatively, the horizontal coordinates can be obtained from the centre of pressure position with the person stationary on the platform, and the vertical coordinate as a fraction of the person's height. Integration can be performed graphically (see Chapter 2) from a hard copy of the relevant traces, or numerically. A third alternative is to represent the appropriate digital data by an analytic function of suitably close representation, such as cubic splines, and then to integrate this function analytically (Bartlett, 1992).

The load rate (Figure 6.10(h)) can be calculated as the rate of change of the contact force (dF/dt) using an acceptable method of numerical differentiation. The load rate has been linked to injury (see, for example, Nigg, 1986). Other calculations that can be performed involve whole body power (Davies, 1971) ($P = \boldsymbol{F} \cdot \boldsymbol{v}$) (Figure 6.10(i)).

All the above variables can be presented graphically as functions of time, as in Figure 6.10. In addition, the forces acting on the performer can be represented as instantaneous force vectors, arising from the instantaneous centres of pressure (a side view is shown in Figure 6.11(a)).

Front, top and three-dimensional views of the force vectors are also possible. The centre of pressure path can also be shown superimposed on the platform surface (Figure 6.11(b)).

(a)
Force (kN)
1.50
0.75
0
−0.75
−1.50
0.3
0.6
0.9
1.2
Time (s)
x
y
z
(b)
Acceleration (m·s−2)
15
7.5
0
−7.5
−15
0.3
0.6
0.9
1.2
Time (s)
x
y
z
(c)
Point of force application (m)
0.2
0.1
0
−0.1
−0.2
0.3
0.6
0.9
1.2
Time (s)
x
z
(d)
Moment about mass centre (N·m)
300
150
0
−150
−300
0.3
0.6
0.9
1.2
Time (s)
x
y
z
(e)
Mass centre velocity (m·s−1)
2.50
1.25
0
−1.25
−2.50
0.3
0.6
0.9
1.2
Time (s)
x
y
z
(f)
Angular momentum (kg·m2·s−1)
150
75
0
−75
−150
0.3
0.6
0.9
1.2
Time (s)
x
y
z

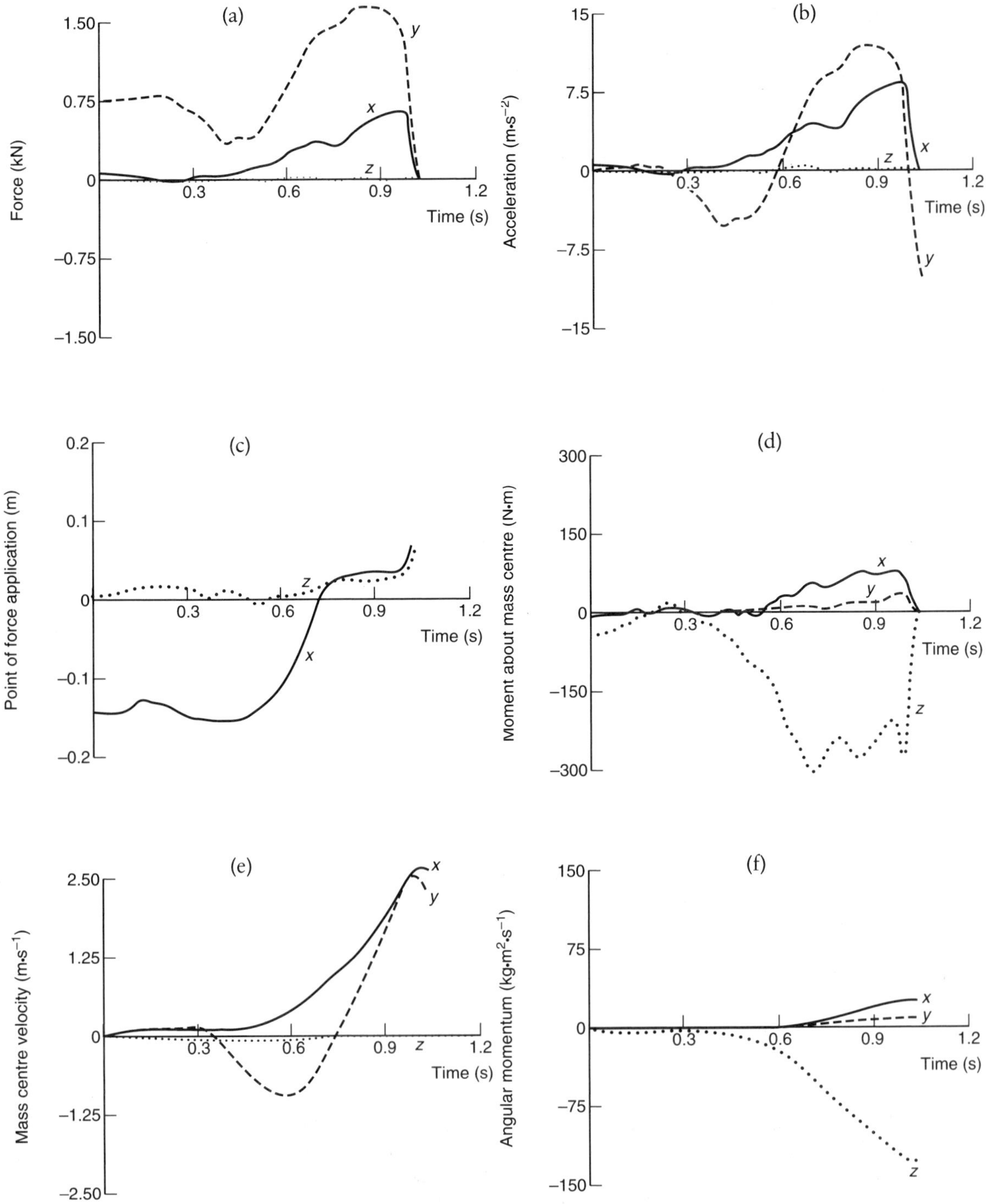

(g)

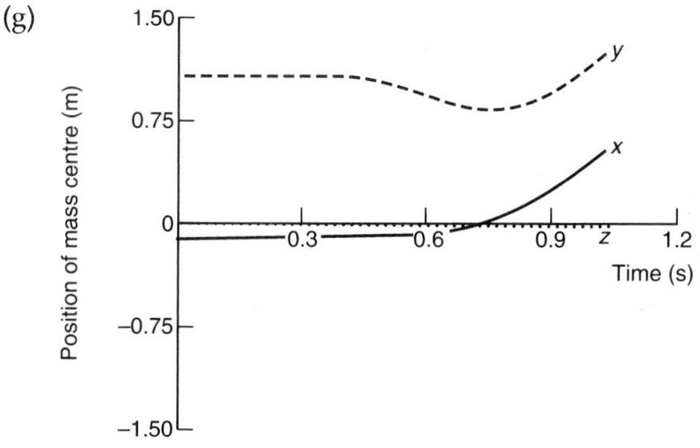

(h)

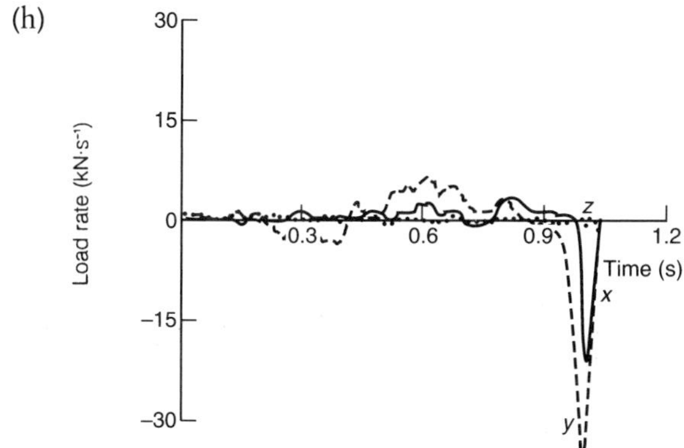

(i)

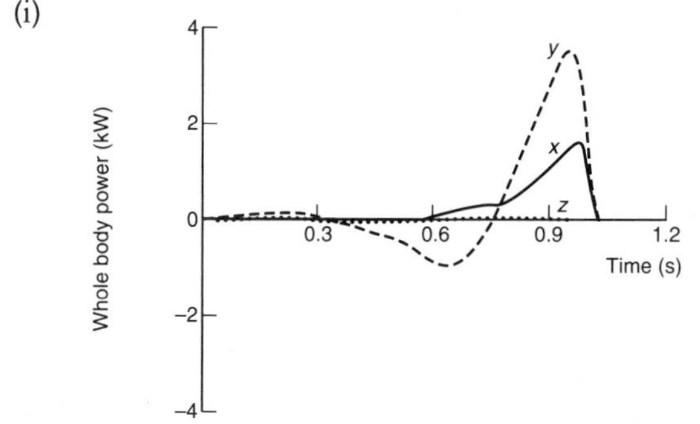

Figure 6.10 Force platform variables (as *x,y,z* components) as functions of time for a standing broad jump: (a) force; (b) centre of mass acceleration; (c) point of force application (*x,z* only); (d) moment; (e) centre of mass velocity; (f) whole body angular momentum; (g) centre of mass position; (h) load rate; (i) whole body power.

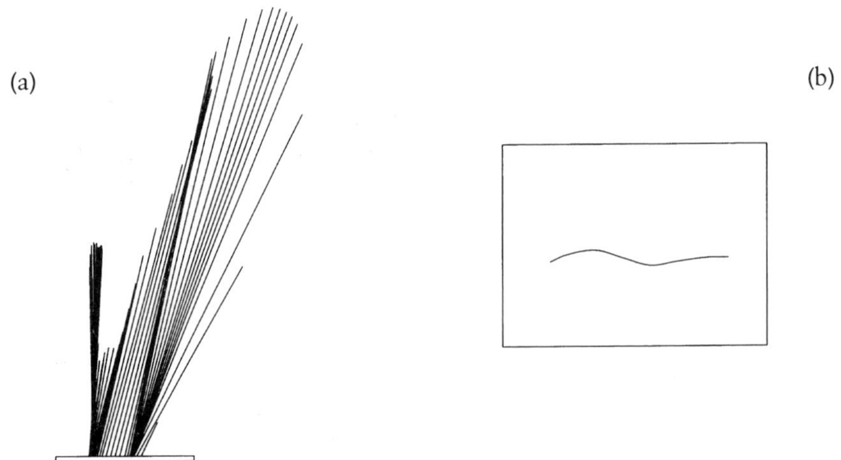

Figure 6.11 (a) Side view of force vectors for a standing broad jump; (b) centre of pressure path from above.

6.4 Examples of the use of force measurement in sports biomechanics

Force platforms have been relatively widely used in sports biomechanics research. Most of this research has involved measurement of ground contact forces, with few examples being reported of the use of force platforms in different sitings. There has also been research into other forces exerted by the sports performer, using various types of force transducer.

Much of the force platform research in sport has focused on various aspects of running. The difference between the initial passive (or impact) loading on the foot and the later active loading are evident in much of this research (for example, Figure 6.12(a)).

The initial contact has high-frequency components (above 30 Hz) with the impact peak force occurring after around 20–30 ms. The active, or propulsive, phase involves low frequencies, below 30 Hz, with the peak propulsive force occurring after about 100 ms. Different patterns of running were identified using force platforms by Cavanagh and Lafortune (1980). They classified runners into rearfoot, midfoot or forefoot strikers, depending on which region of the foot experienced the initial contact. The differences in centre of pressure paths for these three groups of runners can be seen in Figure 6.13.

Differences in ground contact forces have also been identified for different surfaces (Figure 6.12(a)), different types of foot strike (Figure 6.12(b)) and different shoes (Figure 6.12(c)). The variation of maximum force with running speed has also been investigated and was reported, for example, by Nigg *et al.* (1987) to increase from twice body weight at 3 $m \cdot s^{-1}$ to three times body weight at 6 $m \cdot s^{-1}$ wearing running shoes. Force platforms have also been used widely in

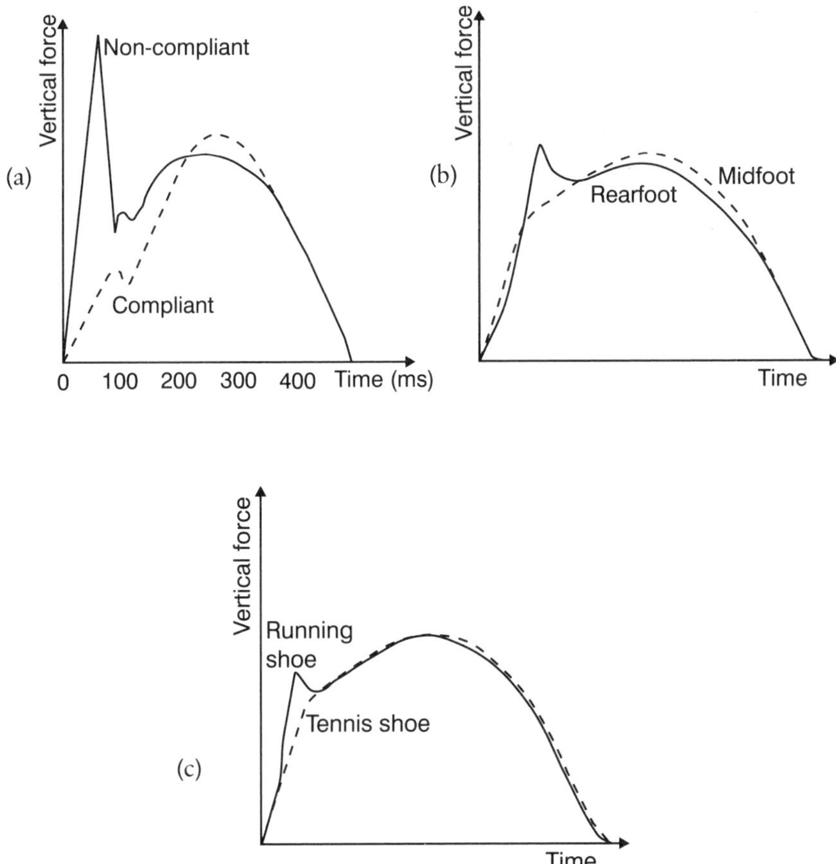

Figure 6.12 Ground contact forces in running: (a) two different surfaces; (b) two types of foot strike; (c) two different shoes (first and last examples adapted from Nigg, 1986).

evaluating running shoes and changes in their construction (see, for example, Nigg, 1986) and in investigating various aspects of sport surfaces (see Nigg and Yeadon, 1987).

Larger vertical impact forces have been recorded from force platforms in other sports. Figures include 8.3 times body weight (BW) for a long jump take-off from an 8.0 $m \cdot s^{-1}$ approach run (Nigg, Denoth and Neukomm, 1981), 9.1 BW for the back foot strike and 6.6 BW for the front foot strike in the delivery stride of the javelin throw (Deporte and van Gheluwe, 1988) and up to 12.3 BW for front foot strike in the delivery stride for cricket fast bowlers (Mason, Wiessensteiner and Spence, 1989). There is an obvious injury potential for such high-impact activities, especially when they are repeated a large number of times, as is the case for fast bowlers in cricket (e.g. Elliott *et al.*, 1992).

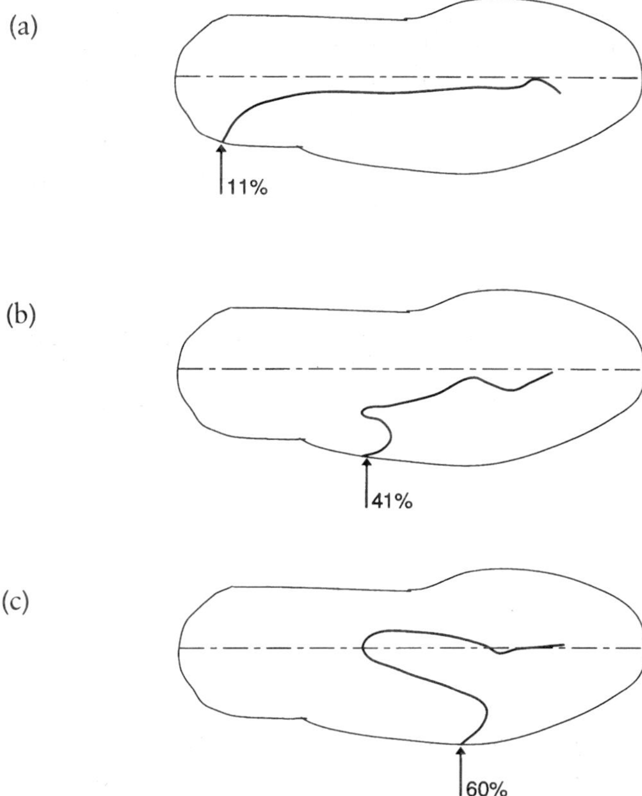

Figure 6.13 Variation of centre of pressure (gait line) patterns for different foot strikes: (a) rearfoot striker; (b) midfoot striker; (c) forefoot striker. The percentage is the strike index, the centre of pressure position from the heel at foot strike as a ratio to shoe length (adapted from Cavanagh and Lafortune, 1980).

An example of the use of a force transducer, other than a force platform, in sports biomechanics is provided by the study of Smith and Spinks (1989). They attached small force transducers to an oar, to investigate how the force applied by the rower varied over the entire rowing stroke. This provided useful information about the proportion of the applied force used for propulsion, particularly when combined with information on the variation of boat speed.

Force platform data are widely used in research into the loading on the various joints of the body. In such research, kinematic data from cinematography, or another motion recording technique, are used with the force and centre of pressure data to calculate the resultant forces and moments at body joints (e.g. Winter, 1983). The two recording systems need to be synchronized for such investigations.

6.5 Summary

In this chapter, the use of the force platform in sports biomechanics was covered, including the equipment and methods used, the processing of force platform data, and some examples of the use of force measurements in sport. The material covered included why the measurement of the contact forces on the sports performer is important. It also considered the advantages and limitations of the two main types of force platform and the important measurement characteristics required for a force platform in sports biomechanics. The procedures for calibrating a force platform were outlined, along with those used to record forces in practice. The different ways in which force platform data can be processed to obtain other movement variables were covered, and some of their uses. The chapter concluded with a consideration of some of the reported research in sports biomechanics which has used force measurement.

6.6 Exercises

1. Explain why the measurement of the contact forces on the sports performer is important in sports biomechanics, with reference to examples from research in this area.
2. List the advantages and disadvantages of the two main types of force platform.
3. Explain the main response characteristics required from a force platform. If you have access to a force platform, obtain its technical specification, and ascertain whether this conforms to the guideline figures of section 6.1.4.
4. A procedure for calibrating the horizontal force components recorded by a force platform was outlined in section 6.2.2. Devise one other way in which such a calibration could be carried out. (This may require some careful thought!)
5. Make a list of sports movements in which you have an interest, other than those covered in this chapter, in which measurement of external forces acting on or exerted by the sports performer would be useful. Then seek to identify those for which a standard force platform could be used, and those for which specialized force transducers would be needed.
6. Describe two ways in which you might be able to synchronize the recording of forces from a force platform with a video recording of the movement. (This may also require some careful thought!)
7. Outline how you would statically calibrate a force platform, how you would check for variability of the recorded force with its point of application on the platform surface, how you would check for cross-talk and how you would check the accuracy of centre of pressure calculations.

8. If you have access to a force platform, and with appropriate supervision if necessary, perform the calibration and other checks on the platform from the previous exercise. You may not be able to load the platform in a horizontal direction.
9. If you have access to a force platform, perform an experiment involving a standing broad jump from the platform, with arm counter movements. From the recorded force components, see if you can obtain the other data that were covered in section 6.3. If your force platform software supports all the processing options for these data, perform this for all three force channels. If you have to do any of the data processing manually, then do so for only the vertical channel. Compare the results you obtain with those of Figure 6.10.
10. If you have access to a force platform, have several people run across the platform to investigate what type of striker each is (from the centre of pressure path). Also investigate one or more of the following: differences in forces recorded for different shoe types or makes; differences between surfaces; effect of running speed. Compare your results, where appropriate, with those of section 6.4, or the references cited.

6.7 References

Bartlett, R. M. (1992) Force platform, in *Biomechanical Analysis of Performance in Sport*, (ed. R. M. Bartlett), British Association of Sports Sciences, Leeds, pp.24–27.

Biewener, A. A. and Full, R. J. (1992) Force platform and kinematic analysis, in *Biomechanics – Structures and Systems: A Practical Approach*, (ed. A. A. Biewener), Oxford University Press, Oxford, pp.45–73.

Bobbert, M. F. and Schamhardt, H. C. (1990) Accuracy of determining the point of force application with piezoelectric force plates. *Journal of Biomechanics*, 23, 705–710.

Cavanagh, P. R. and Lafortune, M. A. (1980) Ground reaction forces in distance running. *Journal of Biomechanics*, 13, 397–406.

Dainty, D. A., Gagnon, M., Lagasse, P.P. *et al.* (1987) Recommended procedures, in *Standardizing Biomechanical Testing in Sport*, (eds D. A. Dainty and R. W. Norman), Human Kinetics, Champaign, IL, pp.73–100.

Davies, C. T. M. (1971) Human power output in exercise of short duration in relation to body size and composition. *Ergonomics*, 14, 245–256.

Deporte, E. and van Gheluwe, B. (1988) Ground reaction forces and moments in javelin throwing, in *Biomechanics XI-B*, (eds G. de Groot, A. P. Hollander, P.A. Huijing and G. J. van Ingen Schenau), Free University Press, Amsterdam, pp.575–581.

Elliott, B.C., Hardcastle, P. H, Burnett, A. F. and Foster, D. H. (1992) The influence of fast bowling and physical factors on radiologic features in high performance young fast bowlers. *Sports Medicine, Training, and Rehabilitation*, 3, 113–130.

Kerwin, D. G. and Chapman, G. M. (1988) Resonant mode shapes of a force plate and mounting, in *Biomechanics XI-B*, (eds G. de Groot, A. P. Hollander, P.A. Huijing and G. J. van Ingen Schenau), Free University Press, Amsterdam, pp.978–83.

Mason, R. B., Weissensteiner, J. R. and Spence, P. R. (1989) Development of a model for fast bowling in cricket. *Excel*, **6**, 2–12.

Nigg, B. M. (1986) *Biomechanics of Running Shoes*, Human Kinetics, Champaign, IL.

Nigg, B. M. (1994) Force, in *Biomechanics of the Human Musculoskeletal System*, (eds B. M. Nigg and W. Herzog), John Wiley, Chichester, pp.200–224.

Nigg, B. M. and Yeadon, M. R. (1987) Biomechanical aspects of playing surfaces. *Journal of Sports Sciences*, **5**, 117–145.

Nigg, B. M., Denoth, J. and Neukomm, P.A. (1981) Quantifying the load on the human body: problems and some possible solutions, in *Biomechanics IX-B*, (eds D. A. Winter, R. W. Norman, R. P. Wells *et al.*), Human Kinetics, Champaign, IL, pp.88–99.

Nigg, B. M., Bahlsen, H. A., Lüthi, S. M. and Stokes, S. (1987) The influence of running velocity and midsole hardness on external impact forces in heel-toe running. *Journal of Biomechanics*, **20**, 951–959.

Ramey, M. R. (1975) Force plate designs and applications, in *Exercise and Sport Sciences Reviews*, Vol. 3, (eds J. H. Wilmore and J. F. Keogh), Academic Press, New York, pp.303–319.

Smith, R. and Spinks, W. (1989) Matching technology to coaching needs: on water analysis, in *Proceedings of the VIIth International Symposium of the Society of Biomechanics in Sports*, (ed. W. Morrison), Footscray Institute of Technology Press, Melbourne, Victoria, pp.277–287.

Winter, D. A. (1983) Moments of force and power in jogging. *Journal of Biomechanics*, 7, 157–159.

6.8 Further reading

Nigg, B. M. (1994) Force, in *Biomechanics of the Human Musculoskeletal System*, (eds B. M. Nigg and W. Herzog), John Wiley, Chichester, pp.200–224: this provides a useful overview of aspects and uses of force measurement in biomechanics, although some readers may find the mathematics a little hard going.

Nigg, B. M. (1986) *Biomechanics of Running Shoes*, Human Kinetics, Champaign, IL: chapter 1, chapter 2, pp.27–38, and chapter 5, pp.139–150 contain much useful, interesting, and generally very readable information relating to the use of force measurement in running shoe research.

7 Electromyography

This chapter is intended to provide an appreciation of the applications of electromyography to the study of muscle activity in sports movements. This includes the equipment and methods used, the processing of electromyographic data and the important relationship between the recorded signal (known as the electromyogram – abbreviation EMG) and muscle tension. After reading this chapter you should be able to:

- appreciate the applications of electromyography to the study of sports skills;
- understand why the recorded EMG differs from the physiological signal;
- evaluate the advantages and limitations of the three types of EMG electrode used in sports biomechanics;
- explain the main characteristics of an EMG amplifier;
- describe the methods of quantifying the EMG signal in the time and frequency domains;
- be aware of the research that has addressed the relationship between the EMG and muscle tension, and understand the importance of this relationship.

7.1 Introduction

Electromyography is the technique for recording changes in the electrical potential of a muscle when it is caused to contract by a motor nerve impulse. Each efferent α-motoneuron (or motor neuron) innervates a number of muscle fibres, which may be less than 10 for muscles used for fine control and more than 1000 for the weight-bearing muscles of the legs. The motor neuron forms a **neuromuscular junction**, or **motor end-plate**, with each muscle fibre that it innervates (see Chapter 1, Figure 1.16). The term **motor unit** is used to refer to a motor neuron and all the muscle fibres it innervates, which can be spread over a wide area of the muscle (Nigg and Herzog, 1994). The motor unit can be considered the fundamental functional unit of neuromuscular control (Enoka, 1994). Each nerve impulse causes all the muscle fibres of the motor unit to contract fully and almost simultaneously. The stimulation of the muscle fibre at the motor end-plate results in a reduction of the electrical potential of the cell and a spread of the action potential through the muscle fibre. The **motor action potential** (MAP), or muscle fibre action potential, is the name given to the waveform resulting from this depolarization wave. This propagates in both directions along each muscle fibre from the motor end-plate before being followed by a repolarization wave. The summation in space and time of MAPs from the fibres of a given motor unit is termed a **motor unit action potential**, MUAP (Figure 7.1).

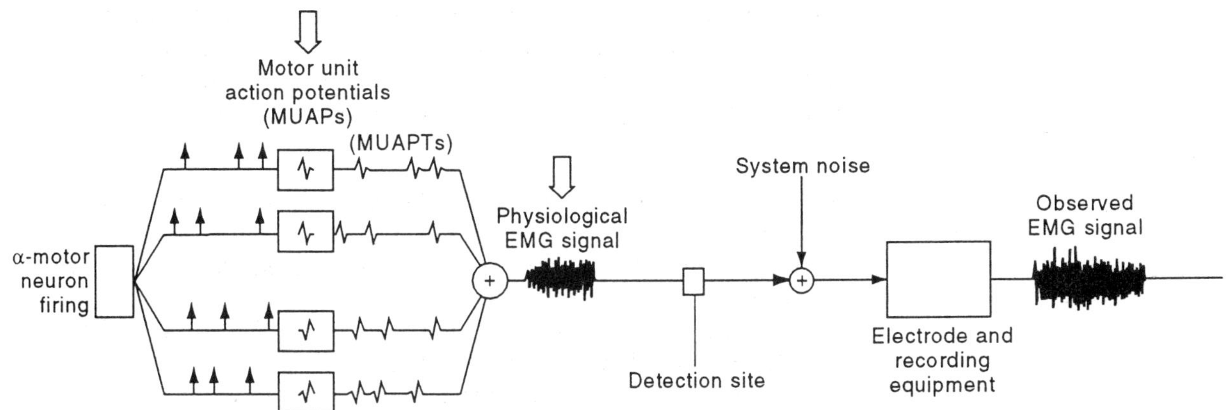

Figure 7.1 Schematic représentation of the generation of the EMG signal.

A sequence of MUAPs, resulting from repeated neural stimulation, is referred to as a **motor unit action potential train** (MUAPT) (Basmajian and De Luca, 1985). The physiological EMG signal is the sum, over space and time, of the MUAPTs from the various motor units.

Electromyography offers the only method of objectively assessing when a muscle is active (Grieve, 1975). It has been used to establish the roles that muscles fulfil both individually and in group actions (see Basmajian and De Luca, 1985). EMG provides information on the timing, or sequencing, of the activity of various muscles in sports movements. By studying the sequencing of muscle activation, the sports biomechanist can focus on several factors which relate to skill level. These include the overlap of agonist and antagonist activity and the onset of antagonist activity at the end of a movement (Lagasse, 1987). EMG also allows the sports biomechanist to study changes in muscular activity during skill acquisition and as a result of training. Clarys and Cabri

(1993) have provided a comprehensive review of all aspects of the use of EMG in the study of sports skills, and this is thoroughly recommended to the reader. EMG can also be used to validate assumptions about muscle activity that are made when calculating the internal forces in the human musculoskeletal system.

7.2 Experimental considerations

7.2.1 RECORDING THE MYOELECTRIC (EMG) SIGNAL

A full understanding of electromyography and its importance in sports biomechanics requires knowledge from the sciences of anatomy and neuromuscular physiology as well as consideration of many aspects of signal processing and recording, including instrumentation. A schematic representation of the generation of the (unseen) physiological EMG signal and its modification to the observed electromyogram is shown in Figure 7.1.

The EMG signal consists of a range of frequencies, which can be represented by the EMG frequency spectrum (Figure 7.2).

The physiological EMG signal is not the one recorded, as its characteristics are modified by the tissues through which it passes in reaching the EMG electrodes used to detect it. These tissues act as a low-pass filter (Chapter 5), rejecting some of the high-frequency components of the signal. The electrode-to-electrolyte interface acts as a high-pass filter, discarding some of the lower frequencies in the signal. The two electrode (bipolar) configuration, which is normal for sports electromyography, changes the two-phase nature of the depolarization–repolarization wave to a three-phase one (Winter, 1990) and removes some low- and some high-frequency signals (i.e. it acts as a bandpass filter). Modern high-quality EMG amplifiers should not unduly affect the fre-

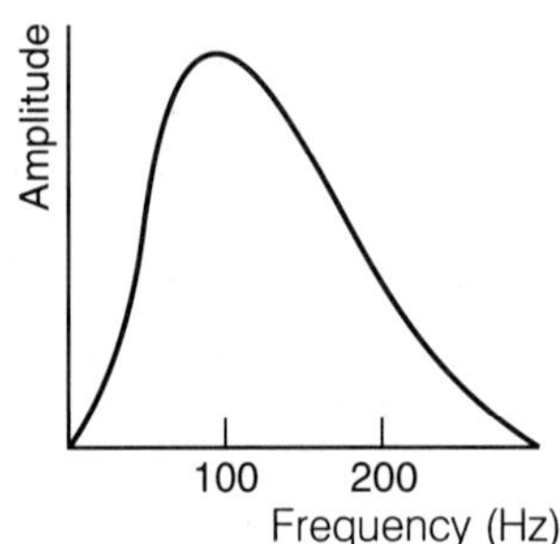

Figure 7.2 Simplified EMG frequency spectrum.

quency spectrum of the signal but the recording device may also act as a bandpass filter.

There are many factors which influence the recorded EMG signal. The intrinsic factors, over which the electromyographer has little control, can be classed (De Luca and Knaflitz, 1990) as:

- **physiological**: these include the firing rates of the motor units; the type of fibre; the conduction velocity of the muscle fibres; and the characteristics of the volume of muscle from which the electrodes detect a signal (the detection volume) – such as its shape and electrical properties;
- **anatomical**: these include the muscle fibre diameters and the positions of the fibres of a motor unit relative to the electrodes – the separation of individual MUAPs becomes increasingly difficult as the distance to the electrode increases.

Extrinsic factors, by contrast, can be controlled. These include the location of the electrodes with respect to the motor end-plates; the orientation of the electrodes with respect to the muscle fibres and the electrical characteristics of the recording system (see below). The use of equipment with appropriate characteristics is vitally important and will form the subject of the rest of this section. The electrical signals that are to be recorded are small – of the order 10 μV to 5 mV. To provide a signal to drive any recording device, there is a requirement for signal amplification. The general requirement is to detect the electrical signals (electrodes), modify the signal (amplifier) and store the resulting waveform (recorder) (compare with the force platform system of Chapter 6). This should be done linearly and without distortion. The following sections summarize the important characteristics of EMG instrumentation (for fuller information, see Loeb and Gans, 1986).

7.2.2 EMG ELECTRODES

These are the first link in the recording chain. Their selection and placement are of considerable importance. Those used in sports biomechanics are usually one of three types, the drawbacks and advantages of which are summarized below.

Passive surface electrodes are placed on the skin and are up to 1 cm in diameter. They are usually silver–silver chloride and are used with conducting gels that contain chloride. Passive surface electrodes are convenient, readily available, require little operator training and cause little or no discomfort. However, it is often necessary to reduce the contact resistance between the skin and electrode, which is relatively time-consuming. Electrode gel is used to avoid poor electrical contact and pressure may need to be applied by the use of adhesive strips. However, applying pressure to an electrode when it is in contact with the skin can cause an artifact voltage indistinguishable from the signal. This can be removed in a similar way to cable artifacts (section 7.2.3).

These electrodes have several limitations (e.g. Grieve, 1975). They are not ideal for fine movement and are mainly used for fairly large groups of muscles or for global pick-up. The separation distance between the detection electrodes is important – the closer they are, the more localized the pick-up. They are not suitable for deep muscles and provide an average measure of the activity of superficial muscles. They do not respond solely to the underlying muscle, but also to neighbouring ones, including deep muscles. The term **cross-talk** is used to describe the interference of EMG signals from muscles other than the ones under the electrode (Basmajian and De Luca, 1985). For example, surface electrodes used to detect activity from triceps brachii might pick up signals from the deltoid during strong abduction at the shoulder. Cross-talk is usually only a serious problem at low or zero activation levels and can be minimized by careful choice of electrode location.

Indwelling (fine wire) electrodes are necessary for recording activity in deep muscles. Fine wire electrodes are the diameter of human hair. They consist of a pair of twisted alloy wires, which are insulated except for the tips. They are inserted with a hypodermic needle after being dry-sterilized (1 h at 1300°C). The hypodermic needle is withdrawn, leaving the wires in place. They are taped at entry to the skin and easily removed after use. The size of the wires (about 0.025 mm diameter) makes them relatively painless. Problems exist for the use of such electrodes in sport. In addition, there are difficulties in locating deep muscles.

There are several limitations to the use of these electrodes (e.g. Basmajian and De Luca, 1985). The signal recorded is a function of the length of the exposed tips. The wires tend to deform or even to fracture – the latter may be undetectable but will affect the signal. They have a higher electrode resistance than surface electrodes, which requires an amplifier with a higher input impedance (see below). There will be some damage to adjacent muscle fibres, which could lead to the recording of abnormal membrane and action potentials. There is no evidence that they represent muscle activity any better than surface electrodes. Finally, there are ethical issues relating to breakage, possible risks of infection and the use of X-rays to locate the electrodes in the muscle. They are not commonly used in sports biomechanics studies in the UK.

Active surface electrodes have now been in use for a decade or so, but have only recently become available commercially. Bipolar active electrodes usually have a standardized electrode spacing of 1 cm (Figure 7.3). They can overcome certain electrical problems (see Basmajian and De Luca, 1985) and a major advantage of such electrodes is that they require no skin preparation or electrode gels. Disadvantages are the need for a

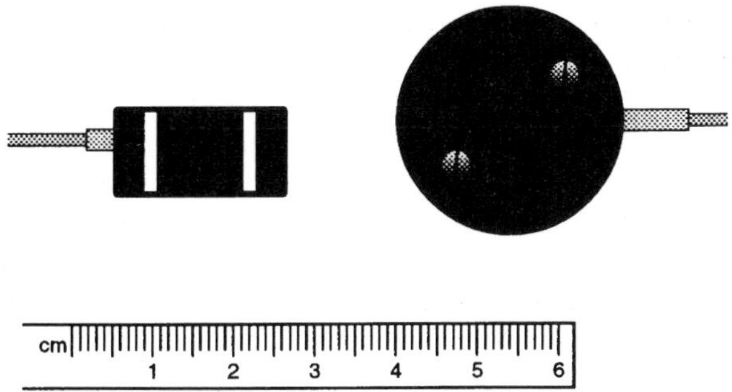

Figure 7.3 EMG electrodes: bipolar configurations of active surface electrodes

power supply to the electrodes, hence the term 'active', which might pose health and safety problems. They also increase the overall noise level in the EMG amplification chain (De Luca and Knaflitz, 1990). It is probable that their advantages outweigh their disadvantages and that they will become increasingly used in sports biomechanics.

In summary, it is worth noting that Grieve (1975), comparing indwelling and passive surface electrodes, stated 'it may be that surface electrodes are not only safer, easier to use and more acceptable to the subject but, for superficial muscles at least, provide a degree of quantitative repeatability that compares favourably with wire electrodes'.

7.2.3 CABLES

Electrical cables are needed to connect the EMG electrodes to the amplifier or pre-amplifier. These can cause problems, known as cable or movement artifacts, which have frequencies in the range 0–10 Hz. The use of high-pass filtering, high-quality electrically shielded cables and careful taping to reduce cable movement can minimize these. The effect of high-pass filtering at 10 Hz is shown in Figure 7.4, where Figures 7.4(a) and 7.4(b) are the unfiltered and filtered signals respectively.

Cable artifacts can be reduced by using pre-amplifiers mounted on the skin near the detection electrodes. The use of active surface electrodes (see above) can virtually eliminate cable artifacts (De Luca and Knaflitz, 1990).

7.2.4 EMG AMPLIFIERS

These are the heart of an EMG recording system. They should provide linear amplification over the whole frequency (Figure 7.2) and voltage range of the EMG signal. Noise must be minimized and interference

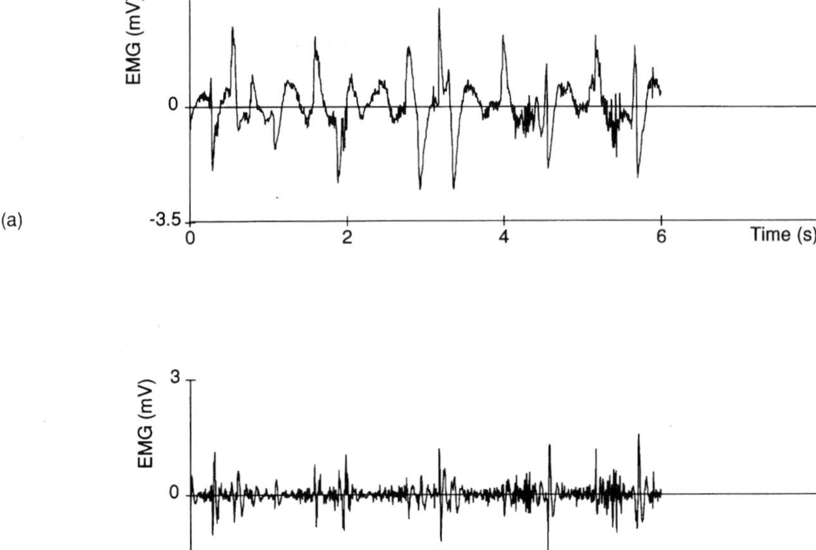

Figure 7.4 Effect of high-pass (10 Hz) filter on cable artifacts: (a) before; (b) after filtering.

from the electrical mains supply (mains hum) must be removed as far as possible. The input signal will be around 5 mV with surface electrodes or 10 mV with indwelling electrodes (Winter, 1990). The most important amplifier characteristics are as follows.

Gain

This is the ratio of output voltage to input voltage and should be high, ideally variable in the range 100–10 000 to suit a variety of recording devices.

Input impedance

The importance of obtaining a low skin resistance can be minimized by using an amplifier with a high input impedance, i.e. a high resistance to the EMG (AC) signal. This impedance should be at least 100 times the skin resistance to avoid attenuation of the input signal. This can be seen from the simplified Equation 7.1, which expresses the relationship between the detected signal (e_{emg}) the amplifier input signal (e_i) and the amplifier input impedance (R_I) and the two skin plus cable resistances (R_1 and R_2).

$$e_i/e_{emg} = R_I/(R_I + R_1 + R_2) \qquad (7.1)$$

Minimum input impedances of 1 MΩ for passive surface electrodes and 5 MΩ for indwelling electrodes have been recommended (Winter, 1990). Input impedances for high performance amplifiers can be as high as 10 GΩ (i.e. 10^{10} Ω).

Frequency response

The ability of the amplifier to reproduce the range of frequencies in the signal is known as its frequency response. The required frequency response depends upon the frequencies contained in the EMG signal (see Figures 7.2 and 7.9). This is comparable with the requirement for audio amplifiers to reproduce the range of frequencies in the audible spectrum. Typical values of EMG frequency bandwidth are 10–1000 Hz for surface electrodes and 20–2000 Hz for indwelling electrodes (Winter, 1990). Most modern EMG amplifiers easily meet such bandwidth requirements. Most of the EMG signal is in the range 20–200 Hz, which unfortunately also contains mains hum at 50 Hz.

Common mode rejection

Use of a single electrode would result in the generation of a two phase depolarization–repolarization wave (Figure 7.5(a)).

It would also contain common mode mains hum. At approximately 100 mV, the hum is considerably larger than the EMG signal. This is overcome by recording the difference in potential between two adjacent electrodes using a differential amplifier. The hum is then largely eliminated as it is picked up commonly at each electrode because the body acts as an aerial – hence the name 'common mode'. The differential wave becomes three-phase (triphasic) in nature (Figure 7.5(b)). The smaller the electrode spacing, the more closely does the triphasic wave approximate to a time derivative of the single electrode wave. In practice, perfect elimination of mains hum is not possible and the success of its removal is expressed by the **common mode rejection ratio** (CMRR). This should be 10 000 or greater.

The overall system CMRR can be reduced to a figure lower than that of the amplifier by any substantial difference between the two skin plus cable resistances. The effective system common mode rejection ratio in this case can be reduced to a value of $R_I/(R_1 - R_2)$. Cables longer than a metre often exacerbate this problem. For passive electrodes, the attachment of the pre-amplifier to the skin near the electrode site reduces noise pick-up and minimizes any degradation of the CMRR arising from differences between the cable resistances.

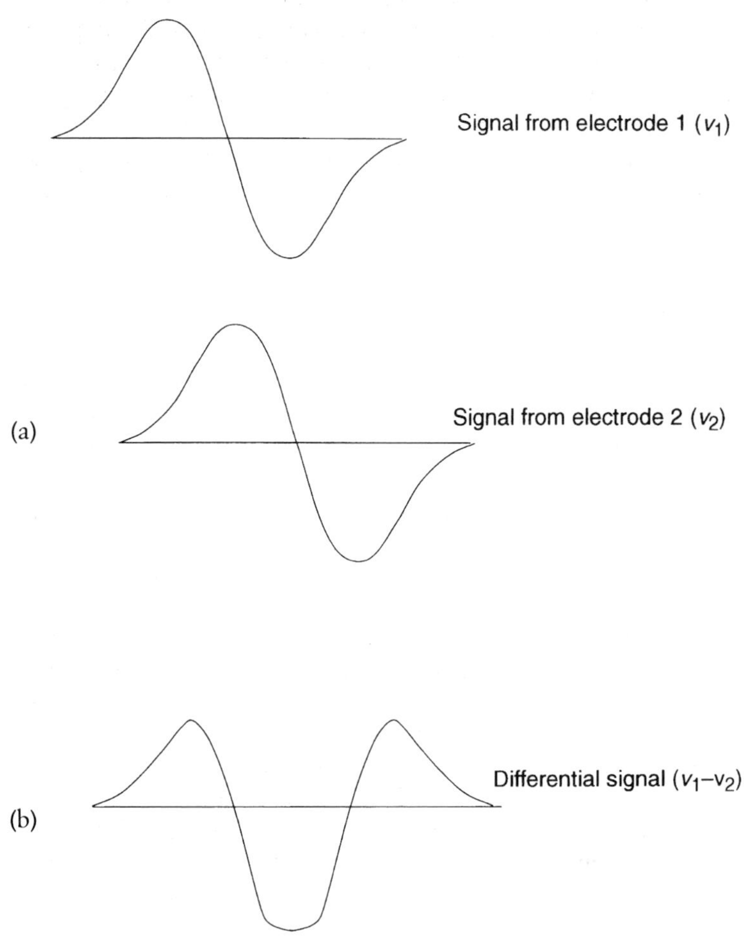

Figure 7.5 EMG signals with no mains hum: (a) biphasic MUAPs recorded from two monopolar electrodes; (b) triphasic differential signal from using the two electrodes in a bipolar configuration.

7.2.5 RECORDERS

A wide variety of devices has been used to record the amplified EMG signal. FM tape recorders are still ideal, permitting, as they do, storage for later processing. Use of UV recorders (oscillographs) permits frequencies of up to 2 kHz to be recorded. Radio telemetry has often had to be used to transmit EMG signals during sporting activity, to avoid excessive cabling. However, large-memory, portable, digital data loggers are now increasingly being used to store EMG signals prior to downloading the data to a computer, and these provide an attractive alternative to telemetry. Currently analog-to-digital conversion and computer processing are by far the most commonly used recording methods. High sampling rates are needed to successfully reproduce the signal in digital form, and telemetered systems do not always provide a sufficiently high sampling rate. Signal aliasing (Chapter 5) will occur if the sampling rate is less than twice the upper frequency limit in the power spectrum of the sampled signal.

7.2.6 EXPERIMENTAL PROCEDURES

It is often considered that the results of an electromyographic investigation are only as good as the preparation of the electrode attachment sites. ISEK (1980) suggested that skin resistance ceases to be a problem if high-performance amplifiers are used and this can also be the case when active electrodes are used. However, as many EMG studies in sports biomechanics currently use passive surface electrodes and amplifiers with input impedances lower than those of high-performance amplifiers, good surface preparation is still beneficial.

The location of the electrodes is the first consideration. It is now generally agreed that, if it is possible to determine the location of the motor end-plate (by an electrical stimulator), then this is definitely not the place to put the electrodes (De Luca and Knaflitz, 1990) contrary to earlier wisdom (e.g. ISEK, 1980). Readers without an electrical stimulator may find the electrode placement positions proposed by Zipp (1982) and Cram (1990) to be useful. However Clarys and Cabri (1993) warned that the placement positions proposed by Zipp (1982) were only valid for isometric contractions. They advised instead that electrodes be placed over the mid-point of the muscle belly, an easily implemented recommendation.

The two detector electrodes should lie at the appropriate place along the muscle belly with the orientation of the electrode pair being on a line parallel to the direction of the muscle fibres (for example Figure 7.6).

If the muscle fibres are neither linear nor have a parallel arrangement, the line between the two electrodes should point to the origin and insertion of the muscle for consistency (Clarys and Cabri, 1993). The separation of the electrodes determines the degree of localization of the signal picked up. Basmajian and De Luca (1985) recommended a standard electrode separation of 1 cm.

The following might be accepted as good experimental practice (adapted from Bartlett, 1992) for investigators not using high-performance amplifiers. The area of the skin on which the electrode is to be placed is firstly shaved and then thoroughly cleaned and degreased using a suitable soap or soap solution. The following procedure (Okamoto et

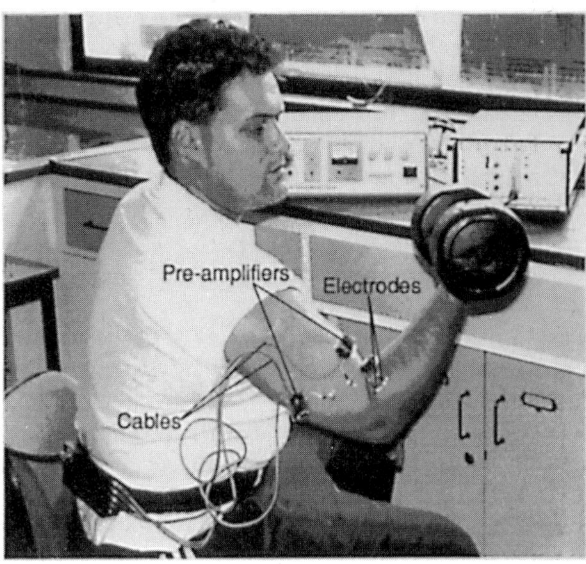

Figure 7.6 EMG system with passive surface electrodes on subject.

al., 1987) is then strongly recommended in preference to earlier techniques of skin abrasion, although even this procedure may require ethics clearance at some institutions. It is easily mastered and is remarkably effective in reducing skin resistance. A sterile lancet should be used to gently mark two lines on the skin in the direction of the line joining the proposed sites of the electrodes. The angle and pressure of the lancet should be carefully adjusted so that only the superficial layer of dead skin is broken. The electrodes are then placed on the faint red scratch lines about 3 cm apart. An earth or ground electrode is also needed. For most EMG systems, the earth electrode can be common for all the active sites or separate for each muscle. In the latter case, it should be placed close to the detector electrodes, preferably over an inactive site.

After the electrode sticker and sterilized electrode have been attached, and electrode gel has been injected into the electrode, any excess gel should be removed. The skin resistance between the two active electrodes should then be checked. In the absence of a device for measuring skin impedance across the signal frequency range, a simple DC ohmmeter reading of skin resistance should be taken. The resistance should be less than 10 kΩ and preferably less than 5 kΩ. If the resistance exceeds the former value, the electrodes should be removed and the preparation repeated. This should rarely be necessary if the procedure of Okamoto *et al.* (1987) is used.

The importance of maintaining a sterile preparation site, using only sterile lancets, electrodes and other equipment, sterilizing or disposing of the electrodes after use and disposing safely of the lancets after use cannot be overstressed (Putnam *et al.* (1992) gives details of electrode sterilization procedures).

7.3 Data processing

Much kinesiological, physiological and neurophysiological electromyography simply uses and analyses the raw electromyogram (for example, see Clarys and Cabri, 1993). The amplitude of this signal, and all other EMG data, should always be related back to the signal generated at the electrodes, not that after amplification. Further EMG signal processing is often performed in sports biomechanics in an attempt to make comparisons between studies. It can also assist in correlating the EMG signal with mechanical actions of the muscles or other biological signals.

While EMG signal processing can provide additional information to that contained in the raw signal, care is needed for several reasons. Firstly, in order to distinguish between artifacts and signal, it is essential that good recordings free from artifacts are obtained. Secondly, the repeatability of the results needs consideration. Grieve (1975) warned of non-exact repetition of EMGs even from a stereotyped activity such as treadmill running. The siting of electrodes, skin preparation and other factors can all affect the results. Even the activity or inactivity of one motor unit near the pick-up site can noticeably change the signal. Thirdly, such experimental factors make it difficult to compare EMG results with those of other studies. However, normalization has been developed to facilitate such comparisons (e.g. Clarys, 1982). This involves the expression of the amplitude of the EMG signal as a ratio to the amplitude of a contraction deemed to be maximal, usually a maximum voluntary contraction (MVC), from the same site. No consensus at present exists as to how to elicit an MVC and it is not always an appropriate maximum (Clarys and Cabri, 1993). Problems associated with the use of MVC as a valid and reliable criterion of maximal force for normalization were summarized by Enoka and Fuglevand (1993). These include the standardization in an MVC of the neural control of muscle coordination and the mechanical factors of joint angle and its rate of change, which could confound interpretation of the results. Furthermore, reported discharge rates of 20–40 Hz and the unsustainable activity in some high-threshold motor units suggest caution in interpreting motor unit activity in the MVC as maximal.

7.3.1 TEMPORAL PROCESSING AND AMPLITUDE ANALYSIS (TIME DOMAIN ANALYSIS)

Temporal processing relates to the amplitude of the signal content or the 'amount of activity'. Such quantification is usually preceded by full-wave rectification as the raw EMG signal $e(t)$ (Figure 7.7(a)) would have an integral of approximately zero, because of its positive and negative deviations.

Full-wave rectification (Figure 7.7(b)) simply involves making negative values positive and the rectified signal is expressed as the modulus of the raw EMG $|e(t)|$. Various terms are used to express EMG amplitude (e.g. Winter, 1990; Bartlett, 1992).

Average rectified EMG (AREMG)

This is the average value of the full-wave rectified EMG over a specified time interval (from t_1 to t_2), such as the duration of the contraction or one running stride:

$$\text{AREMG} = \frac{1}{(t_2 - t_1)} \int_{t_1}^{t_2} |e(t)| \; \mathrm{d}t \tag{7.2}$$

The AREMG is very closely related to integrated EMG (see below). A continuous or moving average may be used. These correspond, respectively, to integrated EMG performed continuously or by resetting after a selected time interval (see below) (ISEK, 1980). A moving average is often used to show the time course of the EMG signal. It is most commonly performed as a window average, where the window is the time span over which the average is taken. Like other time domain descriptors of the electromyogram, the AREMG is affected by: the number of active motor units; the firing rates of motor units; the amount of signal cancellation by superposition; and the waveform of the MUAP. The last of these depends upon electrode position, muscle fibre conduction velocity, the geometry of the detecting electrode surfaces and the detection volume (De Luca and Knaflitz, 1990).

Root mean square EMG (RMSEMG)

This is the square root of the average power (voltage squared) of the signal in a given time. It may be calculated as a continuous or moving value, as above:

$$\text{RMSEMG} = \left[\frac{1}{(t_2 - t_1)} \int_{t_1}^{t_2} e^2(t) \; \mathrm{d}t \right]^{1/2} \tag{7.3}$$

De Luca and Knaflitz (1990) regard this as the preferred temporal descriptor of the EMG signal. It is considered to provide a measure of the number of recruited motor units during voluntary contractions where there is little correlation among motor units (see also Basmajian and De Luca, 1985).

Integrated EMG (IEMG or FWRI)

This is the area under the rectified EMG signal. As for any definite integral, it is expressed as follows, where t_1 and t_2 are the times at the start and end of the integration period.

$$\text{IEMG} = \int_{t_1}^{t_2} |e(t)|\ \mathrm{d}t \tag{7.4}$$

Units are obviously mV·s (often erroneously reported in the literature). Basmajian and De Luca (1985) question the use of the term 'integrated EMG', largely because of its dubious historical provenance. With the linear envelope, it is probably the most widely used EMG time domain descriptor in sports biomechanics. It is closely related to the process of obtaining the AREMG. The time interval $(t_2 - t_1)$ over which the integration is performed is usually one of the following (from Bartlett, 1992):

- **Total activity time**: continuous integration during a contraction so that the integral equals IEMG (Figure 7.7(c)). An average value of the EMG can then be obtained by dividing the total integral by the time of integration. This will give a value identical to the AREMG for the same time period.
- **Resetting at a preselected time**: integrating over a preset time interval (typically 50–200 ms). A series of peaks is obtained (Figure 7.7(d)), the sum of which equals the total IEMG. The trend of the peaks is similar to the result obtained with a linear envelope detector. The choice of the integration period is important in order to follow fluctuations but avoid noise.
- **Voltage level reset**: integrating until a pre-set voltage is reached (Figure 7.7(e)). The IEMG is then equal to the total number of resets multiplied by the voltage reset level. Thus the frequency of the reset pulses is a measure of the intensity of the muscle contraction. This supposedly bears a similarity to the neural action potential rate (Winter, 1990).

Linear envelope

In the past, this was often misleadingly referred to as integrated EMG. It is obtained by the use of a linear envelope detector, which comprises a full-wave rectifier plus a low-pass filter. This is a simple method of quantifying signal intensity and gives an output (Figure 7.7(f)) which, it is claimed, follows the trend of the muscle tension curve (e.g. Winter,

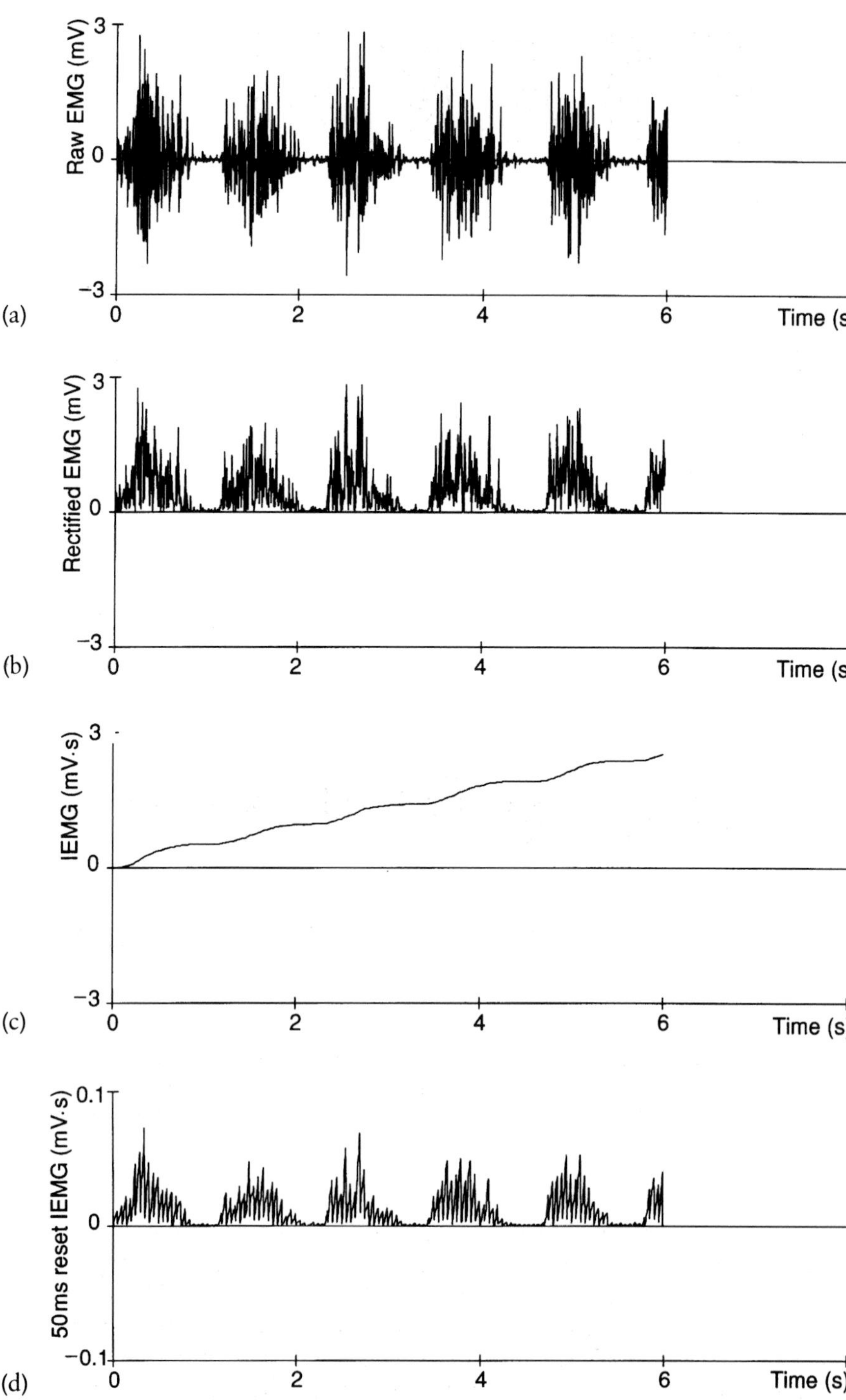
Raw EMG (mV)
3
0
−3
(a)
0
2
4
6
Time (s)
Rectified EMG (mV)
3
0
−3
(b)
0
2
4
6
Time (s)
IEMG (mV·s)
3
0
−3
(c)
0
2
4
6
Time (s)
50 ms reset IEMG (mV·s)
0.1
0
−0.1
(d)
0
2
4
6
Time (s)

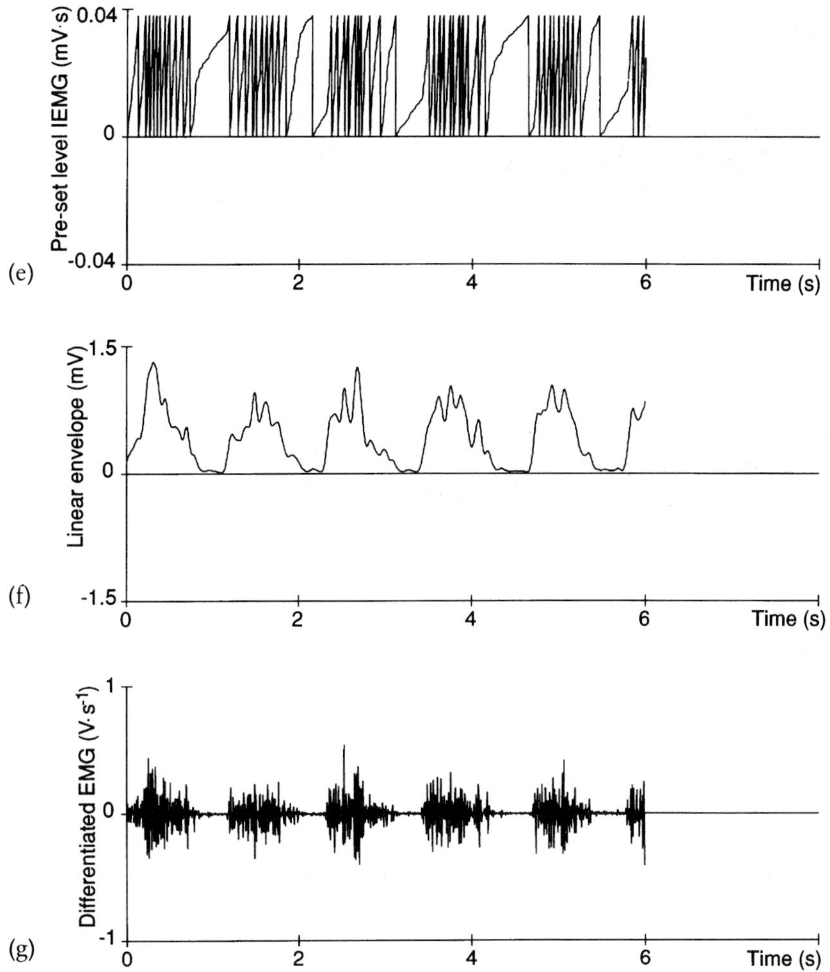

Figure 7.7 Time domain processing of EMG: (a) raw signal; (b) full wave rectified EMG; (c) continuously integrated EMG; (d) integrated EMG resetting after 50 ms; (e) integrated EMG resetting at fixed voltage level; (f) linear envelope using 6 Hz low-pass filter; (g) differentiated EMG.

1990) with no high-frequency components. The choice of cut-off frequency is crucial, and is usually 6 Hz or below, and a second- or higher-order low-pass filter (Chapter 5) is commonly used. Grieve (1975) questioned the effect of this method on the temporal features which are an important part of the EMG description. It is perhaps, at first sight, surprising to find the method so widely used, given the frequency bandwidth of the EMG signal. However, it does appear to isolate the amplitude from the frequency content of the signal.

Differentiated EMG

This is the rate of change of the EMG signal with respect to time and was proposed by Van Leemputte and Willems (1987). It has not received widespread recognition despite the authors' claim that it is better able to predict static muscle torques than the integrated EMG. It has been found to aid analysis of the temporal sequencing of the EMG signal (Figure 7.7(g)). The shape of this is seen to be not dissimilar to that of the raw EMG signal (Figure 7.7(a)) but somewhat more spiky.

7.3.2 FREQUENCY DOMAIN ANALYSIS

Unless the data are explicitly time-limited, as in a single action potential, the EMG signal can be transformed into the frequency domain (as for kinematic data in Chapter 5). The EMG signal is then usually presented as a power spectrum (power equals amplitude squared) at a series of discrete frequencies although it is often represented as a continuous curve. Several measures have been used to define the shape of the power spectrum.

The following so-called discrete measures are often used (Figure 7.8). The **bandwidth** (or half power bandwidth) is the difference between the upper and lower **half power frequencies** (f_a and f_b), that is the frequencies at which power has fallen by half its maximum value. This expresses the spread of the power spectrum. The **centre frequency** is the geometric mean of the upper and lower half power frequencies. This gives a measure of the central tendency of the spectrum. Other discrete measures have also been used (ISEK, 1980). A problem with these discrete measures is that real EMG power spectra (e.g. Figure 7.9) usually contain some noise, unlike the idealized spectrum (Figure 7.8). Determination of some of these discrete measures for a noisy signal can be difficult.

The central tendency and spread of the power spectrum can also be expressed by the use of statistical parameters, which depend upon the distribution of the signal power over its constituent frequencies. Two statistical parameters are often used to express the central tendency of the spectrum. These are comparable to the familiar mean and median in statistics. For a spectrum of *k* discrete frequencies, the **mean frequency** (f_{av}) is obtained by dividing the sum of the products of the power at each frequency $W(f_i)$ and the frequency (f_i) by the sum of all of the powers, $W(f_i)$. That is:

$$fav = (\Sigma\, f_i\, W(f_i))/(\Sigma\, W(f_i)) \tag{7.5}$$

The **median frequency** (f_m) is the frequency that divides the spectrum into two parts of equal power (i.e. the areas under the power spectrum to the two sides of the median frequency are equal). That is (with *fmax* as the maximum frequency in the spectrum):

$$\sum_{0}^{f_m} W(f_i) = \sum_{f_m}^{f_{max}} W(f_i) \qquad (7.6)$$

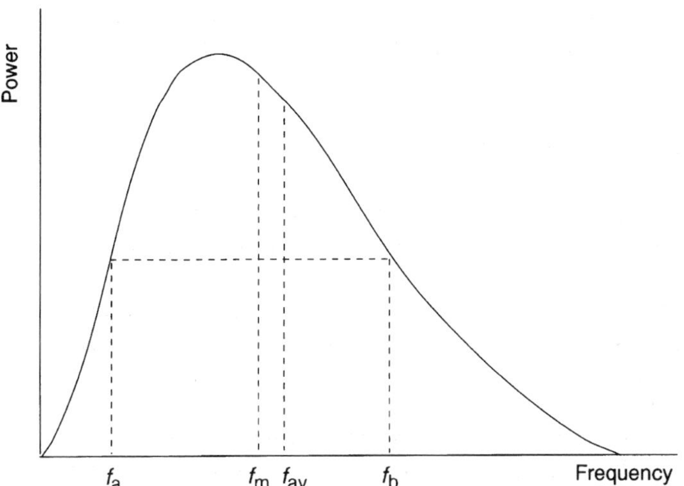

Figure 7.8 Idealized EMG power spectrum.

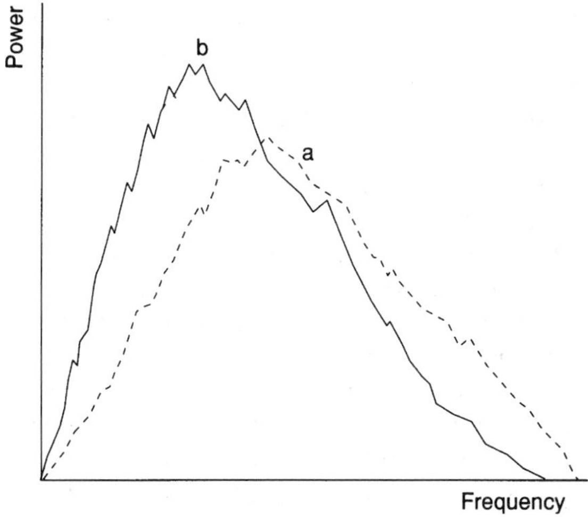

Figure 7.9 EMG power spectra at the start (a) and the end (b) of a sustained, constant force contraction.

The median is less sensitive to noise than is the mean (De Luca and Knaflitz, 1990). The spread of the power spectrum is expressed by the **statistical bandwidth**, which is calculated in the same way as a standard deviation.

The EMG power spectrum can be used, for example, to indicate the onset of muscle fatigue. This is accompanied by a noticeable shift in the power spectrum towards lower frequencies (Figure 7.9) and a reduction in the median (and mean) frequencies. This frequency shift is caused by an increase in the duration of the motor unit action potential. This results either from a lowering of the conduction velocity of all the action potentials or through faster, higher frequency motor units switching off while slower, lower frequency ones remain active.

7.4 EMG and muscle tension

Obtaining a predictive relationship between muscle tension and the electromyogram could solve the problem of muscle redundancy (or muscle indeterminacy). This problem arises because the equations of motion at a joint cannot be solved as the number of unknown muscle forces exceeds the number of equations available. If a solution could be found to this problem, it would allow the calculation of forces in soft tissue structures and between bones. This problem is, arguably, the major issue that confronts biomechanics (see e.g. Norman, 1989). It is not surprising, therefore, that the relationship between the EMG signal and the tension developed by a muscle has attracted the attention of many researchers. Because of the importance of this relationship, this section will briefly summarize some of the research findings in this area.

The electromyogram provides a measure of the level of excitation of a muscle. Therefore, if the force in the muscle depends directly upon its excitation, a relationship should be expected between this muscle tension and suitably quantified EMG. A muscle's tension is regulated by varying the number and the firing rate of the active fibres and the amplitude of the EMG signal depends on the same two factors. It is therefore natural to speculate that a relationship does exist between EMG and muscle tension (De Luca and Knaflitz, 1990). It might further be expected that this relationship would only apply to the active state of the contractile elements, and that the contributions to muscle tension made by the series and parallel elastic elements would not be contained in the EMG.

7.4.1 ISOMETRIC CONTRACTIONS

Lippold (1952) found a linear relationship (Figure 7.10) between IEMG and load up to a maximum voluntary contraction (MVC) for the plantar flexors of the ankle.

Many later studies have supported Lippold's findings. For example Stephens and Taylor (1973) found a linear relationship between mean rectified EMG and the force in the first dorsal interosseus muscle.

Other investigators have found a non-linear or quadratic relationship (Figure 7.10) between IEMG and muscle tension, for example Vredenbregt and Rau (1973) for biceps brachii. Many attempts have been made to resolve the discrepancies between linear and non-linear isometric EMG–tension relationships. The possible causes of the discrepancies can be summarized from the comments of Bouisset (1973), Basmajian and De Luca (1985) and De Luca and Knaflitz (1990), which also highlight other important aspects of the EMG signal, as follows:

- technical factors, such as bandwidths of amplifiers, integrators, transducers and recorders; recording method; electrode spacing and position; duration of integration period; ranges of force; joint position; treatment of results;
- the spread of activity from neighbouring muscles and variation in the pattern of activity between agonists; simultaneous antagonist activity;
- the type of muscle: its elastic characteristics; proportion of different types of motor unit (Basmajian and De Luca (1985) considered fibre type not to be an important factor);
- modalities of contraction, for example the level of motor unit synchronization; with increasing tension beyond the level of a newly recruited motor unit, the latter's firing rate, and hence its contribu-

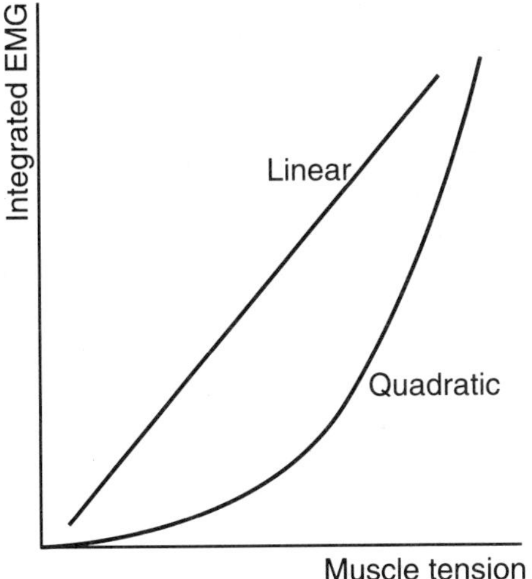

Figure 7.10 Schematic representation of linear and quadratic relationships between EMG and muscle tension in isometric contractions.

tion to the EMG amplitude, will increase while its contribution to muscle tension saturates;

- individual factors relating to age, strength, endurance (Basmajian and De Luca (1985) did not consider training to be a factor);
- the difference between the muscle volume and electrode detection volume; this leads to a contribution to the EMG from a newly recruited motor unit within the detection volume that is not linearly proportional to its contribution to the muscle tension, an effect that depends on the muscle's size;
- the control strategy used by the central nervous system varies between muscles: while the first dorsal interosseus recruits all of its motor units below 50% MVC and has a wide range of firing rates, larger muscles have rapidly increasing and saturating firing rates and recruit motor units across the whole muscle tension range.

7.4.2 NON-ISOMETRIC CONTRACTIONS

Bigland and Lippold (1954) reported a linear relationship between IEMG and Achilles tendon force at a slow constant speed. They also found a linear relationship between speed and IEMG when force was constant but found little change of IEMG with speed for lengthening contractions. Their results were supported by the work of Komi (1973) who found a linear relationship between IEMG and force at each of five contraction speeds for biceps brachii (–45 to +45 $mm \cdot s^{-1}$). He also found that maximal contractions produced constant IEMGs, strongly suggesting that EMG activity relates only to the state of the contractile element of the muscle. This is not generally the same as the tension transmitted to the muscle tendon.

There are difficulties in applying the results of EMG studies of isometric or very slow contractions to dynamic, voluntary movements which typically last only a few tenths of a second, involve less than ten impulses for any given motor unit, and have different patterns of firing than during sustained contractions. Hof and van den Berg (1981(a)–(d)) assumed a linear relationship between mean rectified EMG and muscle force for isometric or quasi-static contractions. For dynamic muscle contractions, their processing of EMG to muscle tension requires that the mean rectified EMG is an adequate measure of muscle activation and that the muscle length is known. The latter can of course be found from cine film or some other technique (Chapter 5) and from appropriate anatomical data. Their processing method also requires a physiological (mathematical) model of the muscle, so that the muscle tension can be computed from the EMG and muscle length data. For this they used the model of Hill (1938), which forms the basis of most muscle models, incorporating the behaviour of the contractile, parallel and series elastic elements (Chapter 1). They reported reasonable agreement between the plantar flexor force predicted by the model compared with values calculated from force platform and visual recordings for walking (see also Hof, 1984). At least part of the discrepancy was caused by the kinetic computations in which the tibialis anterior activity was not accounted for. This caused greater discrepancies between the two estimates of plantar flexor torque in fast walking, where the tibialis anterior activity was more pronounced than in slow walking (Hof, Pronk and van Best, 1987).

The discrepancies that exist between research studies even for isometric contractions should not lead one to expect a simple relationship between EMG and muscle tension for the fast, voluntary contractions that are characteristic of sports movements. For such movements, the relationship between EMG and muscle tension still remains elusive although the search for it continues to be worthwhile.

7.5 Summary

In this chapter the use of electromyography in the study of muscle activity in sports biomechanics was considered, including the equipment and methods used, the processing of EMG data and the important relationship between EMG and muscle tension. Consideration was given to why the EMG signal is important in sports biomechanics and why the recorded EMG differs from the physiological EMG signal. Also covered were the advantages and limitations of the three types of EMG electrodes suitable for use in sports biomechanics, the main characteristics of an EMG amplifier, and other EMG equipment. The processing of the raw EMG signal was considered in terms of its time domain descriptors and the EMG power spectrum and the measures used to define it. Finally, the chapter overviewed the research that has been conducted into the relationship between EMG and muscle tension, and provided an understanding of the importance of this relationship.

7.6 Exercises

1. Explain clearly how the physiological EMG signal is generated.
2. List the factors that cause the recorded electromyogram to differ from the physiological signal.
3. How does the information contained in each of the time domain processed EMG signals differ from that in the raw EMG? What additional information might this provide for the sports biomechanist and what information might be lost?
4. Outline the uses of the EMG power spectrum and the applications and limitations of the various measures used to describe it.

5. List the advantages and disadvantages of the three types of EMG electrode suitable for use in sports biomechanics. Explain, briefly, which type of electrode you would think most suitable to:
   a) record the muscle activity in the pectoralis major of a swimmer;
   b) record the muscle activity in the vastus intermedius during treadmill running.
6. Explain the main characteristics required of an EMG amplifier. If you have access to an EMG amplifier, obtain its technical specification, and ascertain whether this conforms to the recommendations of section 7.2.4.
7. You are to conduct an experiment in which surface electrodes will be used to record muscle activity from biceps brachii, triceps brachii, rectus femoris and biceps femoris. Using a living subject to identify the muscles, the recommendations of Clarys and Cabri (1993) regarding the placement of the detecting electrodes and other relevant information from section 7.2.6, mark the sites at which you would place the detecting and ground electrodes for each of those muscles.
8. If you have access to EMG equipment, and with appropriate supervision if necessary, perform the preparation, electrode siting etc. from the previous exercise according to the recommendations of section 7.2.6. Then carry out experiments to record EMGs as follows:
   a) From biceps brachii and triceps brachii during an isometric curl, the raising and lowering phases of a biceps curl with a dumbbell, a press-up, and chin-ups with wide and narrow grips and with overhand and underhand grips. Check that you are obtaining good results and repeat the preparation if not. Comment on the results you obtain.
   b) From the rectus femoris and biceps femoris during rising from and lowering on to a chair. Again, check that you are obtaining good results, and repeat the preparation if not. Explain the apparently paradoxical nature of the results (this is known as Lombard's paradox).
9. If you have access to EMG equipment, then carry out the following experiment. Prepare and locate electrodes to record activity in biceps brachii. Then have your subject hold a heavy weight in the hand for as long as possible. During this time, record the EMG signal either continuously or for 15 s periods every minute. In either case, integrate the signal for 15 s periods each minute. Plot the resulting integrated EMG against time and comment on the results. If your EMG system allows it and the sampling frequency is 1 kHz or above, obtain the EMG power spectrum and median frequency of each 15 s sample. Plot the median frequency against time. Comment on the results.
10. Summarize the current state of knowledge of the relationship between EMG and muscle tension. If you have access to EMG equipment, perform a simple experiment to investigate the relationship

between EMG and muscle tension. This could involve holding increasing loads in the hand with the elbow flexed at 90°. EMGs could be recorded from biceps brachii for each load and continuous integration performed over a 10 s period. The value of the integral could then be plotted against the load (making allowances for the weight of the forearm plus hand).

7.7 References

Bartlett, R. M. (1992) Electromyography, in *Biomechanical Analysis of Performance in Sport*, (ed. R. M. Bartlett), British Association of Sports Sciences, Leeds, pp.28–37.

Basmajian J. V. and De Luca, C. J. (1985) *Muscles Alive: Their Functions Revealed by Electromyography*, Williams & Wilkins, Baltimore, MD.

Bigland, B. and Lippold, O. C. J. (1954) The relation between force, velocity and integrated electrical activity in human muscles. *Journal of Physiology*, **123**, 214–224.

Bouisset, S. (1973) EMG and muscle force in normal motor activities, in *New Developments in Electromyography and Clinical Neurophysiology*, (ed. J. E. Desmedt), S. Karger, Basel, pp.547–583.

Clarys, J. P. (1982) A review of EMG in swimming: explanation of facts and/or feedback information, in *Biomechanics and Medicine in Swimming*, (ed. A. P. Hollander), Human Kinetics, Champaign, IL, pp.123–135.

Clarys, J. P. and Cabri, J. (1993) Electromyography and the study of sports movements: a review. *Journal of Sports Sciences*, **11**, 379–448.

Cram, J. R. (ed.) (1990) *Clinical EMG for Surface Recordings*, vol. 2, Clinical Resources, Nevada City, CA.

De Luca, C. J. and Knaflitz, M. (1990) *Surface Electromyography: What's New?*, Neuromuscular Research Centre, Boston, MA.

Enoka, R. M. (1994) *Neuromechanical Basis of Kinesiology*, Human Kinetics, Champaign, IL.

Enoka, R. M. and Fuglevand, A. J. (1993) Neuromuscular basis of the maximum voluntary force capacity of muscle, in *Current Issues in Biomechanics*, (ed. M. D. Grabiner), Human Kinetics, Champaign, IL, pp.215–235.

Grieve, D. W. (1975) Electromyography, in *Techniques for the Analysis of Human Movement*, (eds D. W. Grieve, D. I. Miller, D. Mitchelson *et al.*), Lepus Books, London, pp.109–149.

Hill, A. V. (1938) The heat of shortening and the dynamic constants of muscle. *Proceedings of the Royal Society of London, Series B*, **126**, 136–195.

Hof, A. L. (1984) EMG and muscle force: an introduction. *Human Movement Science*, **3**, 119–153.

Hof, A. L. and van den Berg, J. W. (1981a) EMG to force processing I: an electrical analogue of the Hill muscle model. *Journal of Biomechanics*, **14**, 747–758.

Hof, A. L. and van den Berg, J. W. (1981b) EMG to force processing II: estimation of parameters of the Hill muscle model for the human triceps surae by means of a calf ergometer. *Journal of Biomechanics*, **14**, 759–770.

Hof, A. L. and van den Berg, J. W. (1981c) EMG to force processing III: estimation of muscle parameters for the human triceps surae muscle and assessment of the accuracy by means of a torque plate. *Journal of Biomechanics*, **14**, 771–780.

Hof, A. L. and van den Berg, J. W. (1981d) EMG to force processing IV: eccentric–concentric contractions on a spring-flywheel set up. *Journal of Biomechanics*, **14**, 781–792.

Hof, A. L., Pronk, C. N. A. and van Best, J. A. (1987). Comparison between EMG to force processing and kinetic analysis for the calf muscle moment in walking and stepping. *Journal of Biomechanics*, **20**, 167–178.

ISEK (1980) *Units, Terms and Standards in the Reporting of EMG Research*, International Society of Electrophysiological Kinesiology, USA.

Komi, P. (1973) Relationship between muscle tension, EMG and velocity of contraction under concentric and eccentric work, in *New Developments in EMG and Clinical Neurophysiology*, vol. 1, (ed. J. E. Desmedt), S. Karger, Basel, pp.596–606.

Lagasse, P.P. (1987) Neuromuscular considerations, in *Standardizing Biomechanical Testing in Sport*, (eds D. A. Dainty and R. W. Norman), Human Kinetics, Champaign, IL, pp.59–71.

Lippold, O. C. J. (1952) The relation between integrated action potentials in a human muscle and its isometric contraction. *Journal of Physiology*, **117**, 492–499.

Loeb, G. E. and Gans, C. (1986) *Electromyography for Experimentalists*, University of Chicago Press, Chicago, IL.

Nigg, B. M. and Herzog, W. (1994) *Biomechanics of the Musculoskeletal System*, John Wiley, Chichester.

Norman, R. W. (1989) A barrier to understanding human motion mechanisms: a commentary, in *Future Directions in Exercise and Sport Science Research*, (eds J. S. Skinner, C. B. Corbin, D. M. Landers *et al.*), Human Kinetics, Champaign, IL, pp.151–161.

Okamoto, T., Tsutsumi, H., Goto, Y. and Andrew, P. (1987) A simple procedure to attenuate artifacts in surface electrode recordings by painlessly lowering skin impedance. *Electromyography and Clinical Neurophysiology*, **27**, 173–176.

Putnam, L. E., Johnson, R. and Rath, W. T. (1992) Guidelines for reducing the risk of disease transmission in the psychophysiology laboratory. *Psychophysiology*, **29**, 127–141.

Stephens, J. A. and Taylor A. (1973) The relationship between integrated electrical activity and force in normal and fatiguing human voluntary muscle contractions, in *New Developments in Electromyography and Clinical Neurophysiology*, (ed. J. E. Desmedt), S. Karger, Basel, pp.623–627.

Van Leemputte, M. and Willems, E. J. (1987) EMG quantification and its application to the analysis of human movements, in *Current Research in Sports Biomechanics*, (eds B. Van Gheluwe and J. Atha), S. Karger, Basel, pp.177–194.

Vredenbregt, J. and Rau G. (1973) Surface electromyography in relation to force, muscle length and endurance, in *New Developments in Electromyography and Clinical Neurophysiology*, (ed. J. E. Desmedt), S. Karger, Basel, pp.607–622.

Winter, D. A. (1990) *Biomechanics and Motor Control of Human Movement*, John Wiley, New York.

Zipp, P. (1982) Recommendations for the standardization of lead positions in surface electromyography. *European Journal of Applied Physiology*, **50**, 41–54.

7.8 Further reading

Basmajian J. V. and De Luca, C. J. (1985) *Muscles Alive: Their Functions Revealed by Electromyography*, Williams & Wilkins, Baltimore, MD: a classic text in its fifth edition, although the sixth edition is now perhaps a little overdue. Chapters 12–17 provide a vivid description of the action of muscles as revealed by electromyography, and are highly recommended. Other chapters, for example, cover motor control, fatigue and posture.

Clarys, J. P and Cabri, J. (1993) Electromyography and the study of sports movements: a review. *Journal of Sports Sciences*, **11**, 379–448: this is a wonderfully comprehensive review of all matters relating to the use of EMG in the study of sports skills.

Hof, A. L. (1984) EMG and muscle force: an introduction. *Human Movement Science*, 3, 119–153: a reasonably accessible overview of this topic. Students with a weak mathematics background might find some parts a little hard going.

8 Other techniques for the analysis of sports movements

This chapter is intended to provide an understanding of some of the other techniques used in the recording and analysis of sports movements, including their advantages and limitations. After reading this chapter, you should be able to:

- understand the uses and limitations of single-plate multiple image photography for recording movement in sport;
- outline the advantages and limitations, for sports movements, of three types of opto-electronic system, and compare these with the advantages and limitations of conventional cinematography and video analysis;
- appreciate how electrogoniometry can be used to record joint motion;
- define the restrictions on the use of both accelerometry and tendon force measurements in sport;
- understand the value of contact pressure measurements in the study of sports movements, and outline the relative advantages and limitations of pressure insoles and pressure platforms and of the three types of pressure transducer used in sports biomechanics;
- appreciate how and why isokinetic dynamometry is used to record the net muscle torque at a joint.

8.1 Single-plate photography

Historically a wide range of photographic techniques has been used to record human movement (Adrian, 1973; Smith, 1975). These have included methods of recording multiple images on a single frame of film, such as stroboscopic photography, cyclography and rotating-slit shutter photography (chronocyclography). Also, techniques have been used in which the film moved past an open shutter, including streak photography and gliding cyclography and chronocyclography (Atha, 1984).

For recording linear displacements of, and angular range of movement at, a joint or joints it can be advantageous to have the information presented directly, rather than sequentially as with cine and video. A single-frame, multiple-exposure still photograph can then be useful. In such techniques, the camera shutter is left open and the light admitted to the camera is controlled in one of several ways. To obtain reasonably accurate measurements, very good contrast is required and subjects have to wear dark clothing with highly reflective markers. Although much of the equipment required for such techniques is commercially available, few come as complete systems and some parts may have to be made. The techniques are relatively inexpensive and can be made immediate if a Polaroid camera is used. The procedures used when recording movements in this way share similarities with those for cinematography (Chapter 5). A disadvantage of such photographs, for analysis purposes, can be a cluttered image. This can be minimized if essentially linear (Figure 8.1(a)) or non-repetitive rotational movements (Figure 8.1(b)) are studied, with careful planning of the number of exposures to be taken. Subdued background lighting is needed for many of these techniques, which are for laboratory rather than field use.

The images are best suited for display purposes or qualitative analysis, or for basic measurements, such as joint range of movement. With high-quality cameras and film, more accurate linear and angular measurements can be taken, and velocities have been calculated from stroboscopic photography (e.g. Clark, Paul and Dennis, 1977). The resulting images are, of course, only two-dimensional although, in principle, image coordinates from two cameras could be used to reconstruct three-dimensional coordinates. Generally these techniques are not used for detailed quantitative analysis. Two of them – rotating slit shutter and stroboscopic photography – have been used to a limited extent in sports biomechanics (e.g. Maier, 1968; Merriman, 1975).

In rotating slit shutter photography, the lens shutter is kept open while a slotted disc is rotated to control light entry. The speed of rotation of the disc is usually controlled by a variable speed electric motor with good speed stability. The speed of rotation, combined with the width of each slot (or slit) and the number of slots, controls the exposure time and effective 'frame rate'. For example, a disc with four equispaced slots each of 12° width, rotating at 50 Hz provides an exposure time of $12°/(360° \times 50\text{ Hz}) = 1/1500$ s and a 'frame rate' of $4 \times 50\text{ Hz} = 200\text{ Hz}$ (i.e. 1/200 s between frames). This technique can be used in normal lighting.

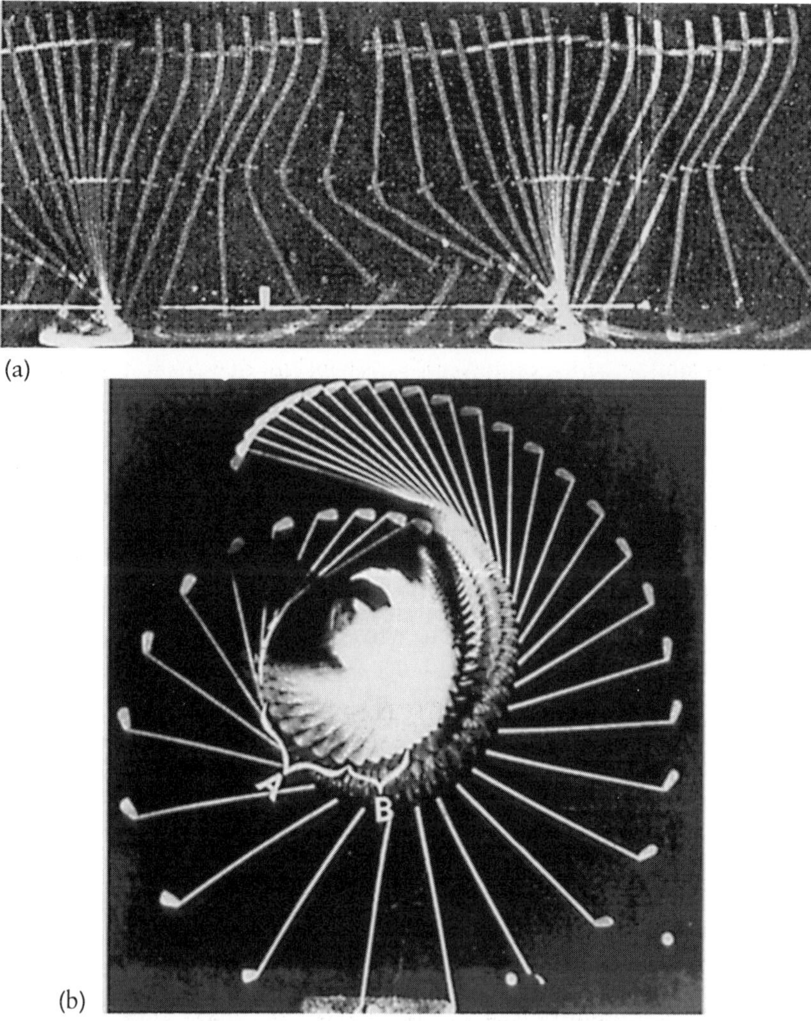

Figure 8.1 Stroboscopic photographs: (a) linear movement – walking (reproduced from Winter, 1990, with the permission of John Wiley & Sons, New York); (b) rotational movement – golf swing (reproduced from Cochran and Stobbs, 1968, with the permission of the Golf Society of Great Britain).

In stroboscopic photography (Figure 8.1), a stroboscope is used. This is a device that emits short and very intense pulses of light at a predetermined flash rate. For biomechanical analysis, the flash rate should be accurate and adjustable and should give around 20–30 exposures in the required study time (a higher rate would normally be used for display purposes). Although the light emitted is intensely bright and of extremely short duration, the illuminated area is restricted to about 1 m^2 for high-quality results using a typical single stroboscope. This area can be increased by using a battery of synchronized stroboscopes. The flickering of the stroboscope can be very distracting for the performer. Furthermore, best results are obtained with no background lighting, which can also cause problems for the performer. The accuracy of such a system should be checked, for example, by filming a disc rotating at a constant, known angular velocity (Dainty *et al.*, 1987).

8.2 Automatic tracking opto-electronic systems

A wide selection of these, generally expensive, devices is commercially available, each with its own measuring principle and performance characteristics. The choice is between video (for example ELITE, Kinemetrics, MacReflex, Motion Analysis, Peak, Vicon), scanning mirror (CODA-3) and light-emitting diode (for example Selspot, IROS, Watsmart) systems. The former two categories use passive markers while the latter category uses active markers (Figure 8.2), which require a power supply and wiring.

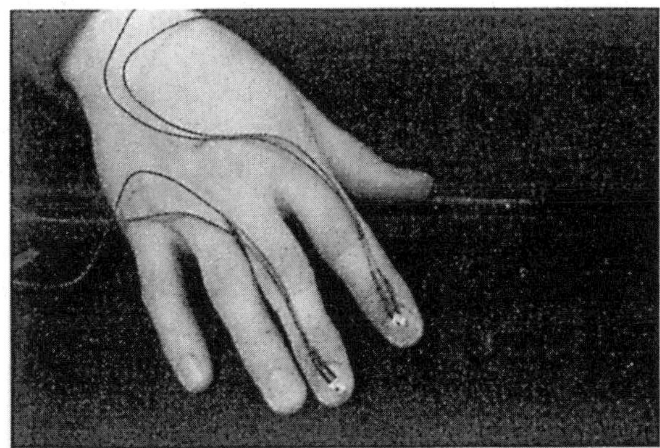

Figure 8.2 LEDs on subject for an active marker opto-electronic system (Selspot) (reproduced from Robertson and Sprigings, 1987, with permission).

These systems all consist, basically, of light sources (markers) mounted on the subject, remote image sensors (cameras), an image processor (marker detector) and a computer to record marker displacements and perform other data processing (Figure 8.3).

These systems are often referred to as on-line systems, as the peripheral devices, such as cameras, are under the control of the processing unit of the system computer. Although based on different technologies, these systems all have the advantage of rapid analysis, facilitated by their on-line operation. They provide more or less immediate coordinate data without manual digitization. They all detect and track the position of the markers placed on the skin. As with all skin-mounted devices, these markers may move relative to deeper tissues. Furthermore, the marker movements do not necessarily relate to movements of underlying, important anatomical features such as joint axes of rotation. This contrasts with manual film or video digitizing, where the operator can seek to intelligently estimate the positions of these axes, although this can be somewhat problematic. Some of these opto-electronic systems have, or are developing, software routines to relate marker positions to joint axes of rotation, which should provide more accurate estimates of the latter. All of these systems claim far higher resolutions than those normally achieved by cine or video digitizing. For example, MacReflex claims a resolution of one-30 000th of the diagonal of the field of view of the camera.

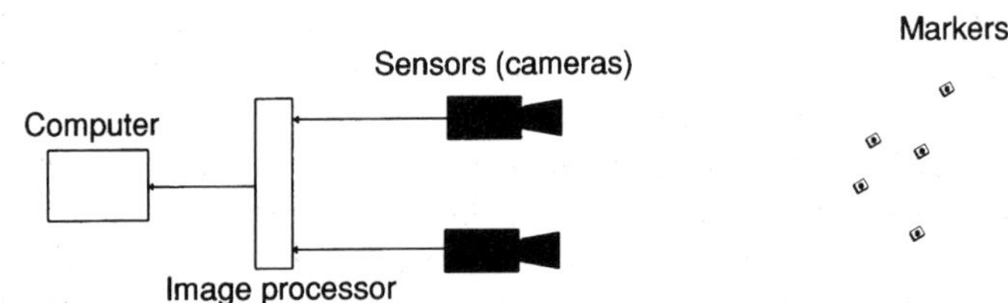

Figure 8.3 Schematic diagram of components of an opto-electronic motion analysis system (adapted from Pedotti and Ferrigno, 1995).

The markers used in infra-red light-emitting diode (LED) systems are active and are flashed sequentially (multiplexed). Their locations are detected by infra-red photo detector cameras and fed to computers for processing. The multiplexing allows the markers to be automatically identified. It does, however, involve non-simultaneous sampling of the marker coordinates, which can lead to problems for fast sports movements (Pedotti and Ferrigno, 1995). These active marker systems provide higher sampling rates and better spatial resolution than devices using passive markers, the resolution of which decreases for sampling rates above 100 Hz (Pedotti and Ferrigno, 1995). However, the leads to power the LEDs from a battery pack or mains supply are a possible impediment for sports movements. LEDs have a light emission angle restricted to a usable range of about 50° (Figure 8.4(a)), creating problems when rotation of the markers occurs, as in many sports movements. Systems using LED markers record the coordinates of the centre of the LED. Stray light sources can cause problems, as their detection may cause the replacement of the marker coordinates by those of the stray light. These systems do not produce a visible image of the movement being recorded.

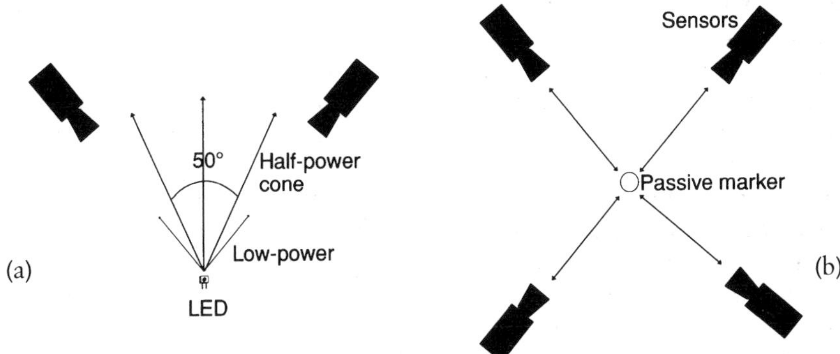

Figure 8.4 Relationship between sensor positions and markers for: (a) active and (b) passive marker systems (adapted from Pedotti and Ferrigno, 1995).

Most passive marker video systems use infra-red light, except peak and motion analysis, which use light in the visible spectrum and which can, therefore, also function as normal video systems (50 or 60 Hz for peak, up to 180 Hz for motion analysis). The light is usually emitted by rings of strobed LEDs located around the camera lenses and reflected by markers covered with retro-reflective tape. Marker shape and size can affect the accuracy of the recorded coordinates (Nigg and Cole, 1994). Spherical markers can be seen, and have the same shape, from any viewing direction (Figure 8.4(b)). They can be difficult to attach to the skin so that they do not move, especially for vigorous movements. Hemispherical markers are often used as a compromise between accuracy, and ease and security of attachment.

Marker detection is by threshold detection or pattern recognition. In the former, a threshold light level is set, which should only be exceeded by light from the markers, not that from the background. This is the less expensive and more used option, but it can result in erroneous detection of other bright objects (such as lights and reflections) in the field of view. Pattern recognition involves comparing the shape of a detected light source with the expected shape. The markers must be symmetrical, and spherical or hemispherical ones are the most suitable. This type of detection is very reliable and rejects erroneous light sources of the wrong shape. It is far more time-consuming than threshold detection and requires greater computer processing speeds. Although the markers are sampled simultaneously, passive marker video systems have the disadvantage that the markers cannot be automatically distinguished from one another. The user will normally be required to initially identify the markers, after which they will be automatically tracked by the system software. For three-dimensional studies two or more cameras are needed. In such studies, tracking of three-dimensional rather than two-dimensional (image) trajectories is becoming more common. Cross-over of markers may result in mis-identification requiring correction or editing by the user.

CODA-3 is an opto-electronic scanning device in which light is swept across the recoding volume by three rotating mirrors (scanners) mounted on a rigid linear frame (Figure 8.5(a)).

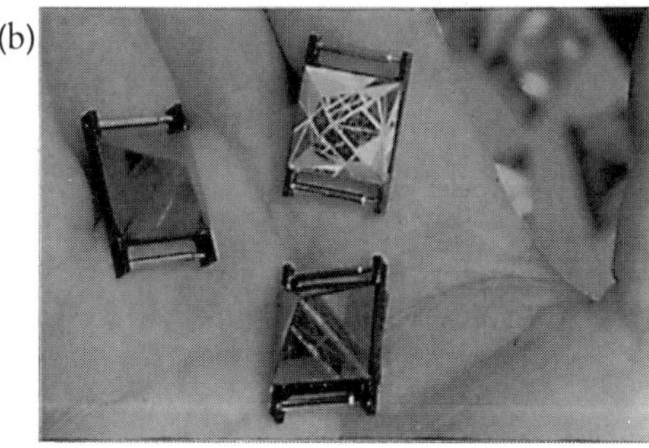

Figure 8.5 CODA-3: (a) scanner; (b) prism marker (reproduced with the permission of Movement Techniques Ltd, Loughborough).

Light is reflected by up to twelve, passive retro-reflective prisms, with a mass of less than 2 g (Figure 8.5(b)). The arrangement of scanners and detectors enables the three-dimensional coordinates of the markers to be captured. CODA-3 is capable of recording at 200 Hz. Unlike the video-based passive marker systems, the markers are distinguished from each other by reflecting different optical spectra (colours). Because the prisms can only reflect within a sector of about 180°, the device is basically limited to movements in a plane, such as walking and running. Other disadvantages include limited scanning range and lack of portability. CODA-3 can be used outdoors, although direct sunlight can produce interference (Atha, 1984). (CODA-3 has now been superseded by CODA mpx30, an active LED system.)

The procedures for use of these systems are, in general, very similar to those for cinematography and video (Chapter 5). Most of these systems require a calibration frame or other calibration structure for three-dimensional analysis, except for CODA-3 and IROS which both have cameras mounted on rigid frames in known locations. Data processing is normally part of the system and fully on-line. All of the systems encounter problems when markers become obscured, whereas in manual digitizing the operator can estimate the positions of such points. While on-line processing is attractive, it demands careful calibration and other accuracy checks, such as recording a known motion. The range of graphical displays of kinematic variables is similar to that for cine and video analysis. Most of these systems provide for synchronization of the recording with force platform, EMG, accelerometer and other analogue inputs. This is very useful in sports biomechanics research.

Most opto-electronic devices were developed for clinical, laboratory use. All the infra-red systems require control over other infra-red light sources, which limits their use outdoors. They all have restrictions on the maximum distance at which markers can be detected. Although these devices are extremely attractive in providing almost immediate feedback, they all require markers on the performer and are not, therefore, appropriate in sports competitions. They have, however, been used in studies of sports movements, including weightlifting, archery and rifle shooting (e.g. Whittle, Sargeant and Johns, 1988; Stuart and Atha, 1990; Zatsiorsky and Aktov, 1990). Some of them, at least, are likely to be increasingly used in sports biomechanics research, and in non-competitive sports environments where rapid processing of repeated trials offers a considerable advantage over manual digitizing techniques.

8.3 Electrogoniometry

If recordings of angular movement are required at only a small number of joints, then electrogoniometry provides a relatively inexpensive, simple, immediate and reliable way of obtaining the information. Electrogoniometers can serve as criterion measures of joint angle (errors can be less than 1°) if well made and correctly mounted (Miller and Nelson, 1973). They have been thoroughly validated (e.g. Adrian, 1973). They have an electrical output signal, which is available for recording or analog-to-digital conversion for computer processing. Early devices were essentially joint protractors with the pivot replaced by a high-quality, wire-wound, rotary resistor (known as a potentiometer), the resistance of which is proportional to the angular displacement of the protractor arms. These devices are suitable for movements in one plane only. Such instruments can still be easily made and used to measure hinge joint angular displacements. Care is needed to ensure that the centre of the potentiometer coincides with the joint axis of rotation and that the two arms lie along the mechanical axes of the bones of the body segments to which they are secured. This problem can be reduced using appropriate linkages but only at the cost of the size of the contact equipment, which is a further limitation of electrogoniometry. For sports movements, light, essentially frictionless and easily fitted devices are needed. Even then, their use in the recording of fast movements is highly problematic, mainly because of the tendency for the potentiometer axis to move relative to the joint axis. Electrogoniometers are, essentially, laboratory instruments although they have been used, for example, in the study of jumping events in sport (Klissouras and Karpovitch, 1967).

Electrogoniometers have also been developed to measure rotations in non-hinge joints about two or three axes, for example Figure 8.6

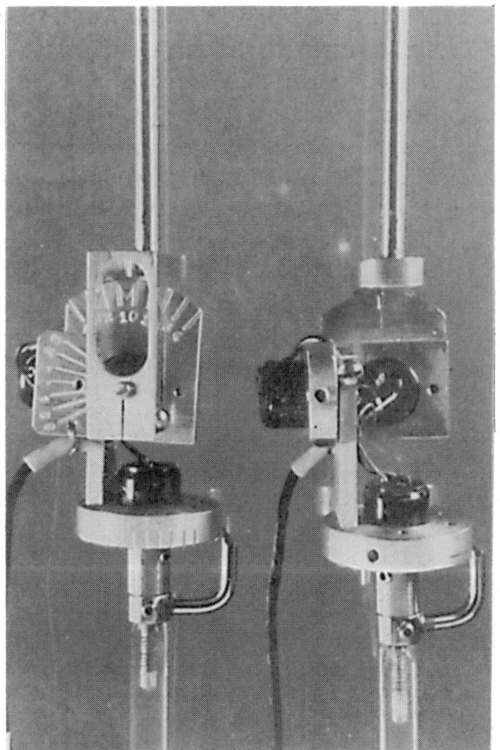

(a)

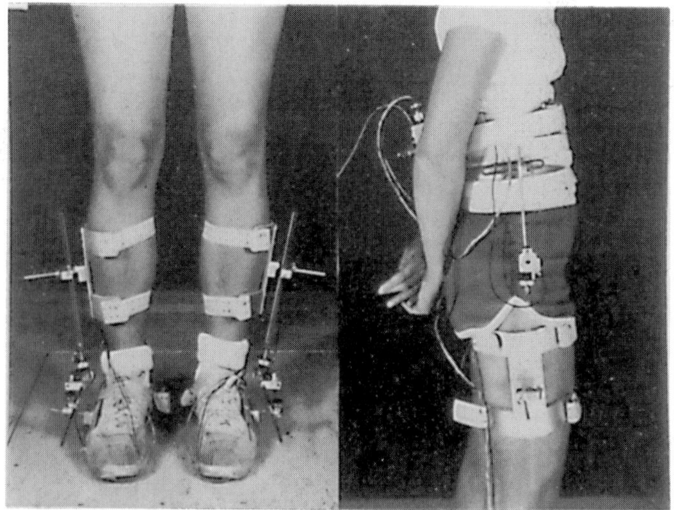

(b)

Figure 8.6 Triaxial electrogoniometer: (a) close up of device; (b) attached to body segments (reproduced from Chao, 1980, with the permission of Elsevier Science Ltd, Kidlington, Oxfordshire).

(Chao, 1980). Such electrogoniometers do, however, restrict movement to some extent and may take a considerable time to fit correctly. They are more suitable for gait analysis than for sports movements.

Flexible electrogoniometers, based on strain gauges, which do not require alignment with a joint axis of rotation, are now available commercially for single plane movements (Nicol, 1987) and movements in two planes (Nicol, 1988). There are, as yet, no reported studies which have fully evaluated their accuracy, validity and reliability for sports movements (Ladin, 1995). However, as they cause little restriction to movement and are easy to attach, they may become increasingly used in laboratory-based studies of sports movements, when only angular kinematic data are to be recorded from a small number of joints.

All electrogoniometers record the angles between body segments rather than the orientation of the segment relative to a global or local coordinate system. This is a major restriction on the use of electrogoniometric data in calculations of body segment dynamics.

8.4 Accelerometry

Accelerometers are used for the direct measurement of acceleration. They consist, in effect, of a mass mounted on a cantilever beam or spring, which is attached to the accelerometer housing (Figure 8.7).

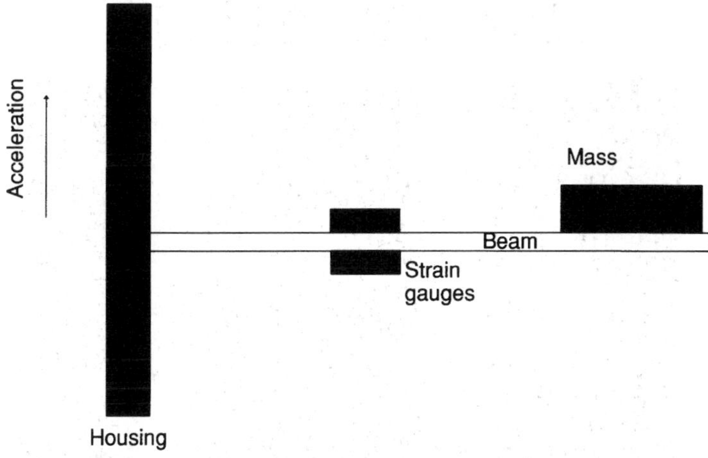

Figure 8.7 Schematic diagram of strain gauge accelerometer.

As the housing accelerates, the mass lags behind because of its inertia, and the beam is deformed. Accelerometers for the study of sports movements usually employ one of the transducer types that were covered for force measurement (Chapter 6). In strain gauge accelerometers, the gauge is usually piezoresistive – in the form of a single silicon crystal or chip. These gauges are gravity-sensitive and their output is the vector sum of the acceleration being measured and that of gravity. Additional information is therefore needed about the orientation of any body segment to which they are attached to correct for gravity. Piezoelectric devices are unaffected by gravity but are also insensitive to slow movements, which can limit their use. They are more robust than piezoresistive devices, which typically break if dropped.

For simple, linear movements, a single accelerometer may be sufficient. However, triaxial accelerometers are often used, consisting of three pre-mounted, mutually perpendicular accelerometers to record the three components of the acceleration vector. These have been used, for example, in the assessment of sports protective equipment, as in Figure 8.8 (Bishop, 1993).

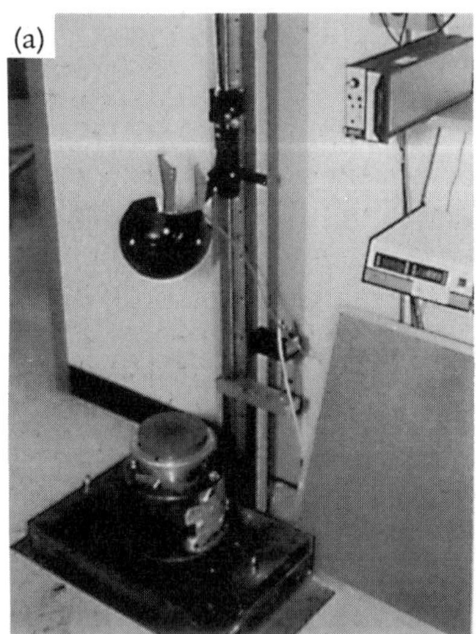

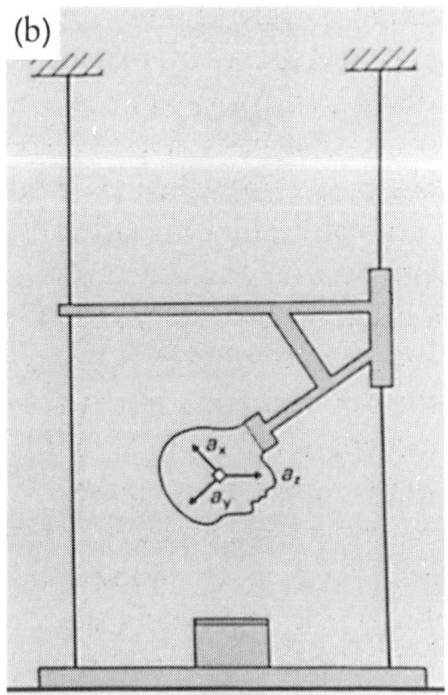

Figure 8.8 Helmet testing (instrumented with accelerometers): (a) in a monorail test rig; (b) in a guidewire test rig with acceleration component directions shown relative to the head (reproduced from Bishop, 1993, with permission).

When mounted on a body segment, the measured acceleration depends on the position of the accelerometers on the segment. The acceleration measured includes terms that arise from both linear and rotational motion of the segment. The rotational terms relate to the angular velocity and angular acceleration of the segment (Chapter 2). The absolute accelerations of a single rigid body segment having six degrees of freedom can be determined by the use of six suitably orientated accelerometers (Morris, 1973), although nine have been recommended to minimize errors (Padgaonkar, Krieger and King, 1975). If the initial conditions and the orientations of the accelerometers are known, then the accelerations can be integrated to obtain velocities and displacements (as in Chapters 2 and 6).

Accelerometers provide a continuous and direct measure of acceleration with very high-frequency response characteristics. Typical figures are 0–1000 Hz for piezoresistive and 0–5000 Hz for piezoelectric devices (Nigg, 1994c). They are usually pre-calibrated and are high-precision instruments giving very accurate measures. The values of the accelerations recorded from such devices are considerably more accurate than those that can be obtained from double differentiation of displacement signals from cine, video or opto-electronic devices. However, for the results to be valid in studying human movements in sport, proper mounting and fixation procedures must be used. Ideally, the devices should be directly mounted to bone (e.g. Lafortune, 1991), although this is not possible for most analyses of sports movements. Errors can arise from skin-mounted accelerometers because of relative movement between the accelerometer and soft tissues (e.g. Gross and Nelson, 1988). Some drift of the signal from piezoelectric accelerometers (as with piezoelectric force transducers, Chapter 6) may also occur with time (Ladin, 1995).

Although most accelerometers used in sports biomechanics are lightweight, cabling and fixation may affect performance. Discomfort to the sports performer may also arise from the attachment of the device (Robertson and Sprigings, 1987). Accelerometers provide better estimates of joint forces, when combined with data from cinematography or similar techniques, than does the use of doubly differentiated displacement data and inverse dynamics alone (Ladin and Wu, 1991). Accelerometers are relatively expensive, though valuable, research tools. They have been used to measure human accelerations in sports as diverse as running, skiing and gymnastics. Accelerometry is, however, probably the most difficult of all of the measuring techniques used in sports biomechanics to use appropriately (Nigg, 1994c).

8.5 Pressure measurement

As was seen in Chapter 6, force platforms provide the position of the point of application of the force (or the centre of pressure) on the platform. This is the point at which the force can be considered to act, although the pressure is distributed over the platform (and foot). Indeed, there may be no pressure acting at the centre of pressure when, for example, it is below the arch of the foot or between the feet during double stance. Information about the distribution of pressure over the contacting surface would be required, for example, to examine the areas of the foot on which forces are concentrated during the stance phase in running in order to improve running shoe design. In such cases, a pressure platform or pressure pad must be used. These consist of a set of force transducers with a small surface (contact) area over the threshold level of 2 N for each sensor, below which the registered force is zero.

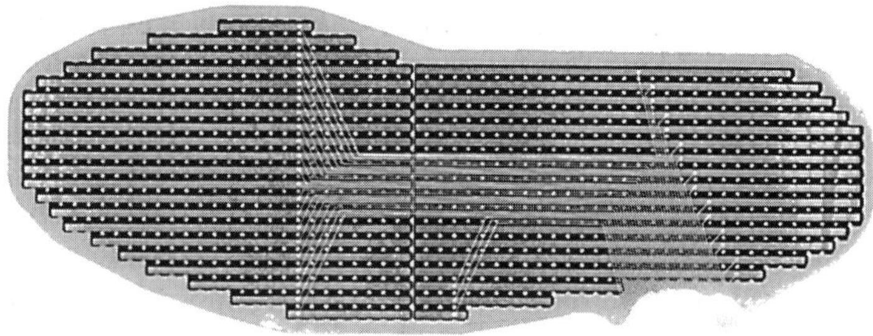

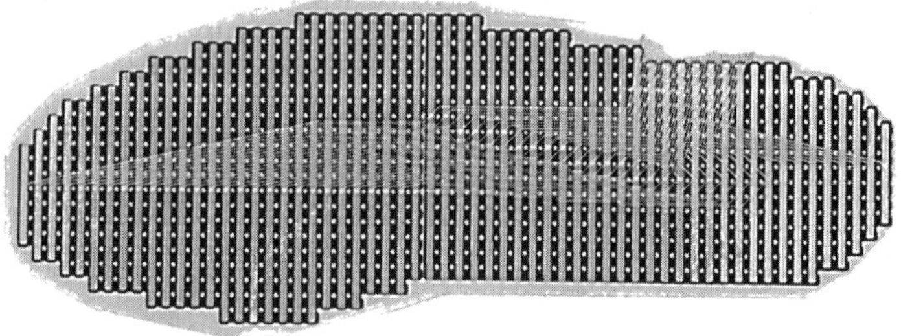

Figure 8.10 Matrix construction of rows and columns on the top and bottom surfaces of a conductive pressure insole (F-Scan).

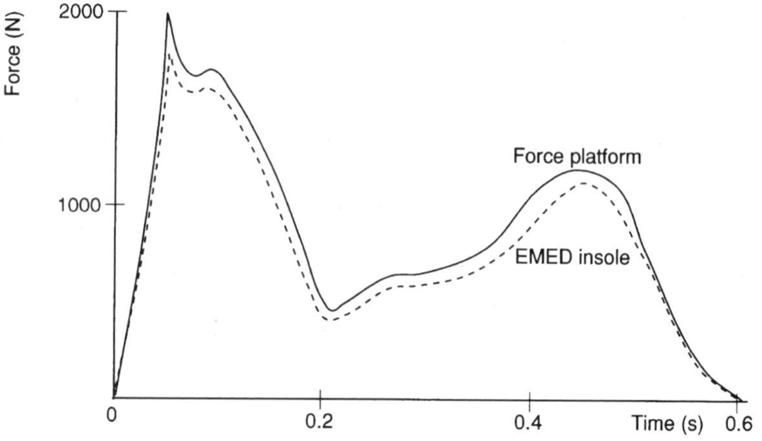

Figure 8.11 Comparison of measured force from an EMED insole with that from a Kistler force platform.

The flexibility of insoles based on capacitive sensors is limited by their thickness, but it has been greatly improved in recent designs. They are expensive. They are not sensitive to temperature. The input force to output signal characteristics are non-linear and the insoles do, therefore, require careful calibration. After this, the calibrated characteristics remain stable and day-to-day reliability is excellent (Cavanagh, Hewitt and Perry, 1992).

Conductive transducers

A very similar matrix principle (Figure 8.10) is used to that for capacitive sensors. The sandwich material is different as it is resistive rather than capacitive. The principle is used both in pressure platforms (e.g. the Musgrave footprint) and insoles (e.g. F-Scan). The devices are not sensitive to temperature. Insoles can be made very thin and flexible but are then fragile and susceptible to wrinkling. The F-Scan insole has 960 sensors each with an area of 5.1 mm^2, and it can be cut to foot size. The sampling rate is up to 100 Hz per sensor and the peak pressure that can be recorded is 1035 kPa. It is much thinner than the EMED insole (0.18 mm) and very inexpensive – it is effectively a disposable item. The calibration is much more linear than for capacitive transducers, but changes in sensitivity during use can cause variations both between insoles and within one insole (Cavanagh, Hewitt and Perry, 1992). Good agreement has been reported in comparing the forces recorded by F-Scan insoles and an AMTI force platform during push-off in walking (Derrick and Hamill, 1992).

Piezoelectric transducers

These were the first sensors used for measurement of pressure distribution inside a shoe. Hennig, Cavenagh and McMillan (1983) developed an insole with 499 sensors, each of which had an area of 23 mm^2, embedded in a 3–4 mm thick layer of very resilient silicone rubber. Sampling rates up to 200 Hz per sensor were possible and individual transducers were claimed to measure peak pressures up to 1500 kPa within 2% linearity and 1% hysteresis. Such transducers do, however, present problems. They are expensive and sensitive to temperature, although the insoles used are flexible. Also, each transducer requires individual connections, resulting in extensive cabling (Nigg, 1994b). No such devices have yet been developed commercially.

Data processing

Many data processing and data presentation options are available to help analyse the results from pressure measuring devices. In principle, all of the data processing options available for force platforms (Chapter

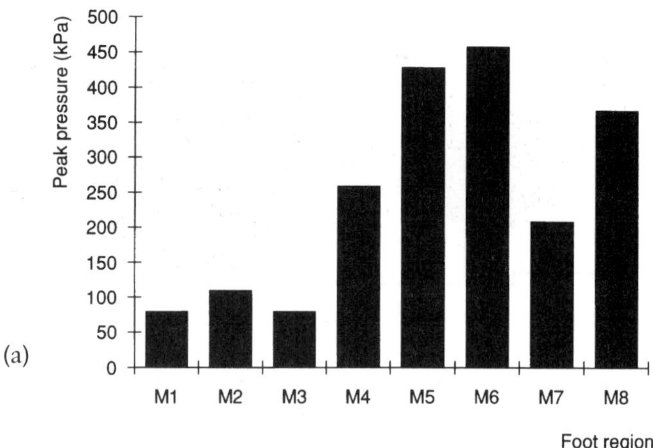

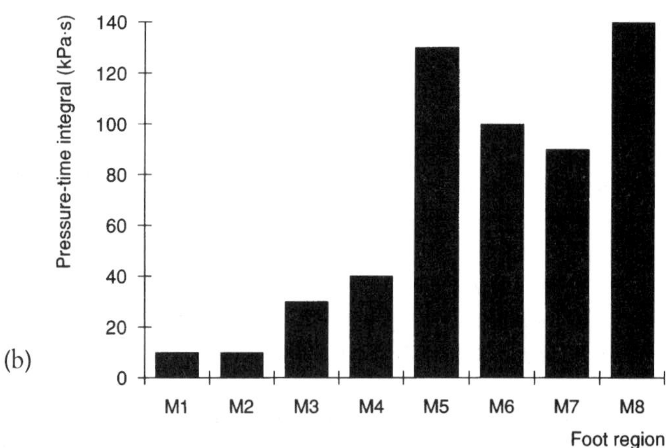

Figure 8.15 Bar chart displays of: (a) peak pressure; (b) pressure–time integral, for eight regions of the foot (M1–M8).

investigate the effect of sports shoe construction and to estimate internal forces in the foot (Morlock and Nigg, 1991). Despite some present limitations of pressure insoles, such as restricted sampling rate and fragility, it is likely that these devices will become increasingly used in sports biomechanics. It is also probable that pressure pads will be more widely used to measure pressure distributions between the hands and sports equipment and other performer–equipment contacts, for example with the seat in wheelchair racing. Future developments of these pressure measuring devices, to include the two shear components of contact force, will further enhance their usefulness in sports biomechanics.

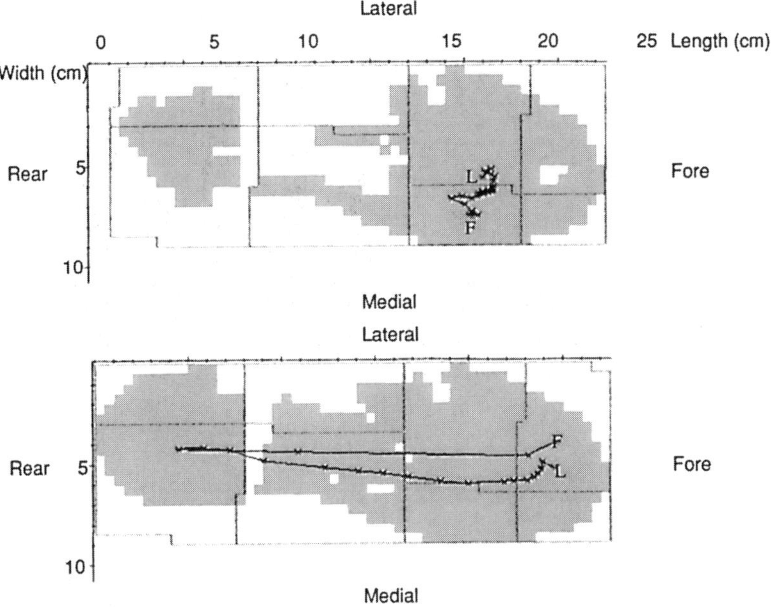

Figure 8.16 Centre of pressure paths (F indicates first and L last contact).

8.6 Measurement of muscle force and torque

8.6.1 DIRECT MEASUREMENT OF MUSCLE FORCE

In Chapter 7, the problem of muscle indeterminacy was touched upon – that is the equations of motion at a joint cannot be solved because the number of unknown muscle forces exceeds the number of equations available for their solution. A possible solution to this problem, through the use of EMG to predict muscle tension, was considered. Another approach would be to directly measure the force exerted by individual muscles, using force transducers attached to the muscle tendons. There have, not surprisingly, been few studies that have directly measured such forces during sports movements. The published studies, of running and cycling (e.g. Komi *et al.*, 1987; Komi, 1990; Gregor *et al.*, 1991), used buckle transducers which typically consist of a stainless steel buckle fitted over the tendon. Strain gauges are mounted on the buckle and wires from these pass through the skin to the recording instrumentation. These transducers are limited to only a few structures (such as the Achilles tendon) as they need a specific length of tendon for attachment and they must be away from any bones (Nigg, 1994a). There are also calibration problems, questions as to whether the forces are affected by the measurement technique, and ethical considerations. This technique is likely to continue to be used only in highly specialized areas of sports biomechanics research.

8.6.2 ISOKINETIC DYNAMOMETRY

External measurements of muscle force, using various strength tests, cannot reveal the force in an individual muscle and do not therefore contribute to a solution of the muscle indeterminacy problem. However, the measurement of the net muscle torque at a joint using isokinetic dynamometry is very useful in providing an insight into muscle function and in obtaining muscle performance data for various modelling purposes. ('Isokinetic' is derived from Greek words meaning 'constant velocity'.)

Isokinetic dynamometry is used to measure the net muscle torque (called muscle torque in the rest of this section) during isolated joint movements (Figure 8.17).

A variable resistive torque is applied to the limb segment under consideration; the limb moves at constant angular velocity once the preset velocity has been achieved, providing the person being measured is able to maintain that velocity in the specified range of movement. This allows the measurement of muscle torque as a function of joint angle and angular velocity. At certain joints, these may then be related to the length and contraction velocity of a predominant prime mover (e.g. quadriceps femoris in knee extension). By adjusting the resistive torque, both muscle strength and endurance can be evaluated. Isokinetic dynamometers are also used as training aids, although they do not replicate the types and speeds of movement in sport (Enoka, 1994).

Passive isokinetic dynamometers operate using either electromechanical or hydraulic components. In these devices, resistance is developed only as a reaction to the applied muscle torque, and they can, therefore, only be used for concentric movements. Electromechanical dynamometers with active mechanisms allow for concentric and eccentric movements with constant angular velocity. The SPARK system can be used for concentric and eccentric movements involving constant velocity, linearly changing acceleration or deceleration, or a combination of these (Baltzopoulos, 1992).

Several problems affect the accuracy and validity of measurements of muscle torque using isokinetic dynamometers. Failure to compensate for gravitational force can result in significant errors in the measurement of muscle torque and data derived from those measurements (e.g. Appen and Duncan, 1986). These errors can be avoided by the use of gravity compensation methods, which are an integral part of the experimental protocol in most computerized dynamometers.

The development and maintenance of a preset angular velocity is another potential problem (e.g. Sapega *et al.*, 1982). In the initial

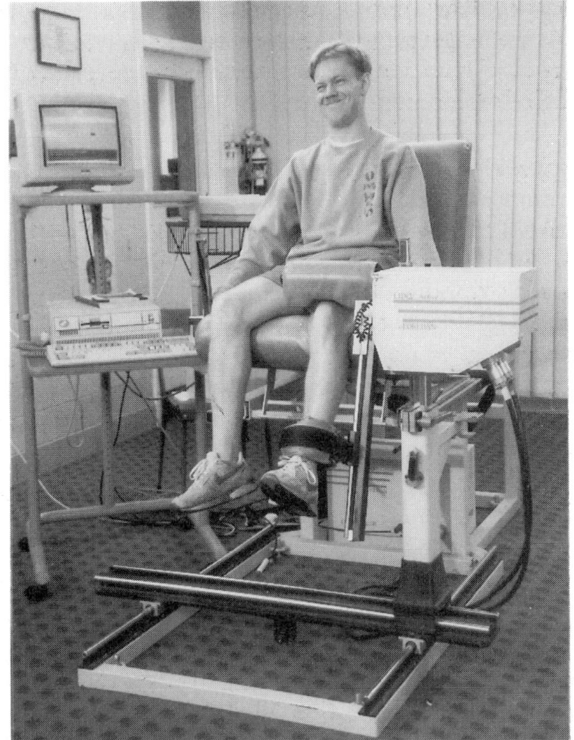

(a)

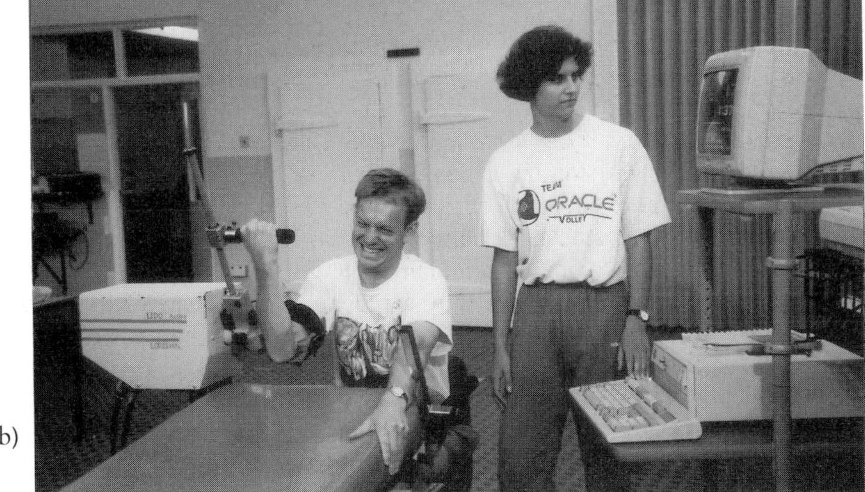

(b)

Figure 8.17 Use of isokinetic dynamometer for: (a) knee extension; (b) elbow flexion.

period of the movement, the dynamometer is accelerated without resistance until the preset velocity is reached. The resistive mechanism is then activated and slows the limb down to the preset level. The duration of the acceleration period, and the magnitude of the resistive torque required to decelerate the limb, depend on the preset angular velocity and the athlete being evaluated. The dynamometer torque during this period is clearly not the same as the muscle torque accelerating the system. If the muscle torque during this period is required, it should be calculated from the moment of inertia and angular acceleration data. The latter should be obtained either from differentiation of the position–time data or from accelerometers if these are available (Baltzopoulos, 1992).

Errors can also arise in muscle torque measurements unless the axis of rotation of the dynamometer is aligned with the axis of rotation of the joint, estimated using anatomical landmarks. For normal subjects and small misalignments, the error is very small and can be neglected (Herzog, 1988). Periodic calibration of the dynamometer system is necessary in respect of both torque and angular position, the latter using an accurate goniometer. Torque calibration should be carried out statically, to avoid inertia effects, under gravitational loading.

Accurate assessment of isokinetic muscle function requires the measurement of torque output while the angular velocity is constant, and computation of angular velocity is therefore essential. Most isokinetic dynamometers output torque and angular position data in digital form. Angular velocity and acceleration can be obtained by differentiation of the angular position–time data after using appropriate noise reduction techniques (Chapter 5). Instantaneous joint power ($P = T \cdot \boldsymbol{\omega}$) can be calculated from the torque (T) and angular velocity ($\boldsymbol{\omega}$) when the preset angular velocity has been reached.

The following parameters can normally be obtained from an isokinetic dynamometer to assess muscle function (Baltzopoulos, 1992).

- **The maximum torque.** The isokinetic maximum torque is used as an indicator of the muscle torque which can be applied in dynamic conditions. It is usually evaluated from two to six maximal repetitions and is taken as the maximum single torque measured during these repetitions. The maximum torque depends on the angular position of the joint. Maximum power can also be calculated (Bloomfield, Ackland and Elliott, 1994).
- **The reciprocal muscle group ratio.** This is an indicator of joint muscle balance, which is affected by age, sex and fitness. It is the ratio of the maximum torques recorded in antagonist movements (usually flexion and extension).
- **The maximum torque position.** This is the joint angular position at maximum torque and provides information about the mechanical properties of the activated muscle group. It is affected by the angular velocity. As the velocity increases, this position tends to occur later in the range of movement and not in the mechanically optimal

joint position. It is therefore crucial to specify the maximum torque position as well as the maximum torque.

- **Muscular endurance under isokinetic conditions.** This is usually assessed in the form of a 'fatigue index'. It provides an indication of the muscle group's ability to perform the movement at the preset angular velocity over a period of time. Although there is no standardized testing protocol or period of testing, it has been suggested that 30–50 repetitions or a total duration of 30–60 s should be used (Baltzopoulos, 1992). The fatigue index can then be expressed as the ratio of the maximum torques recorded in the initial and final periods of the test.

8.7 Summary

In this chapter, some of the other techniques used in the recording and analysis of sports movements were covered, including their advantages and limitations. The uses and limitations of single-plate multiple-image photography for recording movement in sport were considered. The use of automated opto-electronic motion analysis systems was described and the advantages and limitations, for sports movements, of three types of opto-electronic system were considered. The use of electrogoniometry to record joint motion was outlined, along with the use of accelerometry. The limitations of both of these techniques for sports movements were addressed. The value of contact pressure measurements in the study of sports movements was covered, including the relative advantages and limitations of pressure insoles and pressure platforms and of the three types of pressure transducer used in sports biomechanics. Some examples were provided of the ways in which pressure transducer data can be presented to aid analysis. The chapter concluded with a brief consideration of the restrictions on the use of direct tendon force measurement in sport and how isokinetic dynamometry can be used to record the net muscle torque at a joint.

8.8 Exercises

1. List the limitations and possible uses of single-plate multiple image photography for recording movement in sport.
2. If you have access to a stroboscope and good Polaroid camera, photograph a linear movement, such as running, or a non-repetitive rotational movement, such as a tennis serve, following the advice points of section 8.1. (This could be done using any camera, but without the immediate feedback, allowing you to adjust exposure time for example, that Polaroid cameras provide.)
3. List the comparative advantages and disadvantages of the three types of opto-electronic systems covered in section 8.2. Compare these with the advantages and limitations of cinematography and video analysis.

4. Describe how electrogoniometry can be used to record sports movements.
5. If you have access to a single plane electrogoniometer, use it to record knee angles during walking or elbow angles in a simple throwing task. Obtain angular velocities from your recording either using the recording software, if available, or by graphical differentiation (Chapter 2).
6. Outline the restrictions on the use in sport of:
   a) accelerometry;
   b) tendon force measurements.
7. List the relative advantages and disadvantages of pressure insoles and pressure mats. Also, list the relative advantages and disadvantages of the three types of pressure transducer described in section 8.5.
8. Specify two sporting activities in which you consider that the measurement of contact pressure would provide useful information. Explain clearly why the information would be useful.
9. Describe the uses of isokinetic dynamometry in sports biomechanics.
10. If you have access to an isokinetic dynamometer, use it, according to the supplier's guidelines and the recommendations of Baltzopoulos (1992), and with appropriate supervision, to record the maximum muscle torques and time to maximum torque for concentric, isokinetic knee extension and flexion. From these calculate the reciprocal muscle group ratio. Comment on your results and compare them with the hamstrings to quadriceps femoris minimum recommended strength ratio of 40:60, which is often cited.

8.9 References

Adrian, M. (1973) Cinematographic, electromyographic and electrogoniometric techniques for analyzing human movements, in *Exercise and Sport Sciences Reviews*, Vol. 1, (ed. J. H. Wilmore), Academic Press, New York, pp.339–363.

Appen, L. and Duncan, P. W. (1986) Strength relationship of the knee musculature: effect of gravity and sport. *Journal of Orthopaedic and Sports Physical Therapy*, 7, 232–235.

Atha, J. (1984) Current techniques for measuring motion. *Applied Ergonomics*, **15**, 245–257.

Baltzopoulos, V. (1992) Isokinetic dynamometry, in *Biomechanical Analysis of Performance in Sport*, (ed. R. M. Bartlett), British Association of Sports Sciences, Leeds, pp.38–44.

Bartlett, R. M., Müller, E., Raschner, C. and Brunner, F. (1991) Pressure distributions on the plantar surface of the foot during the discus throw. *Journal of Sports Sciences*, **9**, 394.

Bartlett, R. M., Müller, E., Raschner, C. *et al.* (1995) Pressure distributions on the plantar surface of the foot during the javelin throw. *Journal of Applied Biomechanics*, **11**, 163–176.

Bishop, P. J. (1993) Protective equipment: biomechanical evaluation, in *Sports Injuries: Basic Principles of Prevention and Care*, (ed. P. A. F. H. Renström), Blackwell, Oxford, pp.355–373.

Bloomfield, J., Ackland, T.R. and Elliott, B.C. (1994) *Applied Anatomy and Biomechanics in Sport*, Blackwell, Melbourne, Victoria.

Cavanagh, P. R., Hewitt, F. G. and Perry, J. E. (1992) In-shoe plantar pressure measurement: a review. *Foot*, **2**, 397–406.

Chao, E. Y. S. (1980) Justification of triaxial goniometer for the measurement of joint rotation. *Journal of Biomechanics*, **13**, 989–1006.

Clark, F., Paul, T. and Davis, M. (1977) A convenient procedure and computer program for obtaining instantaneous velocities from stroboscopic photography. *Research Quarterly*, **48**, 628–631.

Cochran, A. and Stobbs, J. (1968) *The Search for the Perfect Swing*, Heinemann, London.

Dainty, D. A., Gagnon, M., Lagasse, P.P. *et al.* (1987) Recommended procedures, in *Standardizing Biomechanical Testing in Sport*, (eds D. A. Dainty and R. W. Norman), Human Kinetics, Champaign, IL, pp.73–100.

Derrick, T. R and Hamill, J. (1992) Ground and in-shoe reaction forces when walking, in *Proceedings of NACOB II, The Second American Congress on Biomechanics*, Chicago, IL, 24–28 August.

Enoka, R. M. (1994) *Neuromechanical Basis of Kinesiology*, Human Kinetics, Champaign, IL.

Gregor, R. J., Komi, P. V., Browning, R. C. and Jarvinen, M. (1991) Comparison between the triceps surae and residual muscle moments at the ankle during cycling. *Journal of Biomechanics*, **24**, 287–297.

Gross, T. S. and Nelson, R. C. (1988) The shock attenuation role of the ankle during landing from a vertical jump. *Medicine and Science in Sports and Exercise*, **20**, 506–514.

Hennig, E. M., Cavanagh, P. R. and McMillan, N. H. (1983) Pressure distribution measurements by high precision piezoelectric ceramic force transducers, in *Biomechanics VIII-B*, (eds H. Matsui and K. Kobayashi), Human Kinetics, Champaign, IL, pp.1081–1088.

Herzog, W. (1988) The relation between the resultant moments at a joint and the moments measured by an isokinetic dynamometer. *Journal of Biomechanics*, **18**, 621–624.

Klissouras, V. and Karpovitch, P. V. (1967) Electrogoniometric study of jumping events. *Research Quarterly*, **38**, 41–48.

Komi, P. V. (1990) Relevance of in vivo force measurements to human biomechanics. *Journal of Biomechanics*, **23** (suppl.), 23–34.

Komi, P. V., Solonen, M., Jarvinen, M. and Kokko, O. (1987) *In vivo* registration of Achilles tendon force in man: methodological development. *International Journal of Sports Medicine*, **8** (suppl.), 3–8.

Ladin, Z. (1995) Three-dimensional instrumentation, in *Three-Dimensional Analysis of Human Movement*, (eds P. Allard, I. A. F. Stokes and J.-P. Blanchi), Human Kinetics, Champaign, IL, pp.3–17.

Ladin, Z. and Wu, G. (1991) Combining position and acceleration measurements for joint force estimation. *Journal of Biomechanics*, **24**, 1173–1187.

Lafortune, M. A. (1991) Three-dimensional acceleration of the tibia during walking and running. *Journal of Biomechanics*, **24**, 877–886.

Maier, I. (1968) Measurement apparatus and analysis methods of the biomotor process of sports movements, in *Medicine and Sport, Biomechanics I*, (eds J. Wartenweiler, E. Jokl and M. Hebbelinck), S. Karger, Basel, pp.96–101.

Merriman, J. S. (1975) Stroboscopic photography as a research instrument. *Research Quarterly*, **46**, 256–261.

Milani, T. L. and Hennig, E. M. (1990) Prevention of injuries in triple jump, in *Techniques in Athletics: Conference Proceedings*, Vol. 2, (eds G.-P. Brüggemann and J. K. Rühl), Deutsche Sporthochschule, Cologne, pp.753–760.

Miller, D. I. and Nelson, R. C. (1973) *Biomechanics of Sport: A Research Approach*, Lea & Febiger, Philadelphia, PA.

Morlock, M. and Nigg, B. M. (1991) Theoretical considerations and practical results on the influence of the representation of the foot for the estimation of internal forces with models. *Clinical Biomechanics*, **6**, 3–13.

Morris, J. R. W. (1973) Accelerometry: a technique for the measurement of human body movements. *Journal of Biomechanics*, **6**, 729–736.

Nicol, A. C. (1987) A new flexible electrogoniometer with widespread applications, in *Biomechanics X-B*, (ed. B. Jonsson), Human Kinetics, Champaign, IL, pp.1029–1033.

Nicol, A. C. (1988) A triaxial flexible electrogoniometer, in *Biomechanics XI-B* (eds G. de Groot, A. P. Hollander, P.A. Huijing and G. J. van Ingen Schenau), Free University Press, Amsterdam, pp.964–967.

Nicol, K. (1977) Druckverteilung über den Fuβ bei sportlichen Absprungen und Landungen in Hinblick auf eine Reduzierung von Sportverletzungen. *Leistungsport*, **7**, 220–227.

Nigg, B. M. (1994a) Force, in *Biomechanics of the Musculoskeletal System*, (eds B. M. Nigg and W. Herzog), John Wiley, Chichester, pp.199–224.

Nigg, B. M. (1994b) Pressure distribution, in *Biomechanics of the Musculoskeletal System*, (eds B. M. Nigg and W. Herzog), John Wiley, Chichester, pp.225–236.

Nigg, B. M. (1994c) Acceleration, in *Biomechanics of the Musculoskeletal System*, (eds B. M. Nigg and W. Herzog), John Wiley, Chichester, pp.237–253.

Nigg, B. M. and Cole, G. K. (1994) Optical methods, in *Biomechanics of the Musculoskeletal System*, (eds B. M. Nigg and W. Herzog), John Wiley, Chichester, pp.254–286.

Padgaonkar, A. J., Krieger, K. W. and King, A. I. (1975) Measurement of angular acceleration of a rigid body using linear accelerometers. *Journal of Applied Mechanics*, **42**, 552–556.

Pedotti, A. and Ferrigno, G. (1995) Optoelectronic-based systems, in *Three-Dimensional Analysis of Human Movement*, (eds P. Allard, I. A. F. Stokes and J.-P. Blanchi), Human Kinetics, Champaign, IL, pp.57–77.

Robertson, G. and Sprigings, E. (1987) Kinematics, in *Standardizing Biomechanical Testing in Sport*, (eds D. A. Dainty and R. W. Norman), Human Kinetics, Champaign, IL, pp.9–20.

Sanderson, D. J. and Cavanagh, P. R. (1987) An investigation of the in-sole pressure distribution during cycling in conventional cycling shoes or running shoes, in *Biomechanics X-B*, (ed. B. Jonsson), Human Kinetics, Champaign, IL, pp.903–907.

Sapega, A., Nicholas, J., Sokolow, D. and Sarantini, D. (1982) The nature of torque 'overshoot' in Cybex isokinetic dynamometry. *Medicine and Science in Sports and Exercise*, **14**, 368–375.

Schaff, P., Kulot, M., Hauser, W. and Rosemeyer, B. (1988) Einfluβfaktoren auf die Druckverteilung unter der Fuβsohle in Skischuhen. *Sportverletzungen, Sportschaden*, **2**, 164–171.

Smith, A. J. (1975) Photographic analysis of movement, in *Techniques for the Analysis of Human Movement*, (eds D. W. Grieve, D. Miller, D. Mitchelson *et al.*), Lepus Books, London, pp.3–29.

Stuart, J. and Atha, J. (1990) Postural consistency in skilled archers. *Journal of Sports Sciences*, 8, 223–234.

Whittle, M. V., Sargeant, A. J. and Johns, L. (1988) Computerised analysis of knee moments during weightlifting, in *Biomechanics XI-B* (eds G. de Groot, A. P. Hollander, P.A. Huijing and G. J. van Ingen Schenau), Free University Press, Amsterdam, pp.885–888.

Winter, D. A. (1990) *Biomechanics and Motor Control of Human Movement*, Wiley-Interscience, New York.

Yeadon, M. R. and Challis, J. H. (1994) The future of performance related sports biomechanics research. *Journal of Sports Sciences*, **12**, 3–32.

Zatsiorsky, V. M. and Aktov, A. V. (1990) Biomechanics of highly precise movements: the aiming process in air rifle shooting. *Journal of Biomechanics*, **23** (suppl. 1), 35–41.

8.10 Further reading

Nigg, B. M. and Herzog, W. (eds) (1994) *Biomechanics of the Musculoskeletal System*, John Wiley, Chichester: this contains good chapters on the measurement of pressure (pp.225–236) and acceleration (pp.237–253).

Baltzopoulos, V. (1992) Isokinetic dynamometry, in *Biomechanical Analysis of Performance in Sport*, (ed. R. M. Bartlett), British Association of Sports Sciences, Leeds, pp.38–44: this provides a succinct, yet fairly comprehensive coverage of isokinetic dynamometry, including procedural considerations.

Pedotti, A. and Ferrigno, G. (1995) Optoelectronic-based systems, in *Three-Dimensional Analysis of Human Movement*, (eds P. Allard, I. A. F. Stokes and J.-P. Blanchi), Human Kinetics, Champaign, IL, pp.57–77: this gives a detailed account of the principles and use of these systems, although it is somewhat advanced in places.

Index

UNIVERSITY OF WOLVERHAMPTON
LEARNING RESOURCES

New Book Information *from* E & FN SPON

Sports Biomechanics

Reducing Injury and Improving Performance

***Roger Bartlett*, Sport Science Research Institute, Sheffield Hallam University, UK**

Sports Biomechanics: Reducing Injury and Improving Performance is the companion volume to *Introduction to Sports Biomechanics*, also written by Roger Bartlett. This advanced text focuses on third year undergraduate and postgraduate topics and considers the two key issues of sports biomechanics: why injuries occur, how they can be reduced, and how performance can be improved.

Part One contains a detailed examination of sports injury, including: properties of biological materials, mechanisms of injury occurrence, risk reduction, and the estimation of forces in biological structures. Part Two concentrates on the biomechanical enhancement of sports performance. Subjects covered in detail are: analysis of sports technique, statistical and mathematical modelling of sports movements and the feedback of results to improve performance.

This textbook will prove invaluable for students studying biomechanics in years two and three of a sport science degree, and also for postgraduate students.

- **each chapter contains an introduction, summary, references, example exercises and further reading**
- **mathematical sections are presented in a user-friendly manner**
- **examples from specific sports are used throughout the text to base theory in practice**
- **illustrations reinforce explanations and examples**
- **key course textbook for sports science**
- **takes academic discipline one step further and applies to specific sports, injuries and performance**

Contents: **Section One:** Biomechanical analysis and optimisation of sports techniques. Aspects of biomechanical analysis of sports performance. Biomechanical optimisation of sports techniques. Mathematical models of sports motions. Feedback and communication of model results. **Section Two:** Biomechanics of musculoskeletal injury. Injuries in sport. Properties of materials. Calculating the loads. Biomechanical factors affecting sports injury.

December 1998: 234x156: 304pp: 96 line illustrations
Paperback: 0-419-18440-6: £22.99

Customer Services/Warehouse/Accounts, Cheriton House, Northway, Andover, Hants SP10 5BE, UK. Tel: +44 (0)1264 343071, Fax: +44 (0)1264 343005
Internet: www.routledge.com

Marketing & Editorial, Routledge, 11, New Fetter Lane London EC4P 4EE, UK, Tel: +44 (0)171 583 9855
Email: info@routledge.com, Fax: +44 (0)171 842 2303

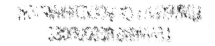